Topology and
Teichmüller Spaces

Topology and Teichmüller Spaces

Katinkulta, Finland   24 – 28 July 1995

Editors

Sadayoshi Kojima
Tokyo Institute of Technology

Yukio Matsumoto
University of Tokyo

Kyoji Saito
Research Institute for Mathematical Sciences, Kyoto

Mika Seppälä
Florida State University

Published by

World Scientific Publishing Co. Pte. Ltd.

5 Toh Tuck Link, Singapore 596224

USA office: 27 Warren Street, Suite 401-402, Hackensack, NJ 07601

UK office: 57 Shelton Street, Covent Garden, London WC2H 9HE

British Library Cataloguing-in-Publication Data
A catalogue record for this book is available from the British Library.

TOPOLOGY AND TEICHMÜLLER SPACES
Proceedings of the 37th Taniguchi Symposium

ISBN-13 978-981-02-2686-2
ISBN-10 981-02-2686-1

PREFACE

The 37th Taniguchi Symposium on "Topology and Teichmüller Spaces", sponsored by the Taniguchi Foundation, was held on July 24-28, 1995, at Katinkulta holiday village in Finland. This volume consists of contributions from the participants, and is dedicated on this occasion to the distinguished founder of the Taniguchi Foundation, the late Mr. Toyosaburo Taniguchi, who passed away on October 26, 1994.

The first article by Shingo Murakami briefly describes the exceptionally outstanding support of Toyosaburo Taniguchi and the Taniguchi Foundation for the Japanese mathematical community to develop international exchanges. All the other articles are research and survey papers essentially based on the talks at the Symposium. They were refereed.

The participants were: Y. Imayoshi (Osaka City U.), N. Kawazumi (Hokkaido), S. P. Kerckhoff (Stanford), S. Kojima (TIT), Y. Matsumoto (Tokyo), Y. Minsky (Stony Brook), J. M. Montesinos-Amilibia (Madrid), S. Morita (TIT), S. Nag (Madras), K. Ohshika (TIT), O. Pekonen (Jyväskylä), R. C. Penner (USC), K. Saito (RIMS, Kyoto), M. Seppälä (Helsinki), T. Soma (Tokyo Denki U.), M. Taniguchi (Kyoto) and S. Murakami (Taniguchi Foundation). The organizing committee and the editorial committee of the Proceedings both consist of Sadayoshi Kojima, Yukio Matsumoto, Kyoji Saito and Mika Seppälä.

On behalf of all participants, first and foremost, we thank the Taniguchi foundation for its exceptionally generous support. We also thank Mika and Cici Seppälä and Osmo Pekonen for making all the local arrangements, which made the Symposium so successful. The production of the present volume could not have taken place without the help of contributers by preparing their papers in TeX, not to mention anonymous referees and the TeXnical work by Masaru Ohta and Shigeru Mizushima. We, the editors, thank all of them.

Sadayoshi Kojima
Yukio Matsumoto
for the editors
July 1996

Proceedings of
THE 37TH TANIGUCHI SYMPOSIUM ON
TOPOLOGY AND TEICHMÜLLER SPACES
held in Finland, July 1995
ed. by Sadayoshi KOJIMA *et al.*
©1996 World Scientific Publishing Co.
pp. 1–4

MR. TOYOSABURO TANIGUCHI AND THE TANIGUCHI FOUNDATION

SHINGO MURAKAMI

(Received December 6, 1995)

1. Mr. Toyosaburo Taniguchi

Mr. Toyosaburo Taniguchi, founder of the Taniguchi Foundation, was born on July 29, 1901 and passed away on October 26, 1994 at the age of 93.

To begin with, I would like to introduce the unforgettable experience I had when I met Mr. Taniguchi for the first time in the spring of 1976. At that time, at the request of Prof. Yasuo Akizuki, an old friend of Mr. Taniguchi, I was preparing an international symposium on differential geometry, to be supported by the Taniguchi Foundation. In the preceding years, three international symposia supported by the Foundation had already been organized with many participants, and Prof. Akizuki notified me that the coming one's scale should be reduced in accordance with Mr. Taniguchi's new policy of managing the Foundation. I therefore planned a symposium of about 40 participants, but Prof. Akizuki then told me by telephone, "I like your plan, but Taniguchi says NO. Because you are in Osaka, go to see him and discuss the number of participants directly with him." Thus I visited Mr. Taniguchi at his office.

From 1959 to 1966 Mr. Taniguchi was the President of Toyobo Co., a big textile company in Osaka, and since 1971 had been the Supreme Adviser of the Japanese Textile Federation. He received the First Class Order of the Sacred Treasure in 1973 from the Japanese Government, and founded the Taniguchi Foundation in 1976.

Mr. Taniguchi received me in a small reception room in Toyobo Co.. He was in his mid-seventies, but looked as though he were in his fifties. He began to speak by explaining why he created the Foundation. He said, "In 1971 I was the head of the delegation sent by the Japanese Government to the United States to negotiate the textile trade problem between the two countries. According to my experience at that time, I firmly believe that the mutual understanding between people of different countries is very important, but really difficult. In order for Japan to survive in

---

This is a modified version of the author's address, delivered at the opening ceremony of the 37th Taniguchi Symposium held at Katin Kulta, Finland on July 23-28, 1995.

1

the world and to develop in the future, it is indispensable for us to have as much contact with foreigners as possible and to have many friends among them. This is why I created a foundation with my personal funds, with the aim of supporting international symposia where young scholars the world over may gather informally to exchange their thoughts. In this way, I want to realize my ideas on an intimate scale. Please cooperate with me. If you organize your symposium respect to my ideas, namely, by inviting only 15 or so prominent young mathematicians to meet together day and night throughout the symposium, I will support you with enough money so that you need not worry about any financial problems."

Indeed, I was inspired by Mr. Taniguchi's ideas, which he enthusiastically explained to me in more detail. When he left me in the reception room, I was very impressed by what he told me and became aware that he had not asked me any questions about the symposium I was going to organize. Shortly thereafter, I gave up my original plan, and I coordinated a Taniguchi Symposium on geometric function theory in 1978. This symposium was organized by T. Ochiai with the help of S. Kobayashi; it had 17 participants whose ages averaged under forty. Fortunately, this style of symposium fitted with Mr. Taniguchi's ideas, and became a standard for organizing Taniguchi Symposia.

2. History of the Taniguchi Foundation

I would like to talk about the history of the Taniguchi Foundation. I said that Mr. Taniguchi founded his Foundation in 1976, but actually this was a renewal of the Foundation that he had created many years before.

In 1929, Mr. Taniguchi established a Foundation, in accordance with his father's will, named "the Taniguchi Technology Promotion Foundation." Its goal was to support scientific activities in the Osaka area. This Taniguchi Foundation contributed exclusively to the development of the Faculty of Science at Osaka University founded in 1931. Among other activities, the Foundation donated a cyclotron to Osaka University which was the first one settled in Japan. By the way, the official name of the new Taniguchi Foundation is "The Foundation Commemorating the 45th Anniversary of the Taniguchi Technology Promotion Foundation."

Now I would like to introduce two mathematicians who were friends of Mr. Taniguchi: Kiyoshi Oka and Yasuo Akizuki. Certainly you know the name Oka from his original work on the theory of functions of several complex variables. Akizuki was an algebraist and, while he was a professor at Kyoto University from 1947 to 1960, he created there a large school of algebraic geometry; Hironaka is one of his students. Mr. Taniguchi and these two mathematicians, Oka and Akizuki, were classmates at the "Sanko", high school in Kyoto. Their friendship continued through their entire lives.

In 1956, supported by reports from Y. Matsushima and K. Nomizu who had just returned from studying abroad, Prof. Akizuki approached Mr. Taniguchi and informed him that the mathematical community in Japan was so short of funds that mathematicians could not afford to organize conferences or meetings to exchange their ideas. Mr. Taniguchi promptly instituted a Mathematics Division in the Taniguchi Foundation, headed by Prof. Akizuki. Since then, the Foundation exclusively supported seminars organized by this Mathematics Division.

The first seminar was held at a ski lodge in Akakura, a small village in the central region of Japan, in the summer of 1956. The seminar was titled "Manifold theory from a Differential Geometric Viewpoint," and was organized by Matsushima and Nomizu. The eleven participants, which fortunately included myself, came from various universities in Japan and were all about thirty years old.

Mr. Taniguchi later told me about the following episode. The ski lodge where the seminar was held is for the employees of Toyobo Co., and Mr. Taniguchi was the vice-president of this company at that time. Now the janitor of the ski lodge, an old man, reported to Mr. Taniguchi about the seminar. He said, "I was very shocked by the behaviour of mathematicians. I couldn't understand what they were doing at all, but they enthusiastically discussed mathematics every day until midnight." Mr. Taniguchi was very impressed by this report, recognizing that what could be done by a small amount of money for him was very useful for mathematicians. Mr. Taniguchi then became a big patron of the mathematical community in Japan.

Until 1972, fifty-two domestic Taniguchi seminars and symposia were held, favoring about seven hundred and fifty mathematicians.

In 1976, Mr. Taniguchi contributed a huge amount of his personal funds to the extant Taniguchi Foundation, and initiated the new Foundation. At this time, Mr. Taniguchi promoted and sponsored small-scale international symposia in various fields of science that are outside of the limelight, ones that might have difficulties in finding financial support from other sources. While his main desire had been to raise the level of scientific thought and research in Japan, at the same time he dreamed of providing a forum where promising young scholars the world over could gather informally to exchange their thoughts, thereby contributing to mutual understanding, peace, and the future development of mankind.

The new Foundation now has eighteen divisions. The titles of these divisions are: Mathematics I and II, Business History, Biophysics, Medical History, Brain Sciences, Ethnology, Neurobiology in Vision, The Theory of Condensed Matter, Philosophy, Art History, Civilization Studies, Catalysis, Life Sciences, Religious Philosophy, Molecular and Cellular Biology, Polymer Chemistry, and Developmental Biology. In every division, an international symposium is held once a year, in accordance with Mr. Taniguchi's original concept.

The Mathematics Divisions were headed by Prof. Akizuki until his death in 1984; since then they have been coordinated by Prof. Kiyoshi Ito, a well-known probabilist, and myself. In these Divisions thirty-six international symposia have been organized, making this symposium the thirty-seventh. Nearly three hundred foreign mathematicians have been invited to these Taniguchi Symposia.

The Taniguchi Foundation will cease to exist by the end of this century, an event that has been scheduled since its beginning. Mr. Taniguchi declared, "Usually a foundation is managed with the interest of the fundamental funds, but I would spend even the fundamental funds for the Foundation to continue its real activity. When all the money has been used up, the Foundation shall cease to exist."

3. Epilogue

Before concluding my talk, I would like to emphasize the fact that the present
Taniguchi Foundation had its origin in the friendship that grew between Mr. Taniguchi
and Prof. Akizuki during their school days. Mr. Taniguchi told me, "I thought of this
style of symposium borrowing the ideas of Akizuki, who organized the Akakura seminar as a small-sale gathering of young mathematicians. If I was able to contribute to
the research of scholars with these symposia, thanks should be addressed to Akizuki."

In 1988 the French Government presented Mr. Taniguchi with the decoration "Of-
ficier dans l'Ordre des Palmes Académiques." Mr. Taniguchi then promptly asked me
to create an exchange program for Japanese and French mathematicians in memory
of the late Prof. Akizuki, and supplied a great amount of capital for this. Thus "The
Akizuki Exchange Program between Japan and France" materialized, enabling five
Japanese and five French mathematicians to cooperate.

Mr. Taniguchi was a man of rare character. Every time a Taniguchi Symposium was
organized, Mr. Taniguchi invited all participants to dinner at a hotel, and explained
his ideas in his welcome speech. The participants were then immediately inspired by
Mr. Taniguchi's thoughts and his fine personality.

The Mathematical Society of Japan awarded the late Mr. Taniguchi the first Kouwa
Seki Prize in the spring of 1995. This prize was newly created in memory of Seki
Kouwa, an eminent 17th-century Japanese mathematician, and recognizes those who
make extraordinary contributions to the Japanese mathematical community.

Finally, I would like to note that Mr. Taniguchi contributed not only to science
but also to art: Born to a wealthy family, he owned many precious Japanese classical
paintings. He donated these to public museums and temples. You can appreciate
some of them in the Taniguchi Collection at the Freer Gallery in the Smithsonian
Institution, Washington, D. C. as well as the Taniguchi Collection in the Kyoto National Museum.

Added on September 1995. It is with deep regret that I mention here the passing
of Mr. Yuichiro Taniguchi, son of Mr. Toyosaburo Taniguchi and successor to the
Foundation, on September 2, 1995 at the age of 65.

MATHEMATICS DIVISION II, TANIGUCHI FOUNDATION, TAKADAI 2-8-11, NAGAOKAKYO, KY-
OTO 617 JAPAN

Proceedings of
THE 37TH TANIGUCHI SYMPOSIUM ON
TOPOLOGY AND TEICHMÜLLER SPACES
held in Finland, July 1995
ed. by Sadayoshi KOJIMA *et al.*
©1996 World Scientific Publishing Co.
pp. 5–30

COMPUTING ON RIEMANN SURFACES

PETER BUSER AND MIKA SEPPÄLÄ

(Received April 1, 1996)

ABSTRACT. These notes are a review on computational methods that allow us to use computers as a tool in the research of Riemann surfaces, algebraic curves and Jacobian varieties.

It is well known that compact Riemann surfaces, projective algebraic curves and Jacobian varieties are only different views to the same object, i.e., these categories are equivalent. We want to be able to put our hands on this equivalence of categories. If a Riemann surface is given, we want to compute an equation representing it as a plane algebraic curve, and we want to compute a period matrix for it.

Vice versa, we want to be able to compute the uniformization for a given algebraic plane curve, or a Riemann surface corresponding to a given Jacobian variety.

In another direction we consider tools that allow us to compute eigenvalues and eigenfunctions of the Laplace operator for Riemann surfaces. The correspondence between the Laplace spectrum of a Riemann surface and the geometry of the surface in general is intriguing. The programs to be described later give us a possibility to explore this correspondence in an explicit manner.

The above mentioned computational problems are hard and most of them are open in the general case. In certain particular cases, like that of hyperelliptic algebraic curves, interesting results are known ([8], [10],[11], [12], [14], [15], [16]). We will review some of these results and consider implementations of programs needed to make practical use of these results. These implementations make use of a larger program, CARS ([5]), currently under development, of which we will describe some features in this paper. CARS stands for *Computer Algebra Riemann Surfaces* and offers a convenient way of defining Möbius transformations and Riemann surfaces for computations.

For the reader's convenience, we also review some of the basic underlying mathematical concepts. Our basic references to the theory of Riemann surfaces are [4], [7], [9] and [13].

The second author wishes to thank the Taniguchi foundation for generous support. This work has also been supported by the European Communities HCM–network "Computational Conformal Geometry", contract nr ERBCHRXCT930408, by the Academy of Finland and by the Swiss National Science Foundation Grant 20-34099.92.

1. Preliminaries

In this section we describe a method for defining Möbius transformations for computations by a computer. We are interested, in particular, in Möbius transformations that map either the upper half–plane or the unit disk onto itself. Such a transformation can be expressed in the form

$$(1) \qquad z \mapsto \frac{az+b}{cz+d}, \quad ad - bc = 1.$$

Transformation (1) maps the upper–half plane U onto itself if and only if all the coefficients a, b, c, and d are real, and it maps the unit disk D onto itself if and only if $c = \bar{b}$ and $d = \bar{a}$, as is well known.

For computational purposes, we should define a Möbius transformation simply by the formula (1), i.e., by giving either exact values (rational or algebraic numbers) or floating point approximations for the coefficients a, b, c, and d. In the applications, that we have in mind, almost always one ends up using simply floating point approximations of the coefficients.

Formula (1) is, of course, sufficient for computing further with the Möbius transformation in question. However, formula (1) is not the geometrically natural way for defining a Möbius transformation.

Geometrically, a hyperbolic transformation

$$(2) \qquad g(z) = \frac{az+b}{cz+d}, \quad ad - bc = 1,$$

is determined by the following three parameters:

- the *attracting* fixed point $a(g) = \lim_{n\to\infty} g^n(z)$
- the *repelling* fixed point $r(g) = \lim_{n\to\infty} g^{-n}(z)$
- the *multiplier*

$$(3) \qquad k(g) = (g(z), z, r(g), a(g)) = \frac{g(z) - r(g)}{g(z) - a(g)} \cdot \frac{z - a(g)}{z - r(g)},$$

where z is any point in $\hat{\mathbb{C}}$ not fixed by g. Especially, $k(g) > 1$. The fixed points are real if $g(U) = U$. The non–Euclidean line through $a(g)$ and $r(g)$, the *axis* of g, is denoted by $ax(g)$. It has natural orientation by $r(g) \to a(g)$.

Any hyperbolic Möbius transformation g having $k = k(g)$, $k > 1$, as the multiplier and $a = a(g)$ and $r = r(g)$ as the attracting and the repelling fixed–points can be written in the form

$$(4) \qquad g(z) = \frac{(ka - r)z - ar(k - 1)}{(k - 1)z + a - kr}.$$

The fixed–points a and r of a hyperbolic Möbius transformation determine and are determined by an oriented hyperbolic line, namely the axis. These are the first two geometric parameters that we will use when defining hyperbolic Möbius transformations for computations.

We still need to specify the multiplier k. This parameter tells us how fast the transformation moves points towards the attracting fixed–point.

A concrete geometric way to define the multiplier is offered by the *isometric circle*. The isometric circle is a circular arc which is perpendicular to ∂D and to the axis of g. Provided that ∞ is not a fixed-point of g, the isometric circle is defined by the condition that the mapping g, when restricted to the isometric circle, is an isometry of the Euclidean metric.

The point where the isometric circle intersects the axis of g is the geometric object that we are looking for.

When defining a hyperbolic Möbius transformation for computations by a computer, the point, where the isometric circle intersects the axis, can be chosen by the mouse. The point will be an interior point of D and a point of the axis of g. The point has to be closer to the repelling fixed-point than to the attracting fixed-point.

Some technical computations are involved here. Using the representation (4), the isometric circle I_g of g can be written as

$$I_g = \{z \in D \mid |(k-1)z + a - kr| = |(ka - r)(a - kr) + ar(k-1)^2|^{\frac{1}{2}}\}.$$

At this point we assume that the parameters a and r are known. By the mouse we choose the point z. The multiplier k is then that solution of the equation

$$(5) \qquad |(k-1)z + a - kr| = |(ka - r)(a - kr) + ar(k-1)^2|^{\frac{1}{2}}$$

which is real and > 1.

Equation (5) leads to awkward computations. In the case where g acts in the unit disk we may assume that $a = -\bar{r}$ since this situation can always be achieved by a conjugation with an elliptic rotation fixing the origin. This rotation maps the isometric circle of the original mapping to that of the conjugated mapping since the rotation is also an isometry of the Euclidean metric. Observe that a general conjugation with a Möbius transformation does not have this property.

Then, if z is the intersection point of the isometric circle and the axis of g, then $g(z) = -\bar{z}$ by symmetry (since $a = -\bar{r}$). This gives us a simple equation from which we can compute k in terms of a, r and z, the intersection point of the isometric circle and the axis of g. Hence the Möbius transformation g becomes uniquely defined by the geometric parameters a, r and z.

Figure 1 illustrates the computer screen that shows the above geometric parameters of a hyperbolic Möbius transformation. This figure is from the display of the program CARS which allows one to define Möbius transformations conveniently and to modify them by mouse actions.

The program CARS allows one also:

- to write words of Möbius transformations,
- to display their axes and isometric circles and
- to see how these geometric parameters of a word of Möbius transformations change as we deform any of the primitive transformations that define this word.

This allows one to study, for instance, the question whether the axis of a certain word of Möbius transformations intersect the axis of some other word of Möbius transformations.

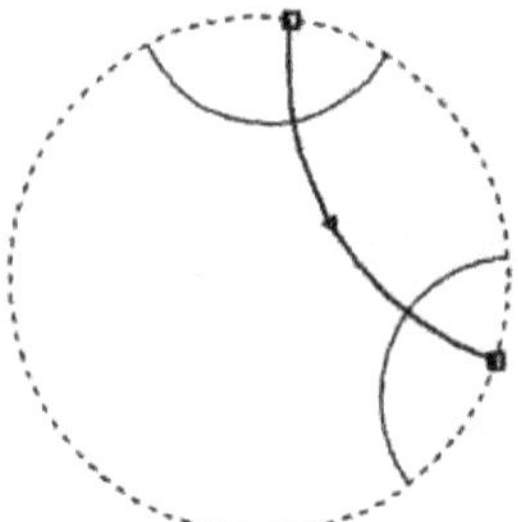

FIGURE 1.   A hyperbolic Möbius transformation as seen by a computer. This picture is captured from the display of the program CARS ([5]). The picture shows: (1) the axis of a hyperbolic Möbius transformation, (2) the orientation of the axis, i.e., giving the attracting and repelling fixed points, and (3) the isometric circle and its image under the mapping in question. All the geometric parameters – fixed-points and the isometric circle – can be modified by convenient mouse actions. The interface CARS then outputs, for computations with other programs, the parameters of the Möbius transformation. From this figure one can also immediately see the action of the defined Möbius transformation in the unit disk. The transformation maps the outside of the isometric circle onto the inside of the image of the isometric circle.

2. Riemann surfaces

We consider Riemann surfaces

$$(6) \qquad\qquad X = D/G$$

where G is a Fuchsian group acting in D. Alternatively we may choose to work in the upper half–plane. We use the representation (6) to define Riemann surfaces for computations.

Hence defining a Riemann surface, on which we compute, reduces to giving a set of Möbius transformations generating the group G. From the computational point of view, these sets of Möbius transformations form simply lists of lists, specifying either the coefficients exactly (rational or algebraic) or giving floating point approximations of the coefficients of the groups in question.

In most of the interesting cases, the group G is not freely generated. Hence, in addition to the floating point approximations of the generators, one should also give the relations satisfied by the generators.

It is an open problem in general to determine when a set of Möbius transformations generates a discontinuous group. In order to ensure that one gets a Fuchsian group, one could proceed as follows. Define first a hyperbolic polygon P in D and then consider Möbius transformations defining a suitable side pairing for P. In such a way one can easily build generators for Fuchsian groups G such that the initial polygon P is a fundamental domain for G and that the Riemann surface D/G is of any desired topological type.

We have, however, chosen an alternative way which builds on the combination theorems of Maskit and others. To that end we start with considering certain simple cases first.

2.1. Elementary cases

Let $G = \langle g, h \rangle$ be a Fuchsian group freely generated by two hyperbolic Möbius transformations whose isometric circles *do not* intersect[1]. We have two possible configurations:

(1) the axis of g and h do not intersect (Fig. (2))
(2) the axis of g and h do intersect (Fig. (3)).

This geometric configuration of the axis of g and that of h is reflected in an interesting way in the formulae regarding the product $g \circ h$. To see this we consider pairs of hyperbolic Möbius transformations in general.

For technical convenience we consider Möbius transformations acting in the upper half plane instead of the unit disk. By a suitable conjugation one can transform all the considerations to the unit disk as well.

Let (g, h) be a pair of hyperbolic transformations fixing the upper half-plane U. Suppose that g and h have no common fixed points and denote

- $t = (r(g), r(h), a(h), a(g))$,

---

[1] This is a mere technical condition which guarantees the discreteness of the corresponding group. Oberserve that this is neither a necessary condition for the discretness nor invariant under conjugation by a Möbius transformation.

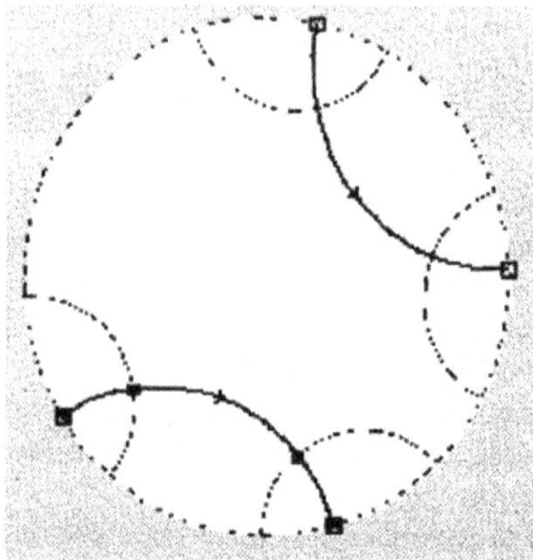

FIGURE 2.   Möbius transformations g and h as in this figure generate
a Fuchsian group corresponding to a Y–piece, i.e., to a pair of pants.

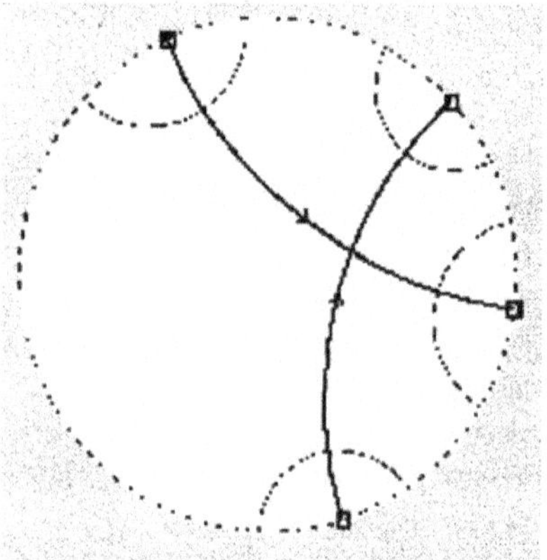

FIGURE 3.   Möbius transformations g and h as in this figure generate
a group corresponding to a Q–piece, i.e., to a handle.

- $k_1 = k(g)$,
- $k_2 = k(h)$,
- $k_3 = k(g \circ h)$.

In order to derive an expression for k_3 in terms of t, k_1 and k_2 we normalize by conjugation such that $r(h) = 1$, $a(h) = 0$ and $a(g) = \infty$. Then $t = r(g)$ and we have by formula (4)

$$g(z) = k_1 z - t(k_1 - 1),$$
$$h(z) = \frac{z}{(1 - k_2)z + k_2}.$$

It follows that

(7) $$f(k_3) = f(g \circ h) = |tf(k_1 k_2) + (1 - t)f(k_1/k_2)|.$$

Here we use the notation

$$f(k) = \sqrt{k + \frac{1}{k} + 2} = \sqrt{k} + \frac{1}{\sqrt{k}}.$$

For a Möbius transformation g, we also write $f(g) = f(k(g))$, where $k(g)$ is the multiplier of g. Observe that if

$$g(z) = \frac{az + b}{cz + d}, \quad ad - bc = 1,$$

then $f^2(g) = (a + d)^2$, i.e., $f(g)$ is the absolute value of the trace of g.

We divide pairs (g, h) of hyperbolic Möbius transformations into three disjoint classes $\mathcal{H}$, $\mathcal{P}$ and $\mathcal{E}$ (cf. formula (7)):

$$
\begin{aligned}
(g, h) \in \mathcal{H} &\iff f(g \circ h) = tf(k_1 k_2) + (1 - t)f(k_1/k_2) \geq 2 \\
&\iff t \geq t_1 = \frac{2 - f(k_1/k_2)}{f(k_1 k_2) - f(k_1/k_2)}, \\
(g, h) \in \mathcal{P} &\iff f(g \circ h) = -tf(k_1 k_2) - (1 - t)f(k_1/k_2) \geq 2 \\
&\iff t \leq t_2 = \frac{-2 - f(k_1/k_2)}{f(k_1 k_2) - f(k_1/k_2)}, \\
(g, h) \in \mathcal{E} &\iff t_2 < t < t_1.
\end{aligned}
$$

Here $\mathcal{H}$ stands for "handle", $\mathcal{P}$ for "pants" and $\mathcal{E}$ for "elliptic". The Möbius transformation $g \circ h$ is either hyperbolic, parabolic or elliptic. Note that $g \circ h$ is elliptic if and only if $(g, h) \in \mathcal{E}$.

All pairs of Möbius transformations corresponding to the setting of Fig. (2) belong to the class $\mathcal{P}$ and all pairs of Fig. (3) belong to the class $\mathcal{H}$. For a detailed proof of this result and a general treatment of the geometry of pairs of Möbius transformations see [13, Section 1.5].

We have the following result ([13, Lemma 1.6.1, p. 40]):

Theorem 1. *The conjugacy class of a pair (g, h) of hyperbolic Möbius transformations belonging either to the class $\mathcal{P}$ or to the class $\mathcal{H}$ is uniquely determined by the multipliers of g, h and $g \circ h$.*

Theorem 1 has a natural geometrical meaning. Consider a pair (g, h) belonging to the class $\mathcal{P}$. The corresponding Fuchsian group $G = \langle g, h \rangle$ is of the second kind and easy considerations show that D/G is a Y–piece, i.e., a pair of pants. The hyperbolic metric of the unit disk defines a complete hyperbolic metric on D/G. The lengths of the geodesic curves freely homotopic to the three boundary curves at infinity are $\log k(g)$, $\log k(g)$ and $\log k(g \circ h)$ (provided that the axis of $g \circ h$ projects onto the third geodesic curve in question, if not, then take $g \circ h^{-1}$ instead). Hence the hyperbolic metric of a Y–piece is determined by the lengths of these three geodesics.

The same applies to the situation where $G = \langle g, h \rangle$ defines a Q–piece, i.e., a handle. Also in this case the metric is defined by the lengths of the geodesic curves corresponding to the axes of g, h and to that of the product of g and h.

These cases are not completly analogous. It can be shown that the three parameters defining a Y–piece can be freely chosen, i.e., that the lengths of the boundary geodesics can be freely given. In the case of a Q–piece, the respective parameters correspond also to the lengths of geodesic curves. But now two of these curves intersect and their lengths must satisfy certain inequalities (see [13, Section 3.13]).

2.2. Building Riemann surfaces out of Y–pieces and of Q–pieces

Cutting a Y– or a Q–piece along the geodesics homotopic to the boundaries at infinity and then taking away the unbounded components of the complement (the *funnels*), we obtain a compact surface of the same signature having geodesic boundaries.

Let now Y and Y' be two such Y– or Q–pieces with geodesic boundaries and assume that the boundary curve α of Y has the same length as the boundary curve α' of Y'. In this case we may identify α with α' using constant speed. The identification depends on one parameter, called the *gluing angle* (see Definition 1 below).

Apart from the gluing angle, the above identification defines a unique surface X. The hyperbolic metric of Y and that of Y' define a smooth hyperbolic metric of X hence making X into a Riemann surface.

To make this unique and our argument general, we consider, for $p \geq 2$, a collection of $p - 2$ Y–pieces, $Y_1, Y_2, \ldots, Y_{p-2}$ and a collection of p Q–pieces $Q_1, Q_2, \ldots, Q_p$.

We assume that the list of Y–pieces is ordered and that the boundary geodesics of each Y–piece, Y_j, $\alpha_1^j, \alpha_2^j, \alpha_3^j$, form also an ordered list. Assume furthermore that the lengths of the boundary geodesics of the Y–pieces satisfy

$$(8) \qquad \qquad \ell(\alpha_3^j) = \ell(\alpha_1^{j+1})$$

Then we can first form a sphere S with p holes by identifying the boundary geodesic α_3^j with α_1^{j+1} for $j = 1, \ldots, p - 3$.

Next assume that the list (g_j, h_j) of generators of each of the Q–pieces Q_j is ordered and that the generators are Nielsen reduced.[2]

---

[2] Nielsen reduced generators are generators whose multipliers are as small as possible. There is an algorithm which finds Nielsen reduced generators for each Y–piece $D/\langle g, h \rangle$ starting from arbitrary generators g and h.

The axes g_l and h_l project onto intersecting geodesic curves on Q_l. Possibly replacing g_l or h_l by their inverses, we may always assume that the commutator

$$(9) \qquad c_l = g_l \circ h_l^{-1} \circ g_l^{-1} \circ h_l$$

corresponds to the single boundary geodesic γ_l of the Q–piece Q_l.

Let α_k^j be a boundary geodesic of Y. We can glue a Q–piece Q_l to α_k^j if and only if

$$(10) \qquad \ell(\alpha_k^j) = \ell(\gamma_l).$$

From the ordering of the Y–pieces Y_j and from the ordering of the boundary curves α_k^j of each Y_j, we get an ordering for the boundary curves of the p–hole sphere S. Call these boundary curves $\delta_1, \delta_2, \ldots, \delta_p$ and identify the boundary curve γ_l of Q_l with δ_l.

In order to parametrize these identifications we do the following:

(1) Use the notation $\alpha_j^4 = \alpha_j^1$. Drop perpendiculars from each α_j^n to α_j^{n+1}, $j = 1, \ldots, p$, $n = 1, 2, 3$. Call y_j^n the end point of the corresponding perpendicular ending at α_j^n.

(2) Drop perpendiculars from each geodesic corresponding to g_l onto the boundary geodesic γ_l of the Q–piece Q_l. Call x_l the end-point of the perpendicular on γ_l.

The points x_l and y_j are called *base points* of their respective boundary geodesics.

In the gluing process we identify pairs of boundary geodesics of Y–pieces and of Q-pieces. On the resulting genus g Riemann surfaces all these pairs of boundary geodesics correspond to a single geodesic curve. Call them $\delta_1, \ldots, \delta_{2g-3}$.

Next assume that the geodesic curves δ_j are all oriented and that the base point sets are also ordered so that we can speak about the first base point ξ_j^1 and of the second base point ξ_j^2 of δ_j.

Definition 1. *The* gluing angle θ_j *associated to the geodesic δ_j is*

$$(11) \qquad \theta_j = 2\pi \frac{\textit{distance from } \xi_j^1 \textit{ to } \xi_j^2 \textit{ along the positive direction of } \delta_j}{\textit{length of } \delta_j}$$

In view of Theorem 1 and the above definition of the gluing angles we have:

Theorem 2. *Let X be a genus p, $p > 1$, Riemann surface obtained by gluing $p - 2$ Y–pieces and p Q–pieces together in the above fashion. Let $\alpha_1, \beta_1, \ldots, \alpha_p, \beta_p$ be the geodesic curves corresponding to the axes of the generators g_j and h_j of the Fuchsian groups of the Q–pieces. Let ρ_j be the geodesic curve corresponding to the axis of the product $g_j \circ h_j$. Let $\delta_1, \ldots, \delta_{2p-3}$ be the curves along which gluing has been performed and let θ_k, $k = 1, 2, \ldots, 2p - 3$ be the gluing angles. Then the $7p - 6$ parameters*

$$(12) \qquad \ell(\alpha_j), \ell(\beta_j), \ell(\rho_j), \ell(\delta_k), \theta_k, j = 1, 2, \ldots, p, \ k = 1, 2, \ldots, 2p - 3$$

define the Riemann surface X uniquely up to an isometry.

Remark. The parameters $\ell(\rho_j)$ in the above theorem are almost superfluous in the sense that once the parameters $\ell(\alpha_j)$ and $\ell(\beta_j)$ are fixed, $\ell(\rho_j)$ can take only two different values.

Let X be a Riemann surface constructed in the above fashion and defined by the parameters of Theorem 2. Then one can *deform* X by cutting X open along any of the curves δ_j and letting one part of X slide and gluing back again. This sliding changes only the respective gluing angle θ_j and the sliding is the *Fenchel–Nielsen twist* of the Riemann surface X along the geodesic curve δ_j.

It is useful to know how the above construction and the Fenchel–Nielsen twist can be carried out on the universal cover.

To that end, consider two Y-pieces Y and Y' and the gluing of these Y-pieces together along a boundary geodesic. The general gluing and the corresponding Fenchel–Nielsen twist is performed in the same way as in this more elementary case.

Represent the Y-piece Y as $Y = U/\langle g, h \rangle$ and Y' as $Y' = U/\langle g', h' \rangle$. Assume that the fixed points of g and h both lie in the positive real axis and those of g' and h' in the negative. Assume furthermore, that the products $g \circ h$ and $g' \circ h'$ are both hyperbolic Möbius transformations fixing 0 and ∞, i.e., that they have the same axis which is the positive imaginary axis. Then $g \circ h = g' \circ h'$ and it turns out[3] that the group $\langle g, h, g', h' \rangle$ uniformizes the Riemann surface X. The Fenchel–Nielsen twist corresponds to conjugating the pair (g', h') by a hyperbolic Möbius transformation fixing 0 and ∞ and leaving the other pair (g, h) intact.

This construction can be generalized to all situations where we glue two Riemann surfaces together along *one* boundary curve ([9]). This construction has also been implemented in the program CARS , which allows one to build any Riemann surface out of Y-pieces and Q-pieces. The only restriction is that, at any step, we glue two Riemann surfaces together along *one* boundary curve only. This means, in particular, that we cannot realize, by the program CARS, all possible decompositions of compact Riemann surfaces into pairs of pants, i.e., into Y-pieces.

CARS allows one to perform Fenchel–Nielsen twists along any of the geodesic curves along which gluing has been performed. CARS allows also one to make these curves shorter or longer in such a generality that one can, in this way, construct any Riemann surface of any given finite type.

3. Equations for Riemann surfaces

In sections 1 and 2 we have seen how one can conveniently define Möbius transformations and Riemann surfaces for computations. The interface CARS, described in those sections, is an input device that can be used to input Riemann surfaces for computations with other programs, most importantly with general purpose mathematics programs such as AXIOM, MAPLE or MATHEMATICA. The communication between CARS and other programs will later be provided by a communication protocol known as OPEN MATH.

Anyway, now that we have defined Riemann surfaces we want to compute with them. An important problem is to compute a polynomial equation representing a given compact Riemann surface as a projective algebraic plane curve. It is classical

---

[3]This statement actually includes two theorems. First of all one has to show that the group generated by g, h, g' and h' is discontinuous and secondly that it uniformizes X. For a detailed treatment of these topics see Combination Theorems in [9].

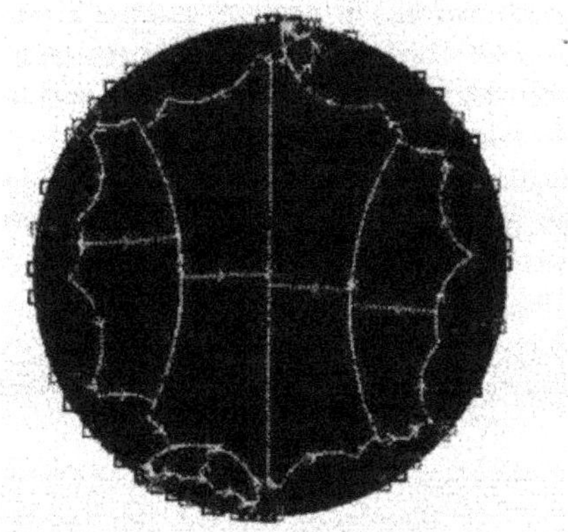

FIGURE 4. A genus 4 Riemann surface as seen by the program CARS . The bright hyperbolic geodesic arcs determine polygons that are fundamental domains for the corresponding Y-pieces or Q–pieces. Each such fundamental domain has also a geodesic arc perpendicular to two of the sides of the polygon. The purpose of these geodesic arcs is simply to fix two base–points on each of the geodesics along which gluing happens. The directed distance of these base–points determines the gluing angle. The program CARS allows one to perform the Fenchel–Nielsen twists by grabbing one of the base–points by mouse and then moving it.

that compact Riemann surfaces are simply projective algebraic curves. To find a polynomial defining a plane algebraic curve that corresponds to a given compact genus p, $p > 1$, Riemann surface X we have to solve two problems:

(1) approximate a *good* mapping of X into $\mathbf{P}^2(C)$ and compute points of the image of X in $\mathbf{P}^2(C)$

(2) solve a system of linear equations to find a polynomial of degree at most $4p - 4$ passing through the points computed above

The above simple plan involves a difficulty: The approximation of a mapping of X into $\mathbf{P}^2(C)$ involves an approximation of a bicanonical mapping of X into a higher dimensional projective space and then projecting down from there. The approximation of the bicanonical mapping in question is done by certain suitable Poincaré series. These series, however, converge slowly and therefore pose a serious numerical problem. We have to be able to distinguish those elements from the Poincaré series that contribute most to the value of the series.

To be more precise, assume that G is a Fuchsian group of the first kind acting in the unit disk D in such a way that D/G is a compact Riemann surface of genus p. Let $g \in G$ be a (primitive in G) Möbius transformation covering the homotopy class of a simple closed geodesic curve $\gamma \in D/G$. Assume that

$$(13) \qquad g(z) = \frac{az + b}{cz + d}, \quad ad - bc = 1.$$

Let

$$(14) \qquad \nu(g) = (\mathrm{tr}^2 g - 4)^{1/2} = \sqrt{k_g} - \frac{1}{\sqrt{k_g}} = ((a + d)^2 - 4)^{1/2}.$$

Here k_g is the multiplier of the hyperbolic Möbius transformation g and $\mathrm{tr}^2 g$ is the square of the trace of a matrix, of determinant $+1$, corresponding to the Möbius transformation g. Observe that such a matrix is well–defined only up to sign. The *square* of the trace is, however, always well–defined.

Using this notation define

$$(15) \qquad \omega_g = \nu(g)(cz^2 + (d - a)z - b)^{-2}.$$

The importance of the formula (15) lies in the fact that

$$(16) \qquad (\omega_g \circ g)\left(\frac{\partial g}{\partial z}\right)^2 = \omega_g$$

as can be verified by a direct computation. Therefore ω_g is a holomorphic 2–form of the cyclic group $\langle g \rangle$. This motivates one to look at the *Petersson–Poincaré series*

$$(17) \qquad \theta_\gamma = \sum_{h \in \langle g \rangle \backslash G} (\omega_g \circ h)\left(\frac{\partial h}{\partial z}\right)^2$$

associated to the simple closed geodesic curve γ.

Wolpert has shown the following result [17, Theorem 3.7, page 521]:

Theorem 3. *Let $\alpha_1, \ldots, \alpha_{3p-3}$ be a decomposition of the Riemann surface D/G into Y-pieces (namely, a maximal set of non-intersecting simple closed curves). Then the Petersson–Poincaré series θ_{α_j} converge, their limits are holomorphic 2-forms of the group G, and $\{\theta_{\alpha_j}\}$ is a basis (over $\mathbb{C}$) for the space of holomorphic 2-forms of the group G.*

This result allows us to approximate the values of quadratic differentials of a compact Riemann surface at as many points as one wants. In particular, this means that we can approximate numerically the corresponding bicanonical mapping

$$(18) \qquad \phi_{2\kappa} : X \to \mathbb{P}^{3g-4}(\mathbb{C}), \ X \ni p \mapsto (\theta_{\alpha_1}(z), \ldots, \theta_{\alpha_{3g-3}}(z)) \in \mathbb{P}^{3g-4}(\mathbb{C}),$$

where $z \in D$ is any point projecting onto $p \in X = D/G$.

This numerical approximation involves computing a finite part of the series (17). For numerical accuracy it is important that the elements to be computed of the series (17) are the ones which contribute most to the sum.

Loris Renggli and Klaus–Dieter Semmler have provided a solution to this problem. Let G_F be a finite part of G. Let us consider that part of the series (17) which corresponds to G_F. Let $A \subset D$ be a fundamental domain for the group G.

Let

$$(19) \qquad \theta_\gamma^F = \sum_{h \in \langle g \rangle \backslash G_F} (\omega_g \circ h)$$

be a finite approximation of θ_γ corresponding to the finite part G_F of G.

Considering the usual proof for the convergence of the Poincaré series, one can estimate the error $|\theta_\gamma(z) - \theta_\gamma^F(z)|$ in terms of the euclidean area of the set

$$(20) \qquad\qquad D \setminus (\cup_{g \in G_F} g(A)),$$

for *any* fundamental domain A of G in D. This leads to the observations that, when selecting elements of the finite part G_F, we should take elements g in the order defined by the Euclidean areas of the domains $g(A)$. This observation has been coded, by Loris Renggli and Klaus–Dieter Semmler, to a program producing good finite parts of a Fuchsian group G.

Further details regarding the error estimates and generalizations of this method will be contained in a forthcoming thesis of Pekka Smolander (University of Joensuu, Finland).

4. Numerical uniformization

The methods described in Section 3 allows one to approximate numerically a polynomial equation defining a given compact Riemann surface as a plane algebraic curve.

The converse problem is more difficult. Starting with a complex algebraic curve C, we would like to find a group G of Möbius transformations such that $C = \Omega(G)/G$, where $\Omega(G)$ is (a component of) the domain of discontinuity of G.

In full generality this is an open problem. We can, however, solve it in certain particular cases. In the following we review the considerations of P. J. Myrberg ([10])

18   PETER BUSER AND MIKA SEPPÄLÄ

as presented in [11], show a new uniqueness result (Lemma 8) and clarify the proof
of Lemma 10 (which contained an error in [11]).

4.1. Opening mapping

We need to consider auxiliary mappings, called *opening mappings*. They map the
exterior of an interval in the complex plane onto the exterior of a disk. The algorithm
is based on iteration of such mappings.

Let

$$(21) \qquad \tau_j \in \mathbb{C},\ j = 1, 2,$$

be two different points in the complex plane. Let I denote the closed interval having
τ_1 and τ_2 as end–points.

Let

$$(22) \qquad m = \frac{\tau_1 + \tau_2}{2} \text{ and } s = \frac{\tau_2 - \tau_1}{4}.$$

We assume that the indices are so chosen that the real part of s is non–negative.
In the applications considered here, the τ's will be real and we simply assume that
$\tau_2 > \tau_1$.

Consider the equation

$$(23) \qquad \frac{z - (m + 2s)}{z - (m - 2s)} = \left(\frac{w - (m + s)}{w - (m - s)} \right)^2.$$

A suitable choice of the square–root gives us, by equation (23), a mapping

$$(24) \qquad \phi : \mathbb{C} \setminus I \to \mathbb{C} \setminus \overline{D},$$

where D is a disk centered at m with radius $|s|$.

This mapping ϕ plays a central role in our construction and it is, therefore, necessary to take a closer look at it.

We observe that the following lemmata hold:

Lemma 4. *The mapping ϕ determined by equation (23) is given by*

$$(25) \qquad \phi(x) = x + \frac{x - m}{2} \left(\sqrt{1 - 4 \frac{s^2}{(x - m)^2}} - 1 \right),$$

for $|x - m| \geq 2|s|$,
or

$$(26) \qquad \phi(x) = x - s \left(t + t^3 + t^5 + O\left(t^7\right) \right),$$

where $t = s/(x - m)$ and $|x - m| \geq 2|s|$.

Lemma 5. *Assume that the τ's are real and $\tau_2 > \tau_1$. Then m is also real and
$s > 0$. In this case the mapping ϕ moves points of the real axis outside the interval
$[m - 2s, m + 2s]$ towards the mid-point m. For $|x - m| \geq 2|s|$, let*

$$(27) \qquad \delta(x) = |x - \phi(x)|$$

or

$$\delta(x) = |x - m| \left(t^2 + t^4 + 2t^6 + O(t^8) \right), \tag{28}$$

where $t = s/(x - m)$.

Then, for $x \geq m + 2s$, $\delta(x)$ is strictly decreasing, and, for $x \leq m - 2s$, $\delta(x)$ is strictly increasing. The function $\delta(x)$ attains its maximum for $x = m + 2s$ and for $x = m - 2s$. The maximum value is

$$\delta(m \pm 2s) = s. \tag{29}$$

Furthermore,

$$\lim_{x \to \pm \infty} \delta(x) = 0. \tag{30}$$

Both of the above lemmata are simple straightforward technical computations and proofs are left to the reader. Observe that the property (30), which is important for our construction, follows from the fact that on the right hand side of the defining equation (23), $2s$ was replaced by s only.

In the above, the mapping ϕ was determined by equation (23) and by requiring that ϕ maps the complement of the interval I onto the complement of the disk D with m as center and $|s|$ as radius.

The other choice of the sign of the square root (when solving w in terms of z from equation (23)) yields a mapping $\tilde{\phi} : \mathbb{C} \setminus I \to D$. Let

$$e(z) = \frac{mz - m^2 + s^2}{z - m} \tag{31}$$

be the elliptic rotation, with angle π, of the complex plane keeping the points $m - s$ and $m + s$ fixed. Then e maps the complement of D onto D and we have

$$\tilde{\phi} = e \circ \phi. \tag{32}$$

In the subsequent considerations we use this mapping ϕ with varying parameters τ_j to perform the desired iteration.

4.2. Opening up the slots

What follows will later be applied to real hyperelliptic M–curves C of genus p defined by the equation

$$y^2 = (x - \lambda_1)(x - \lambda_2) \ldots (x - \lambda_{2p+2}), \tag{33}$$

where all the parameters λ_i are real and no two of them agree. For the considerations of this section, however, the curve C itself is irrelevant.

Here we simply consider, for a positive integer p, a set $\{\lambda_1, \lambda_2, \ldots, \lambda_{2p+2}\}$ of $2p + 2$ distinct *real* numbers. We assume that the points λ_i are numbered in such a way that

$$\lambda_1 < \lambda_2 < \cdots < \lambda_{2p+2}.$$

Then the closed intervals $I_j = [\lambda_{2j-1}, \lambda_{2j}]$, $j = 1, \ldots, p+1$, are disjoint. We call these intervals *slots* or, more precisely, *first generation slots*.

We start now describing our iterative method.

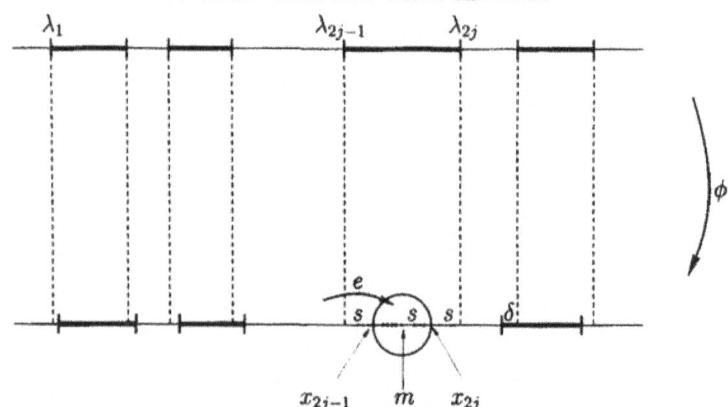

FIGURE 5.   The mapping ϕ opens a slot of length $4s$ and replaces it by a disk D of radius s. The higher generation slots are obtained as images of the other slots under an elliptic Möbius transformation e mapping the complement of the closure of the disk D onto D itself. Also the other slots move towards the mid-point m of the slot that is being opened. The function δ measures this movement. The points x_{2j-1} and x_{2j} are the intersection points of ∂D and the real axis.

Let ϕ_1 be the mapping ϕ defined by the equation (23) with $\tau_1 = \lambda_1$ and $\tau_2 = \lambda_2$. The mapping ϕ_1 *opens up* the first slot I_1 replacing it by a disk D_1 and maps all other slots onto new slots $\phi_1(I_j)$.

The corresponding elliptic rotation e_1, defined by (31), maps the complement of $\overline{D}_1$ onto D_1. In particular, e_1 maps the deformed slots $\phi_1(I_2), \phi_1(I_3), \ldots$ onto new slots $e_1(\phi_1(I_j)), j = 2, 3, \ldots$ in D_1. These are the first 2nd*-generation slots*. Let us denote them by I_j^2. The opening of a slot and the creation of 2nd generation slots is illustrated in figure 5.

Next, let ϕ_2 be the mapping ϕ defined by the equation (23) with $\tau_1 = \phi_1(\lambda_3)$ and $\tau_2 = \phi_1(\lambda_4)$. The mapping ϕ_2

- opens up the slot $\phi_1(I_2)$ and replaces it by a disk D_2
- maps the boundary of the disk D_1 onto a Jordan curve going through the points $\phi_2(\phi_1(\lambda_1))$ and $\phi_2(\phi_1(\lambda_2))$
- maps the deformed slots $\phi_1(I_j)$ onto new slots $\phi_2(\phi_1(I_j))$.

Let e_2 be the corresponding elliptic rotation, defined by (31) with

$$(34) \qquad m = \frac{\phi_1(\lambda_3) + \phi_1(\lambda_4)}{2} \text{ and } s = \frac{\phi_1(\lambda_4) - \phi_1(\lambda_3)}{4}.$$

The rotation e_2 maps the slots $\phi_2(\phi_1(I_3)), \phi_2(\phi_1(I_4)), \ldots$ onto new 2nd-generation slots $I_j^2, j = p + 1 \ldots 2p$ inside the disk D_2. The images $e_2(\phi_2(I_j^2))$ of the first 2nd-generation slots $I_j^2, j = 1, \ldots p$ inside D_2 are referred to as 3rd-generation slots I_j^3.

We repeat this procedure until all the first generation slots have been opened and new higher generation slots have been formed. For the moment our main interest lies in following what happens to the points x_{2j-1}^0 and x_{2j}^0 which are defined as the intersection points of the boundaries of the disks D_j, $j = 1, \ldots, p + 1$, and the real axis. Choose the indices in such a way that $x^0_{2j-1} < x^0_{2j}$ for all values of j.

For all indices $j = 1, \ldots, p + 1$, define the sequences x^k_{2j-1} and x^k_{2j} by setting

$$(35) \qquad x^1_{2j-1} = \phi_{j+1}(x^0_{2j-1}) \text{ and } x^1_{2j} = \phi_{j+1}(x^0_{2j})$$

and

$$(36) \qquad x^2_{2j-1} = \phi_{j+2}(x^1_{2j-1}) \text{ and } x^2_{2j} = \phi_{j+2}(x^1_{2j})$$

and so on.

Repeating the above construction for all the openings of all generations of slots we get $2p + 2$ infinite sequences x^k_i. The next two lemmata are taken directly from [11]. For the reader's convenience we reproduce also the proofs here.

Lemma 6. *For all values $i = 1, 2, \ldots, 2p+2$, the corresponding sequences x^k_i converge as $k \to \infty$.*

Proof. Assume that the slots are ordered according to their generation and, within each generation, from left to right. Let us fix i and consider the sequence x^k_i, which depends on the choice of the ordering of the slots. We have to show that the limits

$$(37) \qquad \lim_{k \to \infty} x^k_i = x^\infty_i$$

exist.

Let ϕ_m denote the opening of the slot I_m.

For a given sequence x^k_i where $i = 2j$ or $2j - 1$ observe that the convergence of the sequence itself is equivalent to the convergence of the series

$$(38) \qquad x^0_i + (\phi_{j+1}(x^0_i) - x^0_i) + (\phi_{j+2}(\phi_{j+1}(x^0_i)) - \phi_{j+1}(x^0_i)) + \ldots$$

This series simply keeps track on how much the initial point x^0_i moves as we continue opening up the slots. It is enough to show that the series (38) converges. That immediately implies the convergence of the sequence x^k_i. Furthermore, the limit point x^∞_i is simply the sum of the series (38).

To prove the convergence of the series (38) observe that it is some kind of alternating series in the sense that some of its terms are positive and some are negative. By taking only positive terms or only negative terms we get two series. It suffices to show that both of them converge.

Let $\Delta x^0_i = x^0_i$ and use the notation Δx^{k-1}_i to denote the k:th term of series (38), $k = 2, 3, \ldots$. Let $\Delta x^{k_n}_i$, $n = 1, 2, \ldots$ be the negative terms only.

All the initial slots I_j are contained in the closed interval $[\lambda_1, \lambda_{2p+2}]$. By Lemma 5 all openings of slots I_j move the exterior points towards the mid-point of the slot I_j. Therefore all the various generation slots will be contained in the same closed interval $[\lambda_1, \lambda_{2p+2}]$.

Taking the negative terms of the series (38) only we add up the movements of the initial point to left. Such movements occur only when opening up slots that are on the left hand side of the point in question. Since, by the construction, the point x^0_i and all its iterates are always in the interval $[\lambda_1, \lambda_{2p+2}]$, it follows that the negative terms of the series (38) must form a convergent series.

The same argument shows that the positive terms of the series (38) form also a convergent series. We deduce therefore that the original series (38) converges absolutely. $\square$

Lemma 7. *For $i \neq i'$, $x_i^\infty \neq x_{i'}^\infty$.*

Proof. Observe first that two limit points x_i^∞ and $x_{i'}^\infty$ certainly do not agree if the have been obtained from the end–points of *different* first generation slots. This follows from the properties of the function $\delta(x)$ (see Lemma 5).

Consider limit points x_i^∞ and $x_{i'}^\infty$ that have been obtained from the end–points of the same first generation slot I_j.

Let x_{2j-1} and x_{2j} be the intersection points of the real axis and the disk D_j obtained by opening up the slot I_j. Let y_1 and y_2 be the end–points of the 2nd generation slots in the interval $[x_{2j-1}, x_{2j}]$ closests to x_{2j-1} and to x_{2j}, respectively. The intervals (x_{2j-1}, y_1) and (y_2, x_{2j}) do not contain any 2nd–generation slots. Nor will their deformations contain any higher generation slots. Let

$$\epsilon = |x_{2j-1} - y_1| + |y_2 - x_{2j}|$$

be the combined Euclidean length of these intervals.

Using again Lemma 5 we infer that, when opening any other slots, the intervals (x_{2j-1}, y_1) and (y_2, x_{2j}) *will get longer*. The distance between x_{2j-1} and x_{2j} and any of their deformations will, therefore, always be $> \epsilon$. This implies that the distance between the limit points x_{2j-1}^∞ and x_{2j}^∞ is also $> \epsilon$. Therefore $x_i^\infty \neq x_{i'}^\infty$ whenever $i \neq i'$. $\square$

Let g be a Möbius transformation such that

$$(39) \qquad\qquad g(x_i^\infty) = i - 1 \text{ for } i = 1, 2, 3.$$

Replacing each of the x_i^∞ by the corresponding $g(x_i^\infty)$ we get the *normalized set* of limit points x_i^∞.

The sequences which give us the limit points x_i^∞ depend on the order in which the slots are opened. We have, however, the following result:

Lemma 8. *The normalized limit points x_i^∞ are independent on the ordering of the slots.*

The proof will be based on the considerations of the next section and will therefore be postponed to Subsection 4.3.1.

4.3. Hyperelliptic M–curves

Let $p \geq 2$ be an integer and consider an algebraic plane curve C defined by the affine equation

$$(40) \qquad\qquad y^2 = (x - \lambda_1)(x - \lambda_2) \ldots (x - \lambda_{2p+2})$$

where all the parameters λ_i are distinct, real and numbered in such a way that $\lambda_1 < \lambda_2 < \cdots < \lambda_{2p+2}$. Such an equation determines a real algebraic M–curve of genus p. We wish to use the considerations of the previous sections to find generators $g_1, g_2, \ldots$ for a discontinuous group G of Möbius transformations such that

$$(41) \qquad C = \Omega(G)/G,$$

where $\Omega(G)$ denotes the domain of discontinuity of G.

To that end it is useful to recall the usual construction for the curve C. It is the following:

- take first two copies of the Riemann sphere
- cut them both open along the slots $I_j = [\lambda_{2j-1}, \lambda_{2j}], j = 1, \ldots, p+1$.
- identify the boundary components of the first Riemann sphere, that was cut open, with the boundary components of the second one in such a way that you get a double cover of the original Riemann sphere ramified at the points λ_i.

The first important observation is that the complement Ω_0 of the union of the slots I_j in the Riemann sphere $\hat{\mathbb{C}}$, $\hat{\mathbb{C}} \setminus \cup_{j=1}^{p+1} I_j$ can be identified, conformally, with a region C_u of the original algebraic curve C. Let $\epsilon : C \to C$ denote the hyperelliptic involution. Then $C_u \cup \epsilon(C_u)$ is the complement of $p+1$ simple closed geodesic curves[4] of the algebraic curve C.

Consider the first opening of the slot I_1. It is a conformal mapping

$$(42) \qquad \phi_1 : \hat{\mathbb{C}} \setminus I_1 \to \hat{\mathbb{C}} \setminus \overline{D}_1.$$

Here we are using the notation of section 4.2. Via this mapping,

$$(43) \qquad U_1 = \hat{\mathbb{C}} \setminus \left(\overline{D}_1 \cup \left(\cup_{j=2}^{p+1} I_j \right) \right)$$

is conformally equivalent to $\mathbb{C}_u$. The elliptic rotation e_1 maps the outside of D_1 onto its inside. It follows that $e_1(U_1) \cup U_1$ is conformally equivalent to the complement of the above mentioned simple closed geodesic curves on the algebraic curve C. In other words, $e_1(U_1) \cup (U_1)$ is almost all of the Riemann surface in question. Furthermore, the mapping e_1 corresponds to the hyperelliptic involution.

Let now Ω_2 denote the complement of the union of the slots $I_2, I_3, \ldots, I_{p+1}$ and the 2nd-generation slots $e_1(I_j)$, $j > 1$. The above remarks imply that we have an analytic local homeomorphism $\pi_1 : \Omega_2 \to C$.

We continue this procedure and keep up opening slots in a certain given order, for instance, by generation and in each generation from left to right. Let Ω_n denote the complement of all the various generation slots after having opened n slots (which have disappeared since they have been replaced by disks).

Since each opening of slots is a conformal mapping, we have, at each level an analytic local homeomorphism $\pi_n : \Omega_n \to C$. Once we have opened all the 1st-generation slots, i.e., for $n > p+1$, the mapping π_n can already be extended to a surjective mapping. (The extension refers to the extension of the mapping π_n to the boundaries of the disks that have replaced the opened slots. This is straightforward.)

---

[4] When speaking of geodesic curves, we refer to the hyperbolic metric of the corresponding Riemann surface and to curves that are geodesics in that metric.

Define the set Ω^∞ by setting

$$(44) \qquad \hat{\mathbf{C}} \setminus \Omega^\infty = \{ z \in \hat{\mathbf{C}} | \exists z_n \in \hat{\mathbf{C}} \setminus \Omega^n \text{ such that } \lim_{n \to \infty} z_n = z \}.$$

Theorem 9. *The mappings π_n form a convergent sequence and the limit mapping π_∞ is an analytic regular covering mapping*

$$(45) \qquad \pi_\infty : \Omega^\infty \to C.$$

The proof of the theorem relies on Lemma 6 and will be contained in a paper, by Semmler and Seppälä, which is presently under preparation. The main point to be shown here is, that for any $z \notin \mathbf{R}$, the sequence $\phi_k(z)$ converges. This is simply Lemma 6 extended to the whole complex plane. The details are technically involved and will appear elsewhere later. P. J. Myrberg has another proof for this ([10]).

Consider the limits x_j^∞ of Lemma 6. Let e_j^∞ denote that elliptic Möbius transformation which keeps the point x_{2j-1}^∞ and x_{2j}^∞ fixed and rotates the infinite complex plane by angle π.

Let G be the group generated by all products $e_j^\infty e_{j'}^\infty$, $j, j' = 1, 2, \ldots, p+1$.

The following result follows immediately from the definitions.

Lemma 10. *The group G is the cover group of the analytic cover (45). Hence G is a Kleinian group whose domain of discontinuity is Ω^∞.* □

4.3.1. *Proof of Lemma 8.* Let x_i^∞ and y_i^∞, $i = 1, 2, 3, \ldots 2p+2$ be two sets of normalized limit points corresponding to the same initial data $\lambda_1, \ldots, \lambda_{2p+2}$. Let $\pi_\infty : \Omega^\infty \to C$ and $\pi'_\infty : \Omega'^\infty \to C$ be the corresponding coverings of the original curve C (given by equation (40)). Now the identity mapping of the curve C induces a Möbius transformation $h : \Omega^\infty \to \Omega'^\infty$ with the property

$$(46) \qquad h(x_i^\infty) = y_i^\infty \text{ for all } i = 1, 2, \ldots, 2p+2.$$

since the sets $\{x_i^\infty\}$ and $\{y_i^\infty\}$ are normalized, $x_i^\infty = y_i^\infty$ for $i = 1, 2, 3$. Therefore the Möbius transformation h has three fixed points implying that h *is* the identity. Hence $x_i^\infty = y_i^\infty$ for all values of $i = 1, 2, \ldots, 2p+2$. This argument proves Lemma 8.

5. Period matrices

The Torelli theorem states that a compact Riemann surface is determined by its Jacobian variety. Jacobian varieties are p–dimensional complex torii having certain special properties. This theorem provides a fasinating new point of view to Riemann surfaces and algebraic curves. It is, therefore, natural to try to produce the result of this theorem in a computable way, i.e., given a compact Riemann surface $X = \Omega(G)/G$, we would like to be able to compute the corresponding Jacobian variety. Such a variety is determined by a period matrix of the Riemann surface X. Hence the computation of the Jacobian reduces to the computation of periods of differentials on X.

5.1. Poincaré series

In this section we review certain considerations of [12].

Let $D = \{z \in C \mid |z| < 1\}$ be the unit disk and G a Fuchsian group acting in D. Let $Y = D/G$.

Assume that G *is of the second kind*. This means that the domain of discontinuity of G, $\Omega(G)$, is connected. Then $X = \Omega(G)/G$ is a symmetric Riemann surface, the symmetry $\sigma : X \to X$ being the mapping induced by the reflection in the unit circle ∂D. All so called orthosymmetric[5] Riemann surfaces can be expressed in this way.

Use the notation $D^1(X)$ for the complex vector space of holomorphic Abelian differentials on X. Holomorphic differentials ω on X lift to automorphic holomorphic functions $\omega : D \to \mathbb{C}$ satisfying:

$$(47) \qquad \omega(g(z)) \left(\frac{\partial g(z)}{\partial z} \right) = \omega(z)$$

for all $z \in D$ and for all $g \in G$. This is, of course, quite standard.

The limit set, $L(G)$, of a Fuchsian group G of the *second kind* is a subset of ∂D of measure 0. The Riemann surface $Y = D/G$ is an open Riemann Y surface which is the interior of a Riemann surface $Y \cup \partial Y$ *with boundary components*.

Let $f : D \to \hat{\mathbb{C}}$ be a meromorphic function. Let $m \geq 1$ be an integer. Consider *the Poincaré series*

$$(48) \qquad \Theta^m(f)(z) = \sum_{g \in G} f(g(z)) \left(\frac{\partial g(z)}{\partial z} \right)^m.$$

If this series converges, then its limit is a meromorphic function ω satisfying

$$(49) \qquad \omega(g(z)) \left(\frac{\partial g(z)}{\partial z} \right)^m = \omega(z).$$

Use the notation $\Theta(f)(z) = \Theta^1(f)(z)$.

Standard arguments imply the convergence of $\Theta^1(f)(z)$ in the case of Fuchsian groups of the second kind provided that the function f is not too bad ([6, Proposition 3, page 205]):

Lemma 11. *Let f be a rational function with no poles in $L(G)$. Then $\Theta(f)(z)$ is a meromorphic differential of the group G.*

This is an interesting result which allows us to approximate Abelian differentials of groups of the second kind numerically. The above result was contained already in a paper by W. Burnside ([2]) published in 1891.

Assume now that the group G is freely generated by the hyperbolic Möbius transformations $g_1, g_2, \ldots, g_p$ mapping the unit disk onto itself. For $j = 1, 2, \ldots, p$ let

$$(50) \qquad h_j(z) = \frac{\bar{\zeta}_j}{1 - \bar{\zeta}_j z}, \quad \zeta_j = g_j(0), \quad 1 \leq j \leq p.$$

---

[5]An orthosymmetric Riemann surface is a Riemann surface X equipped with an antiholomorphic involution $\sigma : X \to X$ such that the fixed–point set of σ decomposes X into two parts.

Lemma 12. $\{\Theta(h_j) \mid 1 \leq j \leq p\}$ *is a basis for holomorphic differentials on* $X = \Omega(G)/G$.

Proof. ([6, Corollary 1, page 208]).

5.2. Estimating the periods

Using Lemma 12 we can rather easily estimate periods of differentials. To that end we have to integrate the base differentials along simple closed curves on $\Omega(G)/G$. These curves can, furthermore, be taken to be geodesic curves. Consider such a curve α. Choose a lifting of α to the cover $\Omega(G)$. Denote this lifting also by α. It is now an arc in $\Omega(G)$, which is disjoint from the closure of the limit set $L(G)$ of G.

Lemma 13. *The integral of a base differential* $\Theta(h_j)$ *along* α *can be computed by the convergent series:*

$$(51) \qquad \int_\alpha \Theta(h_j)dz = \sum_{g \in G} \log\left(\frac{1 - \overline{\zeta}_j g(\alpha(0))}{1 - \overline{\zeta}_j g(\alpha(1))}\right).$$

Proof. [12, Lemma 4].

In estimating the sum (51) we have the following problems:

(1) Since we can use finite precision only, we can deal with a small part of G at one time only. The part available for our computations depends on the choice of the basis of G. Long words of elements of G get, as their numerical approximations, constant mappings. They do not contribute anything to the sum (51).

(2) Rounding errors.

We can, however, improve our algorithm using the following construction. The arc α, along which we integrate, can often be taken to be an arc on the axis A_{g_α} of some Möbius transformation $g_\alpha \in G$. We may, furthermore, suppose that the orientation of A_{g_α}, as determined by the attracting and the repelling fixed-points a_{g_α} and r_{g_α}, agrees with that of α. For otherwise we replace g_α by its inverse. We may suppose also, without restricting the generality, that:

- g_α is primitive in G, i.e., the cyclic subgroup $G_\alpha = \langle g_\alpha \rangle$ of G generated by g_α contains all elements of G which have A_{g_α} as their axis.
- $g_\alpha(\alpha(0)) = \alpha(1)$. This follows from the fact that, in our applications, the projection of α on the Riemann surface $\Omega(G)/G$ is a simple closed curve. We may, therefore, suppose that g_α maps the starting-point of α onto its end-point.

Having made all the above assumptions, we observe now that

(1) $m(g(\alpha) \cap g'(\alpha)) = 0$ for all $g, g' \in G_\alpha$, $g \neq g'$. Here m denotes the 1–dimensional Lebesgue measure.

(2) $A_{g_\alpha} = \cup_{g \in G_\alpha} g(\alpha)$.

By the above properties we have now:

$$\sum_{g \in G_\alpha} - \int_{g(\alpha)} \frac{-\bar{\zeta_j}}{1 - \bar{\zeta_j}\omega} d\omega = - \int_{A_{g_\alpha}} \frac{-\bar{\zeta_j}}{1 - \bar{\zeta_j}\omega} d\omega$$

$$(52) \qquad\qquad\qquad = \log \frac{1 - \bar{\zeta_j} r_{g_\alpha}}{1 - \bar{\zeta_j} a_{g_\alpha}}.$$

Therefore, by equation (52), we may compute an infinite part of the series (51) by a single expression only. This improves the accuracy of our computations and reduces rounding errors considerably.

Let now $g \in G \backslash G_\alpha$. The arc $\beta = g(\alpha)$ lies on the axis of the Möbius–transformation $g_\beta = g \circ g_\alpha \circ g^{-1}$.

The subgroup $G_\beta = \langle g_\beta \rangle$ of G contains all transformations of G whose axis agrees with that of g_β. Also in this case we have

(1) $A_{g_\beta} = \cup_{h \in G_\beta} h(\beta)$.
(2) $m(h(\beta) \cap h'(\beta)) = 0$ for all h, $h' \in G_\beta$, $h \neq h'$.

Hence

$$\sum_{h \in G_\beta} - \int_{h(g(\alpha))} \frac{-\bar{\zeta_j}}{1 - \bar{\zeta_j}\omega} d\omega = - \int_{A_{g_\beta}} \frac{-\bar{\zeta_j}}{1 - \bar{\zeta_j}\omega} d\omega$$

$$(53) \qquad\qquad\qquad = \log \frac{1 - \bar{\zeta_j} r_{g_\beta}}{1 - \bar{\zeta_j} a_{g_\beta}}.$$

By equations (52) and (53) we conclude that periods of differentials can be computed by the formula

$$(54) \qquad \int_\alpha \Theta(h_j) dz = \sum_{h \in G/G_\alpha} \log \frac{1 - \bar{\zeta_j} h(r_{g_\alpha})}{1 - \bar{\zeta_j} h(a_{g_\alpha})},$$

which is, in numerical computations, much more accurate and faster than formula (51).

5.3. Period matrices of general compact Riemann surfaces

A general compact genus p, $p > 1$, Riemann surface $X = D/G$ is defined for computations by giving generators for the corresponding Fuchsian group G of the first kind. These Fuchsian groups can be produced by the input device CARS described in Sections 1 and 2.

In order to be able to compute a period matrix for X we have to do the following:

(1) Find a basis for the homology of X.
(2) Find a basis for Abelian differentials of X.
(3) Integrate the base differentials along curves in the homology basis.

From a given presentation $X = D/G$ of X in terms of its Fuchsian group (of the first kind) one can rather easily find Möbius transformations $g_1, h_1, \ldots, g_p, h_p$

corresponding[6] to a homology basis of X.

The main problem is to find differentials on X. In a recent manuscript ([8]) this problem has been solved by representing X first as a plane algebraic curve and then using methods of algebraic geometry to find differentials on that curve. This allows one to approximate numerically period matrices of any compact Riemann surface X. Since the manuscript [8] is still being revised, we will not go into any more details here.

6. Riemann surfaces and the Laplacian

Many of the analytic and geometric properties of the Riemann surfaces are reflected in the properties of the Laplace operator. This operator arises, for instance, in the computation of the period matrices by means of conformal capacities. It also arises in connection with the closed geodesics.

Since the early sixties, the relationship between the Laplacian and the geometry of the underlying surface has been studied extensively by many authors. The best known connection, and nowadays a field of research of its own right, is the Selberg trace formula which relates the sequence of the eigenvalues of the Laplacian to the sequence of the closed geodesics.

It is a peculiar fact that even today no compact Riemann surface of genus $g \geq 2$ is known where one would be able to compute the eigenvalues in a closed form. This is quite in contrast to the case in genus 1 where the eigenvalue equation is separable and all eigenfunctions are given explicitly by means of trigonometric functions.

An obvious approach to the computation of the eigenvalues is to compute the lengths of the closed geodesics and to use the Selberg trace formula. Some experiments in this direction, however, yielded completely unsatisfactory results due to the exponential growth of the length spectrum and the slow convergence of the series involved in the trace formula.

We started, therefore, another approach using methods from numerical analysis. For this it was necessary to find a new triangulation algorithm which works on closed surfaces with non–trivial topology. A well known difficulty here is that, as the automatic triangulation proceeds around a non–trivial closed loop, it gets into collision with parts of the surface triangulated at earlier steps. This difficulty is sometimes described as the problem of colliding front waves. To overcome it we used techniques from the theory of sphere packings which also proved to be useful for proving theoretical results. For the details we refer to [3].

A special feature of our triangulation is that it may be adapted to the decompositions of a Riemann surface into building blocks, i.e., into Y–pieces and into Q–pieces. This makes it possible to *input* a Riemann surface using the Fenchel–Nielsen coordinates.

The numerical computation of the eigenfunctions and eigenvalues is based on the finite element method and may be carried out on a workstation yielding quite accurate

---

[6]This simply means that the axes of the transformations g_j and h_j project onto simple closed geodesic curves homotopic to curves defining a basis for the homology and that each g_j and each h_j is primitive in G.

results. Details with a number of explicit examples will appear in a forthcoming paper (by Buser, Renggli and Semmler).

F. Steiner (Physics Department, University of Hamburg) has used another numerical approach in genus 2. His approach is based on the use of the so–called boundary method on fundamental polygons ([1]).

This method is particularly accurate if no small closed geodesics occur. A joint paper with F. Steiner, which is presently under preparation, will bring these two methods together. This paper will provide an interface allowing one to go from the Fenchel–Nielsen description to the description based on symmetric octagons and inversely.

References

1. R. Aurich and F. Steiner. Quantum eigenstates of a stongly chaotic system and the scar phenomenon. *Chaos, Solitons & Fractals*, 5(2):229 – 255, 1995.
2. W. Burnside. On a Class of Automorphic Functions. *Proc. London Math. Soc.*, XXIII:49 – 88, 1891.
3. P. Buser, M. Seppälä, and R. Silhol. Triangulations and moduli spaces of Riemann surfaces with group actions. *Manuscripta Mathematica*, 88(2):209 – 224, 1995.
4. Peter Buser. *Geometry and Spectra of Compact Riemann Surfaces*. Birkhäuser Verlag, Basel–Boston–New York, 1992.
5. CARS. Computer Algebra Riemann Surfaces. Input device for Möbius transformations and Riemann surfaces presently (1996) being coded by Loris Renggli at the Florida State University. For up-to-date information see http://klein.math.fsu.edu/.
6. C. J. Earle and A. Marden. On Poincaré Series with Application to h^p Spaces of Bordered Riemann Surfaces. *Ill. J. Math.*, 13:202 – 219, 1969.
7. Hershel M. Farkas and Irwin Kra. *Riemann Surfaces*, volume 71 of *Graduate Texts in Mathematics*. Springer–Verlag, Berlin–Heidelberg–New York, 1980.
8. Patrizia Gianni, Mika Seppälä, Robert Silhol, and Barry Trager. Riemann surfaces, plane algebraic curves and their period matrices, 1995. Manuscript.
9. Bernard Maskit. *Kleinian Groups*, volume 287 of *Grundlehren der mathematischen Wissenschaften*. Springer–Verlag, Berlin–Heidelberg–New York, 1987.
10. P. J. Myrberg. Über die Numerische Ausführung der Uniformisierung. *Acta Soc. Sci. Fenn.*, XLVIII(7):1 – 53, 1920.
11. Klaus-Dieter Semmler and Mika Seppälä. Numerical uniformization of hyperelliptic curves. In *Proceedings ISSAC95*, 1995.
12. Mika Seppälä. Computation of period matrices of real algebraic curves. *Discrete Comput Geom*, 11:65 – 81, 1994.
13. Mika Seppälä and Tuomas Sorvali. *Geometry of Riemann Surfaces and Teichmüller Spaces*. Number 169 in Mathematics Studies. North–Holland, 1992.
14. Robert Silhol. Normal forms for period matrices of real algebraic curves of genus 2 and 3. *J. Pure and Applied Algebra*, 87:79 – 92, 1993.
15. Robert Silhol. Numerical approximation of period matrices of real algebraic curves, 1994. To appear.
16. Robert Silhol. A numerical approach to defining a legendre map in genus 2, 1995. To appear in Contemporary Mathematics.
17. Scott Wolpert. The Fenchel–Nielsen deformation. *Ann. Math., II. Ser.*, 115:501–528, 1982.

DÉPARTEMENT DE MATHÉMATIQUES, ECOLE POLYTECHNIQUE FÉDÉRALE DE LAUSANNE, CH-1015 LAUSANNE

E-mail address: buser@dpma.epfl.ch

DEPARTMENT OF MATHEMATICS, FLORIDA STATE UNIVERSITY, TALLAHASSEE, FL 32306-3027

E-mail address: seppala@math.fsu.edu

Proceedings of
THE 37TH TANIGUCHI SYMPOSIUM ON
TOPOLOGY AND TEICHMÜLLER SPACES
held in Finland, July 1995
ed. by Sadayoshi KOJIMA *et al.*
©1996 World Scientific Publishing Co.
pp. 31–55

VOLUMES AND CHERN-SIMONS INVARIANTS OF CYCLIC COVERINGS OVER RATIONAL KNOTS

HUGH M. HILDEN, MARÍA TERESA LOZANO*, AND
JOSÉ MARÍA MONTESINOS–AMILIBIA*

(Received March 1, 1996)

0. Introduction

The volume and the Chern-Simons invariant ([C-S]) are among the most important geometric invariants of a hyperbolic 3-manifold, and, after the Mostow Rigidity Theorem [Mos], they are also topological invariants. In this paper we compute the volume and the Chern-Simons invariant of the hyperbolic 3-manifold $M_n(p/q)$, obtained as an n-cyclic covering of S^3 branched over a 2-bridge knot p/q. We adapt the general method introduced in [HLM$_5$] for cyclic coverings of S^3 branched over hyperbolic knots to the particular case of 2-bridge knots. The general method considers a one parameter family of cone-manifold structures in S^3 having the knot K as a singular curve with angle α, such that the angle $\alpha(t)$, the jump $\beta(t)$ and the length $\delta(t)$ are differentiable functions of a parameter t. The orbifold $(S^3, K_{\frac{2\pi}{n}})$ with a cyclic isotropy group of order n in the knot K is a member of the cone manifold family. Then, Schläffli formulas and initial conditions allow the computation of the volume and the Chern-Simons invariant of this orbifold, and therefore those of the cyclic covering manifold $M_n(p/q)$.

In the case of a 2-bridge knot p/q we take the parameter to be $x = 2\cos\alpha$, where α is the cone angle of the singular curve p/q, and we can obtain the functions jump $\beta(x)$ and length $\delta(x)$ from the excellent component $\mathcal{E}[p/q]$ of $C[p/q]$, the curve of representations (or of characters) of the knot group in $SL(2,\mathbb{C})$, (see [Ri]).

In Sec.1 we recall some concepts and results, and we prove Theorem 1.6 which gives the formulas for $CS(M_n(p/q))$ and $V(M_n(p/q))$, using as starting points, the spherical cone-manifold of angle π, $(S^3, (p/q)_\pi)$, and the euclidean cone-manifold $(S^3, (p/q)_{\alpha_h})$, respectively.

In Sec.2 we present an easy method to find the excellent component $\mathcal{E}[p/q]$ of $C[p/q]$, the curve of representations (or of characters) of the knot group in $SL(2,\mathbb{C})$.

---

*This research was supported by grant PB92-0236

We obtain formulas for the jump β and the lengh δ of the knot p/q in each geometric cone manifold structure of S^3 with singularity of angle α in the knot p/q, using $\mathcal{E}[p/q]$. And then we apply Theorem 1.6. Observe that we obtain some invariants of the knot: the functions volume $V(S^3, K_\alpha)$, length $\delta(\alpha)$ and jump $\overline{\beta}(\alpha)$; and the numbers jump and twist of the knot in the euclidean cone-manifold structure, and the jump for the complete hyperbolic structure of the complement of the knot. We also obtain the Chern-Simons invariants of the orbifolds $(S^3, p/q, n)$.

We have implemented the computations using the program Mathematica [W]. Since for amphicheiral knots, like 5/3 or 13/5, the Chern-Simons invariants vanish, we tabulate in Sec.3 the results of our calculations for the knots 7/3, 9/5, 11/3 and 15/11.

1. The cone-manifold family $(S^3, (p/q)_\alpha)$

Let p/q be the 2-bridge knot determined by the pair of relatively prime odd integers p, q, with $1 < q < p$. It is a hyperbolic knot in S^3 ([T]), i.e. the exterior of the knot p/q has a complete hyperbolic structure of finite volume. We may consider that this structure is a hyperbolic cone-manifold structure in S^3 with cone angle 0 around the singularity. On the other hand the double covering of S^3 branched over the knot p/q is the lens space $L(p, q)$, which has a spherical structure, i.e., a Riemannian metric of constant curvature $+1$. This spherical structure induces, via the covering, a spherical cone manifold structure in S^3 with cone angle π around the knot p/q. These two structures are members of the family of cone-manifold structures on S^3 having the 2-bridge knot p/q as a singular curve with angle α. We denote this family by $(S^3, (p/q)_\alpha)$, $\pi \leq \alpha \leq 0$. There exists an angle α_h, such that the cone manifold $(S^3, (p/q)_\alpha)$ is hyperbolic for $\alpha < \alpha_h$, euclidean for $\alpha = \alpha_h$, and spherical for $\pi \geq \alpha > \alpha_h$ [HLM$_4$]. (See [HLM$_3$] for a detailed construction of this family for the knot 5/3, and [S] for a more general study.)

We consider the standard orientation in the 3-sphere, $\overrightarrow{S}^3$, and we choose an orientation on the knot, $\overrightarrow{p/q}$. Let $\overrightarrow{m}$ be an oriented meridian, such that the orientation of $\overrightarrow{m}$ followed by the orientation of $\overrightarrow{p/q}$ coincides with the orientation of $\overrightarrow{S}^3$. Let $\overrightarrow{l}_c$ be the canonical longitude of $\overrightarrow{p/q}$, and $o = \overrightarrow{m} \cap \overrightarrow{l}_c$. The curves $\overrightarrow{m}$ and $\overrightarrow{l}_c$ represent elements of the fundamental group $\pi_1(S^3 \setminus p/q, o)$, which will be denoted by the same letters, $\overrightarrow{m}$ and $\overrightarrow{l}_c$. The geometric cone manifold structure $(S^3, (p/q)_\alpha)$ is given by an holonomy map $h_\alpha : \pi_1(S^3 \setminus p/q, o) \longrightarrow X$, where $X = Iso^+(H^3) = PSL(2, \mathbf{C})$ if the geometry is hyperbolic, $\alpha < \alpha_h$; $X = Iso^+(E^3)$ if the geometry is euclidean, $\alpha = \alpha_h$; and $X = Iso^+(H^3) = SO(4)$ if the geometry is spherical, $\alpha_h < \alpha \leq \pi$. In all three cases $h_\alpha(l_c)$ is an isometry with a *length* $\delta(\alpha)$ and an angle $\beta(\alpha)$. Then the *jump* of $(p/q)_\alpha$ is the point in $\mathbf{R}/4\pi\mathbf{Z}$ represented by the angle $\beta(\alpha)$. It will be denoted by $\overline{\beta}$. The *twist* of $(p/q)_\alpha$ is $tw((p/q)_\alpha) = \beta\frac{\alpha}{2\pi}$, where $-2\pi \leq \beta < 2\pi$, and $\overline{tw}((p/q)_\alpha) = \overline{\beta}\frac{\alpha}{2\pi}$ is a point in $\mathbf{R}/2\alpha\mathbf{Z}$. The length, jump and twist are invariants of the cone manifold.

The real number $I(S^3, (p/q)_\alpha)$ defined in [HLM$_5$] is associated to a orthonormal frame field s in the Riemannian manifold $S^3 \setminus (p/q \cup m)$, having special singularities at $p/q \cup m$. Recall ([Y, Def1.3], and compare [Me]) that a frame field on a 3-manifold N is *special* on a link J if it is an orthonormal frame field on $N \setminus J$ which has the following behavior near each component K of J. Let U_K be an open neighborhood of radius ϵ of K. For each point x of $U_K \setminus K$

i) $e_3(x)$ is tangent to $\gamma(x, y)$ and it has direction opposite to y, where $\gamma(x, y)$ is the unique geodesic in U_K such that $d(x, y) = d(x, K) = length(\gamma(x, y)) = \delta$,

ii) $e_2(x)$ is tangent to $\vec{S}_\delta(y)$, where $\vec{S}_\delta(y) = \{z \in U_K | d(z, K) = d(z, y) = \delta\}$

It follows that, in the limit, $e_1(x)$ is tangent to $\vec{K}$.

Following an idea of Yoshida in [Y], for the Figure Eight knot, it is proved in [HLM$_5$, Pror 3.1], that for each hyperbolic knot in S^3 and for the complete hyperbolic structure on $S^3 \setminus (p/q)$, there exists an orthonormal frame field s_0 on $S^3 \setminus (p/q \cup m)$, having special singularities at $p/q \cup m$. Applying the Schmidt orthonormalization process to s_0 with respect to the geometric structure of the cone manifold $(\vec{S}^3, \overrightarrow{(p/q)_\alpha})$, one obtains an orthonormal frame field s_α on $(\vec{S}^3 \setminus \overrightarrow{(p/q)_\alpha} \cup m)$. By deforming it in a neighborhood of $\overrightarrow{(p/q)_\alpha} \cup m$ if necessary, we may assume that s_α has special singularities at $\overrightarrow{(p/q)_\alpha} \cup m$.

We follow the notation of [Y] for Q, the Chern-Simons form, which is defined on the positively-oriented orthonormal frame bundle, $F(\vec{S}^3 \setminus (p/q)_\alpha)$ by

$$Q = \frac{1}{4\pi^2}(\theta_{12} \wedge \theta_{13} \wedge \theta_{23} + \theta_{12} \wedge \Omega_{12} + \theta_{13} \wedge \Omega_{13} + \theta_{23} \wedge \Omega_{23})$$

where (θ_{ij}) is the connection 1-form and (Ω_{ij}) is the curvature 2-form of the Riemannian connection on the 3-manifold $\vec{S}^3 \setminus (p/q)_\alpha$.

Let $s' = (f_1, f_2, f_3)$ be an orthonormal framing defined on a subset of $\vec{S}^3 \setminus (p/q)_\alpha$ containing m such that $f_1(y)$ is the tangent vector to m at each $y \in m$ having the same direction as the e_1-vectors of s near y, and $f_2(y)$ is tangent to the meridian disc of p/q bounded by m. Then $\tau(m, s') = -\int_{s'(m)} \theta_{23}$. In fact $\tau(m, s') \equiv \tau(m)$ (mod 2π) where $\tau(m)$ is the torsion of the curve m.

Definition 1.1. We define

$$(1) \qquad I(\vec{S}^3, \overrightarrow{(p/q)_\alpha}) = \frac{1}{2}\int_{s(S^3 - p/q - m)} Q - \frac{1}{4\pi}\tau(m, s') - \frac{1}{4\pi}tw((p/q)_\alpha)$$

which depends only on the frame field s, when the value of β is chosen to be $-2\pi \leq \beta < 2\pi$.

We denote by a subscript r, the class (mod r) of $I(\vec{S}^3, \overrightarrow{(p/q)_\alpha})$. Some classes have special properties. For example the class $I_1(\vec{S}^3, \overrightarrow{(p/q)_\alpha})$ is independent of the frame field s; the class $I_{\frac{\alpha}{2\pi}}(\vec{S}^3, \overrightarrow{(p/q)_\alpha})$ depends on the frame field s, but is independent of the representative in the equivalence class $\overline{\beta}$. In the case of $\alpha = \frac{2\pi}{n}$ the cone-manifold is an orbifold. Then $\frac{\alpha}{2\pi} = \frac{1}{n}$ and

$$I_{\frac{\alpha}{2\pi}}(\vec{M}, \vec{\Sigma}_{\frac{2\pi}{n}}) = I_{\frac{1}{n}}(\vec{M}, \vec{\Sigma}_{\frac{2\pi}{n}}) \equiv I(\vec{M}, \vec{\Sigma}_{\frac{2\pi}{n}}) \quad (\text{mod } \frac{1}{n})$$

is both independent of the frame field s and of the representative in the equivalence class $\overline{\beta}$. Then $I_1(\overrightarrow{M}, \overrightarrow{\Sigma}_{\frac{2\pi}{n}})$ (mod $\frac{1}{n}$), which is an invariant of the orbifold $(\overrightarrow{M}, \overrightarrow{\Sigma}, n)$, should be called the *Chern-Simons invariant of the orbifold*.

Given a family of cone manifold structures, there exist Schläffli formulas for the functions $I_1(\overrightarrow{S}^3, \overrightarrow{(p/q)}_\alpha)$, and $I(\overrightarrow{S}^3, \overrightarrow{(p/q)}_\alpha)$, if for each α the integral of Q is taken along the orthonormalization of the same frame field, for instance the field s_α.

Theorem 1.2 ([HLM$_5$]). *Let $(S^3, K_{\alpha(t)})$ be a family of geometric cone-manifold structures in S^3, where $\alpha(t)$ and $\beta(t)$ are differentiable functions of t. Then the following equation ("the Schläffli formula") between differential forms holds*

$$(2) \qquad dI_1(S^3, K_\alpha) = -\frac{1}{4\pi^2}\beta d\alpha$$

To obtain the value of the Chern-Simons invariant of the orbifold $(\overrightarrow{S}^3, \overrightarrow{(p/q)}_{\frac{2\pi}{n}})$ we are going to use the above Schläffli formula taking the orbifold $(\overrightarrow{S}^3, \overrightarrow{(p/q)}_\pi)$ as the starting point. Therefore we first compute the Chern-Simons invariant of $(\overrightarrow{S}^3, \overrightarrow{(p/q)}_\pi)$.

The double cover of S^3 branched over the knot p/q is the Lens space $L(p,q)$.

We adopt the following definition of the *oriented* Lens space $L(p,q)$, which corresponds to Dehn surgery of type (p,q) in the trivial knot in the oriented S^3 (see [Ro]). The Lens space $L(p,q)$ is the quotient space of $\overrightarrow{S}^3$ by the cyclic action

$$\mathbb{Z}_p \times \overrightarrow{S}^3 \longrightarrow \overrightarrow{S}^3$$

$$(exp(2\pi i/p), (z_1, z_2)) \to (exp(2\pi i/p)z_1, exp(-2\pi qi/p)z_2).$$

The standard Riemannian metric of constant sectional curvature 1 in S^3 induces a Riemannian metric of constant sectional curvature 1 in the quotient $L(p,q)$. With this metric, $L(p,q)$ has the structure of a spherical 3-manifold.

Theorem 1.3. *Let (p,q) be a coprime pair of odd integers, $1 < q < p$. Let r be the integer such that $0 < r < p$ and $qr \equiv 1$ (mod p). The value of the Chern-Simons invariant of the spherical manifold $L(p,q)$ is the following*

$$CS(L(p,q)) \equiv \begin{cases} \frac{r+q}{2p} & (mod\,1) \text{ if } r \text{ is odd} \\ \frac{r+q-p}{2p} & (mod\,1) \text{ if } r \text{ is even.} \end{cases}$$

Proof. We use the torsion formula of Meyerhoff-Ruberman, [M-R, Theorem 3.1], and results of Yoshida, [Y, Lemmas 5.6 and 5.7] to compute $CS(L(p,q))$. Note that the definition of Lens space in [Y] differs from our definition; actually we understand for $L(p,q)$ the manifold $L(p, p-r)$ in [Y]. We need to have a link L in $(L(p,q))$ and a frame field on $(L(p,q)) \setminus L$ having special singularities at L, as well as the torsion of each component of L (mod 4π). Let L be $\mu(S_1) \cup \mu(S_2)$, where $S_1 = \{(z_1,0)||z_1| = 1\}$, $S_2 = \{(0,z_2)||z_2| = 1\}$ and $\mu : S^3 \longrightarrow L(p,q)$ is the natural quotient map. Lemma

5.6 of [Y] gives a frame field s on $(L(p,q)) \setminus L$ having special singularities at L, and such that $s^*Q = 0$. Lemma 5.8 of [Y] gives the torsion of $\mu(S_1)$ and $\mu(S_2)$ (mod 4π):

$$\tau(\mu(S_1)) = (\frac{p-q}{p} + \lambda)2\pi, \quad \tau(\mu(S_2)) = (\frac{p-r}{p} + \nu)2\pi$$

where $\lambda = \frac{1+(-1)^{p-q}}{2}$, $\nu = \frac{1+(-1)^{p-r}}{2}$. Since q and p are always odd, $\lambda = 1$. Then the torsion formula is

$$CS(L(p,q)) \equiv \frac{1}{2}\int_{s(L(p,q)\setminus L)} Q - \frac{1}{4\pi}((\tau(\mu(S_1)) - 2\pi) + (\tau(\mu(S_2)) - 2\pi))$$

$$= 0 - \frac{1}{2}(\frac{p-q}{p}) - \frac{1}{2}(\frac{p-r}{p} + \nu - 1)$$

$$\equiv -\frac{1}{2}(\frac{2p-r-q}{p} + \nu - 1) \quad (mod\, 1)$$

The value of $CS(L(p,q))$ depends on the parity of r. If r is odd, $\nu = 1$ and $CS(L(p,q)) \equiv -\frac{1}{2}(\frac{2p-r-q}{p}) = -\frac{2p-r-q}{2p} \equiv \frac{r+q}{2p}$ (mod 1). If r is even, $\nu = 0$ and $CS(L(p,q)) \equiv -\frac{1}{2}(\frac{2p-r-q}{p} - 1) = -\frac{p-r-q}{2p} = \frac{r+q-p}{2p}$ (mod 1). Observe that in both cases the numerator is even. Therefore $pCS(L(p,q)) \equiv 0$ (mod 1), which is the Chern-Simons invariant of $\overrightarrow{S}^3$, the p-cyclic covering of $L(p,q)$. $\square$

It follows that the Chern-Simons invariant of the spherical manifold $L(p,q)$ is $2I_{\frac{1}{2}}(S^3, (p/q)_\pi)$ (mod 1). Then

$$(3) \qquad I_{\frac{1}{2}}(S^3, (p/q)_\pi) \equiv \frac{1}{2}CS(L(p,q)) \quad (mod\, \frac{1}{2}).$$

The above value and the Schläffli formula permit us to compute $I_{\frac{1}{2}}(S^3, (p/q)_\alpha)$ (mod $\frac{1}{2}$) for any α between π and 0, if we write α and β as a differentiable funtion of a parameter. The following theorem states this result for the hyperbolic orbifolds in the family.

Theorem 1.4. *Let p/q be the 2-bridge knot determined by the pair of relatively prime odd integers p, q, with $1 < q < p$. The Chern-Simons invariant $I_{\frac{1}{n}}(\overrightarrow{S}^3, \overrightarrow{(p/q)}_{\frac{2\pi}{n}})$ (mod $\frac{1}{n}$) of the hyperbolic orbifold $(\overrightarrow{S}^3, \overrightarrow{(p/q)}_{\frac{2\pi}{n}})$ is*

$$(4) \qquad I_{\frac{1}{n}}(\overrightarrow{S}^3, \overrightarrow{(p/q)}_{\frac{2\pi}{n}}) \equiv \frac{1}{2}CS(L(p,q)) - \frac{1}{4\pi^2}\int_\pi^{\frac{2\pi}{n}} \beta d\alpha \quad (mod\, \frac{1}{n}),$$

for n even. For n odd, we actually obtain a weaker result, namely:

$$(5) \qquad I_{\frac{1}{n}}(\overrightarrow{S}^3, \overrightarrow{(p/q)}_{\frac{2\pi}{n}}) \equiv \frac{1}{2}CS(L(p,q)) - \frac{1}{4\pi^2}\int_\pi^{\frac{2\pi}{n}} \beta d\alpha \quad (mod\, \frac{1}{2n}),$$

instead of (mod $\frac{1}{n}$), *where β is any differentiable lift of the jump to the universal covering $\mathbb{R}$ of $\mathbb{R}/(4\pi\mathbb{Z})$.*

Proof. Let β be a differentiable lift of $\overline{\beta(t)}$ to $\mathbf{R}$, then

$$
(6) \qquad -\frac{1}{4\pi^2} \int_\pi^{\frac{2\pi}{n}} \beta d\alpha = \tilde{I}(\vec{S}^3, \overrightarrow{(p/q)}_{\frac{2\pi}{n}}) - \tilde{I}(\vec{S}^3, \overrightarrow{(p/q)}_\pi)
$$

where $\tilde{I}(\vec{S}^3, \overrightarrow{(p/q)}_\alpha)(t)$ is a differentiable lift of $I_{\frac{\alpha}{2\pi}}(\vec{S}^3, \overrightarrow{(p/q)}_\alpha)(t)$ to $\mathbf{R}$.

As $\tilde{I}(\vec{S}^3, \overrightarrow{(p/q)}_\alpha)(t)$ and $I(\vec{S}^3, \overrightarrow{(p/q)}_\alpha)(t)$ are two lifts of $\mathbf{R}/(\frac{\alpha}{2\pi})$ to the universal covering $\mathbf{R}$, for the same function $I_{\frac{\alpha}{2\pi}}(\vec{M}, \overrightarrow{(p/q)}_\alpha)(t)$, they differ by a multiple of $\frac{\alpha}{2\pi}$. In particular for $\alpha = \frac{2\pi}{n}$, they differ by $\frac{k}{n}$

$$
\tilde{I}(\vec{S}^3, \overrightarrow{(p/q)}_{\frac{2\pi}{n}}) = I(\vec{S}^3, \overrightarrow{(p/q)}_{\frac{2\pi}{n}}) + \frac{k}{n}.
$$

Then

$$
\tilde{I}_1(\vec{S}^3, \overrightarrow{(p/q)}_{\frac{2\pi}{n}}) \equiv I_1(\vec{S}^3, \overrightarrow{(p/q)}_{\frac{2\pi}{n}}) + \frac{k}{n} \pmod 1
$$

$$
\tilde{I}_1(\vec{S}^3, \overrightarrow{(p/q)}_\pi) \equiv I_1(\vec{S}^3, \overrightarrow{(p/q)}_\pi) \pmod 1.
$$

From (6) we have

$$
\tilde{I}(\vec{S}^3, \overrightarrow{(p/q)}_{\frac{2\pi}{n}}) = \tilde{I}(\vec{S}^3, \overrightarrow{(p/q)}_\pi) - \frac{1}{4\pi^2} \int_\pi^{\frac{2\pi}{n}} \beta d\alpha,
$$

$$
\tilde{I}_1(\vec{S}^3, \overrightarrow{(p/q)}_{\frac{2\pi}{n}}) \equiv I_1(\vec{S}^3, \overrightarrow{(p/q)}_{\frac{2\pi}{n}}) + \frac{k}{n}
$$

$$
\equiv I_1(\vec{S}^3, \overrightarrow{(p/q)}_\pi) - \frac{1}{4\pi^2} \int_\pi^{\frac{2\pi}{n}} \beta d\alpha \pmod 1.
$$

And therefore

$$
I_{\frac{1}{2}}(\vec{S}^3, \overrightarrow{(p/q)}_{\frac{2\pi}{n}}) + \frac{k}{n} \equiv I_{\frac{1}{2}}(\vec{S}^3, \overrightarrow{(p/q)}_\pi) - \frac{1}{4\pi^2} \int_\pi^{\frac{2\pi}{n}} \beta d\alpha
$$

$$
\equiv \frac{1}{2} CS(L(p,q)) - \frac{1}{4\pi^2} \int_\pi^{\frac{2\pi}{n}} \beta d\alpha \pmod{\frac{1}{2}}.
$$

If n is even, $n = 2m$, we work $\pmod{\frac{1}{m}}$ obtaining

$$
I_{\frac{1}{n}}(\vec{S}^3, \overrightarrow{(p/q)}_{\frac{2\pi}{n}}) \equiv \frac{1}{2} CS(L(p,q)) - \frac{1}{4\pi^2} \int_\pi^{\frac{2\pi}{n}} \beta d\alpha \pmod{\frac{1}{n}}.
$$

If n is odd, we work $\pmod{\frac{1}{n}}$

$$
I_{\frac{1}{n}}(\vec{S}^3, \overrightarrow{(p/q)}_{\frac{2\pi}{n}}) \equiv \frac{1}{2} CS(L(p,q)) - \frac{1}{4\pi^2} \int_\pi^{\frac{2\pi}{n}} \beta d\alpha \pmod{\frac{1}{2n}}.
$$

$\square$

The key to compute the volume of the cone manifolds $(S^3, (p/q)_\alpha)$ is the Formula of Schläffli, (See [Vi], [C], [M]). In a one parameter family of polytopes in a space of constant curvature K, $KdV = (1/2)\sum \ell_i d\alpha_i$, where V is volume, and where the sum is taken over all the edges; ℓ_i is the length of the ith edge and α_i is its dihedral angle. As in [H], the volume of a cone manifold is the volume of the polyhedron, from which it is constructed. If several edges of a polytope are identified and the resulting identified edge is <u>not</u> part of the singular set then the sum of the corresponding dihedral angles is 2π and, since the differential of the constant 2π equals zero, these edges make no contribution to dV. Hence, in our case (See also [HLM$_3$]),

$$(7) \qquad K dV(S^3, (p/q)_\alpha) = (1/2)\delta d\alpha.$$

where $K = 1$ if $\pi \geq \alpha > h$, and $K = -1$ if $0 \leq \alpha < \alpha_h$.

Theorem 1.5. *Let p/q be the 2-bridge knot determined by the pair of relatively prime odd integers p, q, with $1 < q < p$. The volume of the geometric cone manifold $(\overrightarrow{S}^3, (\overrightarrow{p/q})_\alpha)$, can be obtained by the following formulas. For $0 \leq \alpha < \alpha_h$*

$$(8) \qquad V(\overrightarrow{S}^3, \overrightarrow{(p/q)}_\alpha) = -\frac{1}{2}\int_{\alpha_h}^{\alpha} \delta d\alpha.$$

For $\alpha_h < \alpha \leq \pi$

$$(9) \qquad V(\overrightarrow{S}^3, \overrightarrow{(p/q)}_\alpha) = \frac{1}{2}\int_{\alpha_h}^{\alpha} \delta d\alpha.$$

where α_h is the cone angle of the Euclidean cone manifold in S^3 with the knot p/q as singular set.

Proof. We compute the volume of a spherical or hyperbolic cone manifold with the assumption that the curvature is plus or minus one. But the spherical, euclidean and hyperbolic geometries on the cone manifolds over a hyperbolic knot form a continuous family in which curvature changes from positive to negative and the Euclidean case is a point in a continuum. Under this point of view, the normalized volume of the Euclidean cone manifold is zero and it can be used as a starting point in the integral. □

Theorem 1.6. *The volume and the Chern-Simons invariant of the hyperbolic 3-manifold $M_n(p/q)$, obtained as n-cyclic covering of S^3 branched over a 2-bridge knot p/q, where p, q are a pair of relatively prime odd integers, with $1 < q < p$, can be obtained by the following formulas.*

$$
\begin{aligned}
V(M_n(p/q)) &= -\frac{n}{2}\int_{\alpha_h}^{\frac{2\pi}{n}} \delta d\alpha \\
CS(M_n(p/q)) &\equiv \frac{n}{2}CS(L(p,q)) - \frac{n}{4\pi^2}\int_{\pi}^{\frac{2\pi}{n}} \beta d\alpha \quad \begin{cases} (\bmod\ 1) \textit{ if } n \textit{ is even} \\ (\bmod\ \frac{1}{2}) \textit{ if } n \textit{ is odd} \end{cases}
\end{aligned}
$$

(10)

where α_h is the cone angle of the Euclidean cone manifold in S^3 with the knot p/q as singular set, and δ, β are the length and the jump of the knot in the cone manifold structure $(\overrightarrow{S}^3, (\overrightarrow{p/q})_\alpha)$.

Proof. This is a consequence of the above theorems and the fact that volume and Chern-Simons invariant of orbifolds are multiplicative funcions under coverings. □

2. Results from the curve of representations $C[p/q]$

We shall express α, δ, β as differentiable funtions of a parameter x by using the excellent component $\mathcal{E}[p/q]$ of the curve of representations $C[p/q]$, of the group $\pi_1(S^3 \setminus (p/q))$ into $SL(2, \mathbb{C})$ ([Ri])). In fact, the excellent component $\mathcal{E}[p/q]$ is the one containing the representations corresponding to the geometric structures of the cone-manifold family that we consider. Then, the first problem to solve after computing the polynomial defining $C[p/q]$ is to recognize the excellent component $\mathcal{E}[p/q]$.

Next we explain an easy method to find the excellent component $\mathcal{E}[p/q]$ of the curve of representations $C[p/q]$, which was announced in [HLM$_2$].

Theorem 2.1. *The excellent component $\mathcal{E}[p/q]$ of the curve of representations $C[p/q]$ is the component containing the points $(-2, 2 \cos \frac{\pi(q+1+kp)}{p})$, $(-2, 2 \cos \frac{\pi(q-1+kp)}{p})$ for some integer k.*

Proof. The double covering of the cone manifold $(S^3, (p/q)_\pi)$ branched along p/q is the Lens space $L(p, q)$, and S^3 is the p-cyclic covering of $L(p, q)$. Then, the dihedral group D_{2p} acts on S^3 with $(S^3, (p/q)_\pi)$ as orbit space. To understand this action, consider the following presentation of the dihedral group

$$D_{2p} = |g_1, g_2 : g_1^2, g_2^2, (g_1 g_2)^p|.$$

Each generator g_i acts on S^3 by π-rotation around an axis such that the element $g_1 g_2$ acts on S^3 as the generator of the cyclic action defining $L(p, q)$, namely $((g_1 g_2)(z_1, z_2)) \to (exp(2\pi i/p)z_1, exp(-2\pi q i/p)z_2)$, where $S^3 = \{(z_1, z_2) \in \mathbb{C} \times \mathbb{C} | |z_1|^2 + |z_2|^2 = 1\}$. Take, for instance, $A_1 = S^3 \cap (\mathbb{R} \times \mathbb{R}) = \{(r_1, r_2))|r_1, r_2 \in \mathbb{R}, r_1^2 + r_2^2 = 1\}$ as axis for the action of g_1, and $A_2 = \{(r_1 e^{(\pi/p)}, r_2 e^{(-\pi q/p)})|r_1, r_2 \in \mathbb{R}, r_1^2 + r_2^2 = 1\}$ as the axis for the action of g_2.

For $x = \pi$, there exit two real values $z_1(\pi), z_2(\pi)$ such that $(\pi, z_1(\pi))$ and $(\pi, z_2(\pi))$ are in $\mathcal{E}[p/q]$ and they define two different representations l, d, into $S^3 = SU(2) \subset SL(2, \mathbb{C})$. The holonomy ω is the composition

$$\pi_1(S^3 \setminus (p/q)) \xrightarrow{(l,d)} SU(2) \times SU(2) \xrightarrow{\lambda} SO(4),$$

where $\lambda(A, B) = \{P \to A^t P \overline{B}\}$, where $P = (x, y, z, t)$ is the element of $\mathbb{R}^4$ represented by the matrix $P = \begin{pmatrix} x + iy & z + it \\ -z + it & x - iy \end{pmatrix}$. This representation gives the cone manifold structure $(S^3, (p/q)_\pi)$. Then, up to conjugation on $SU(2)$, the image of a is the action of g_1, the image of b is the action of g_2, and the image of ab, which is the following:

$$\omega(ab) = (y \to \begin{bmatrix} e^{i\mu} & 0 \\ 0 & e^{-i\mu} \end{bmatrix} y \begin{bmatrix} e^{-i\nu} & 0 \\ 0 & e^{i\nu} \end{bmatrix}),$$

where $z_1(\pi) = tr(l(ab)) = 2 \cos \mu$ and $z_2(\pi) = tr(d(ab)) = 2 \cos \nu$, must be the action of $g_1 g_2$. Then $\mu - \nu = \frac{2\pi}{p}$, and $\mu + \nu = \frac{2\pi q}{p} + 2k\pi$. Therefore

$$\{z_1(\pi), z_2(\pi)\} = \left\{ 2 \cos \frac{\pi(q + 1 + kp)}{p}, 2 \cos \frac{\pi(q - 1 + kp)}{p} \right\}$$

for some k. We have proved that the points $(-2, z_1(\pi))$, $(-2, z_2(\pi))$ in the excellent component $\mathcal{E}[p/q]$ are the points $(-2, 2\cos\frac{\pi(q+1+kp)}{p})$, $(-2, 2\cos\frac{\pi(q-1+kp)}{p})$. As a consequence the minimal polynomial of $2\cos\frac{\pi}{p}$ divides the polynomial $r(-2, z)$ where $r(x, z)$ defines the curve of characters $\mathcal{C}[p/q]$ (see [HLM$_2$]). This observation was independently made by Heusener (private communication). $\square$

Next we recall some notation and we refer to [HLM$_1$] and [HLM$_2$] for full details.

The group $[p/q] = \pi_1(S^3 \setminus (p/q))$ has the following presentation $|a, b : aw = wb|$, where $w = b^{\epsilon_1}a^{\epsilon_2}b^{\epsilon_3}\cdots a^{\epsilon_{p-1}}$ and ϵ_i is the sign (plus or minus 1) of iq reduced mod $2p$ in the interval $(-p, p)$. The generators a, b are represented by meridians and the canonical longitude l_c represents the element of the group $ww^*a^{-2\epsilon}$ where $\epsilon = \sum_{i=1}^{p-1}\epsilon_i$ and $w^* = a^{\epsilon_1}b^{\epsilon_2}b^{\epsilon_3}\cdots b^{\epsilon_{p-1}}$. Notice that $ww^*a^{-2\epsilon}$ is cyclically palindromic since the sequence $\epsilon_1, ..., \epsilon_{p-1}$ is palindromic. Therefore the order of the letters in $ww^*a^{-2\epsilon}$, if reversed, gives $a^{-2\epsilon}ww^* = ww^*a^{-2\epsilon}$, since ww^* and a commute. This shows that the holonomy of the canonical longitude can be calculated both as the product of a sequence of matrices or as the product of the reverse sequence as well.

$\mathcal{C}[p/q] = \{(x(\rho), z(\rho))|\rho$ is a non abelian representation of $[p/q]$ into $SL(2, \mathbb{C})\}$, where $x(\rho)$ is $tr(\rho(a^2))$ and $z(\rho)$ is $tr(\rho(ab))$. If the point $(x, z) \in \mathcal{E}[p/q]$ corresponds to the cone manifold $(S^3, (p/q)_\alpha)$, then $x = 2\cos\alpha$. (i.e. $\alpha(x) = \arccos(\frac{x}{2}) \cap [0, \pi]$.)

The following Lemma defines a representation $\rho : [p/q] \longrightarrow SL(2, \mathbb{C})$ with holonomy point (x, z).

Lemma 2.2. *Let (x, z) be a point in $\mathcal{E}[p/q]$. Define the representation $\rho : [p/q] \longrightarrow SL(2, \mathbb{C})$ by*

$$\rho(a) = \begin{bmatrix} a_{11} & 0 \\ 0 & a_{22} \end{bmatrix} \quad \rho(b) = \begin{bmatrix} b_{11} & b_{12} \\ 1 & b_{22}. \end{bmatrix}$$

where

$$a_{11} = \frac{\sqrt{x+2} + \sqrt{x-2}}{2}, \qquad a_{22} = \frac{\sqrt{x+2} - \sqrt{x-2}}{2},$$

$$b_{11} = \frac{2z - x - 2 + \sqrt{x^2-4}}{2\sqrt{x-2}}, \qquad b_{22} = \frac{-2z + x + 2 + \sqrt{x^2-4}}{2\sqrt{x-2}}$$

$$b_{12} = b_{11}b_{22} - 1.$$

Then $tr(\rho(a^2)) = x$ and $tr(\rho(ab)) = z$. If $x = 2\cos\alpha$ then $a_{11} = e^{\frac{\alpha i}{2}}$ and $a_{22} = e^{-\frac{\alpha i}{2}}$.

Proof. The matrices $\rho(a)$ and $\rho(b)$ belong to $SL(2, \mathbb{C})$, because $\det\rho(a) = a_{11} \cdot a_{22} = \frac{\sqrt{x+2} + \sqrt{x-2}}{2} \cdot \frac{\sqrt{x+2} - \sqrt{x-2}}{2} = \frac{x+2-x+2}{4} = 1$, and clearly $\det\rho(b) = 1$.

$$tr(\rho(a^2)) = a_{11}^2 + a_{22}^2 = \left(\frac{\sqrt{x+2} + \sqrt{x-2}}{2}\right)^2 + \left(\frac{\sqrt{x+2} - \sqrt{x-2}}{2}\right)^2$$

$$= \frac{\left(\sqrt{x+2} + \sqrt{x-2}\right)^2 + \left(\sqrt{x+2} - \sqrt{x-2}\right)^2}{4}$$

$$= \frac{x+2+x-2+x+2+x-2}{4} = x,$$

$$tr(\rho(ab)) = a_{11}b_{11} + a_{22}b_{22} = \left(\frac{\sqrt{x+2} + \sqrt{x-2}}{2}\right)\left(\frac{2z - x - 2 + \sqrt{x^2 - 4}}{2\sqrt{x-2}}\right)$$

$$+ \left(\frac{\sqrt{x+2} - \sqrt{x-2}}{2}\right)\left(\frac{-2z + x + 2 + \sqrt{x^2 - 4}}{2\sqrt{x-2}}\right)$$

$$= \frac{2\sqrt{x+2}\sqrt{x^2-4} + 2\sqrt{x-2}(2z - x - 2)}{4\sqrt{x-2}} = \frac{x + 2 + 2z - x - 2}{2} = z.$$

If $x = 2\cos\alpha$, then $x + 2 = 2(2\cos^2\frac{\alpha}{2} - 1) + 2 = 4\cos^2\frac{\alpha}{2}$ and $x - 2 = 2(1 - 2\sin^2\frac{\alpha}{2}) - 2 = -4\sin^2\frac{\alpha}{2}$. Therefore $a_{11} = \frac{\sqrt{x+2} + \sqrt{x-2}}{2} = \frac{\sqrt{4\cos^2\frac{\alpha}{2}} + \sqrt{-4\sin^2\frac{\alpha}{2}}}{2} = \cos\frac{\alpha}{2} + i\sin\frac{\alpha}{2} = e^{\frac{\alpha i}{2}}$ and $a_{22} = \frac{\sqrt{x+2} - \sqrt{x-2}}{2} = \frac{\sqrt{4\cos^2\frac{\alpha}{2}} - \sqrt{-4\sin^2\frac{\alpha}{2}}}{2} = \cos\frac{\alpha}{2} - i\sin\frac{\alpha}{2} = e^{-\frac{\alpha i}{2}}$. $\square$

The above Lemma allows us to compute the trace of the image of any element of the group $[p/q]$ by any representation associated to (x, z) as a polynomial with integer coefficients in the variables x, z. Let us call $q(x, z)$ the polynomial $tr(\rho(l_c))$, where ρ is the representation with holonomy (x, z) defined in Lemma 2.2, and l_c is the canonical longitude.

We distinguish two cases in order to compute the jump β and the length δ of the singular curve (p/q) as a function of x.

Case 1. $\pi \geq \alpha > h$. The cone manifold $(S^3, (p/q)_\alpha)$ is spherical. The holonomy ω, in this case, is a representation of $\pi_1(S^3 \setminus (p/q))$ in $SO(4)$. It is obtained as follows: For each value of $x = 2\cos\alpha$ there exist two real values $z_1(x), z_2(x)$ such that (x, z_1) and (x, z_2) are in $\mathcal{E}[p/q]$ and they define two different representations l, d, into $S^3 = SU(2) \subset SL(2, \mathbb{C})$. The holonomy ω is the composition

$$\pi_1(S^3 \setminus (p/q)) \xrightarrow{(l,d)} SU(2) \times SU(2) \xrightarrow{\lambda} SO(4)$$

Up to conjugation on $SU(2)$ the image of the meridian a is the following rotation of angle α:

$$\omega(a) = \left(y \rightarrow \begin{bmatrix} e^{\frac{i\alpha}{2}} & 0 \\ 0 & e^{-\frac{i\alpha}{2}} \end{bmatrix} y \begin{bmatrix} e^{-\frac{i\alpha}{2}} & 0 \\ 0 & e^{\frac{i\alpha}{2}} \end{bmatrix}\right)$$

The element in $\pi_1(S^3 \setminus (p/q))$ represented by the canonical longitude l_c of the knot (p/q) conmutes with the meridian a, then $l(l_c)$ and $d(l_c)$ must be also diagonal matrices,

$$l(l_c) = \begin{bmatrix} e^{i\gamma} & 0 \\ 0 & e^{-i\gamma} \end{bmatrix}, \quad d(l_c) = \begin{bmatrix} e^{i\phi} & 0 \\ 0 & e^{-i\phi} \end{bmatrix}$$

Then $\omega(l_c)$ is an helicoidal turn with the same axis as $\omega(a)$ (see [Mon]). The length δ of (p/q) is $\gamma - \phi$ and the jump β is $\gamma + \phi$. The fact that $\gamma - \phi > 0$ allows us to fix which one of the two representations of $\pi_1(S^3 \setminus (p/q))$ into $SL(2, \mathbb{C})$ is the left one, l, and which one is the right one, d. As $tr(l(l_c)) = 2\cos\gamma$ and $tr(d(l_c)) = 2\cos\phi$, one obtains

$$\beta = \arccos \frac{tr(l(l_c))}{2} + \arccos \frac{tr(d(l_c))}{2}$$

$$= \arccos \frac{q(x, z_1)}{2} + \arccos \frac{q(x, z_2)}{2},$$

$$\delta = \arccos \frac{tr(l(l_c))}{2} - \arccos \frac{tr(d(l_c))}{2}$$

$$= \arccos \frac{q(x, z_1)}{2} - \arccos \frac{q(x, z_2)}{2}.$$

If the angle α decreases to α_h, $(x \to h)$, the real values (z_1, z_2) go to a common real value z_h.

Case 2. $0 \leq \alpha < \alpha_h$. The cone manifold $(S^3, (p/q)_\alpha)$ is hyperbolic. The holonomy ω, in this case, can be lifted to two representations $\pm\rho$ of $\pi_1(S^3 \setminus (p/q))$ into $SL(2,\mathbb{C})$. For each value of $x = 2\cos\alpha$ there exist two complex conjugate values $z_1(x), z_2(x)$, $z_2 = \bar{z}_1$, such that (x, z_1) and (x, z_2) are in $\mathcal{E}[p/q]$ and they define two representations $\pm\rho_1, \pm\rho_2$, in $SL(2,\mathbb{C})$, associated to the same hyperbolic structure, up to orientation. Consider for instance $\pm\rho_1$. Up to conjugation in $SL(2,\mathbb{C})$ the image of the meridian a is the following rotation of angle α:

$$\rho_1(a) = \pm \begin{bmatrix} e^{i\frac{\alpha}{2}} & 0 \\ 0 & e^{-i\frac{\alpha}{2}} \end{bmatrix}$$

As in the spherical case $\rho_1(l_c)$ must be also a diagonal matrix,

$$\rho_1(l_c) = \begin{bmatrix} e^{\frac{v}{2}} & 0 \\ 0 & e^{-\frac{v}{2}} \end{bmatrix}$$

(well defined, since the length of the word $ww^*a^{-2\epsilon}$ is even), where $v = \delta + i\beta$, δ is length of $\vec{\Sigma}_\alpha$, and β is the jump. If $\delta < 0$, then we interchange the roles of ρ_1 and ρ_2 in what follows. Then $tr(\rho_1(l_c)) = e^{\frac{v}{2}} + e^{-\frac{v}{2}} = e^{\frac{i(\beta-i\delta)}{2}} + e^{-\frac{i(\beta-i\delta)}{2}} = 2\cos\frac{\beta - i\delta}{2}$, therefore

$$\beta - i\delta = 2\arccos \frac{tr(\rho_1(l_c))}{2}$$

To express β as a funcion of x, recall that $tr(\rho(l_c))$ can be written as a polynomial $q(x, z)$ with integer coefficients for any representation $\pm\rho$. Observe that $tr(\rho_2(l_c)) = q(x, z_2) = q(x, \bar{z}_1) = \overline{q(x, z_1)} = e^{\frac{\bar{v}}{2}} + e^{-\frac{\bar{v}}{2}} = e^{\frac{i(\beta+i\delta)}{2}} + e^{-\frac{i(\beta+i\delta)}{2}} = 2\cos\frac{\beta + i\delta}{2}$, and therefore

$$\beta + i\delta = 2\arccos \frac{tr(\rho_2(l_c))}{2}$$

This gives the same formula for β as in case 1:

$$\beta = \arccos \frac{tr(\rho_1(l_c))}{2} + \arccos \frac{tr(\rho_2(l_c))}{2}$$

$$= \arccos \frac{q(x, z_1)}{2} + \arccos \frac{q(x, z_2)}{2}$$

and

$$\delta = i\left(\arccos\frac{tr(\rho_1(l_c))}{2} - \arccos\frac{tr(\rho_2(l_c))}{2}\right)$$
$$= i\left(\arccos\frac{q(x,z_1)}{2} + \arccos\frac{q(x,z_2)}{2}\right).$$

We resume the above computation in the following theorems.

Theorem 2.3. *The invariant $I_{\frac{1}{2}}(S^3,(p/q)_\alpha)$ for the 2-bridge knot p/q, where (p,q) is a coprime pair of odd integers, $1 < q < p$, and $\pi \geq \alpha \geq 0$ is given by the following formulas: Let r be the integer $0 < r < p$, such that $qr \equiv 1 \pmod{p}$). Then if r is odd*

$$I_{\frac{1}{2}}(S^3,(p/q)_\alpha) \equiv$$
$$\frac{r+q}{4p} - \int_{-2}^{2\cos\alpha} \frac{1}{4\pi^2}(\arccos\frac{q(x,z_1)}{2} + \arccos\frac{q(x,z_2)}{2})d(\arccos\frac{x}{2}) \quad (\mathrm{mod}\ \frac{1}{2})$$

and if r is even

$$I_{\frac{1}{2}}(S^3,(p/q)_\alpha) \equiv$$
$$\frac{r+q-p}{4p} - \int_{-2}^{2\cos\alpha} \frac{1}{4\pi^2}(\arccos\frac{q(x,z_1)}{2} + \arccos\frac{q(x,z_2)}{2})d(\arccos\frac{x}{2}) \quad (\mathrm{mod}\ \frac{1}{2}).$$

Theorem 2.4. *The volume $V(S^3,(p/q)_\alpha)$ for the 2-bridge knot p/q, where (p,q) is a coprime pair of odd integers, $1 < q < p$, and $\pi \geq \alpha \geq 0$ is given by the following formulas: If $\pi \geq \alpha \geq \alpha_h$*

$$V(S^3,(p/q)_\alpha) = \int_{2\cos\alpha_h}^{2\cos\alpha} \frac{1}{2}(\arccos\frac{q(x,z_1)}{2} - \arccos\frac{q(x,z_2)}{2})d(\arccos\frac{x}{2})$$

If $\alpha_h \geq \alpha \geq 0$

$$V(S^3,(p/q)_\alpha) = \int_{2\cos\alpha}^{2\cos\alpha_h} \frac{i}{2}(\arccos\frac{q(x,z_1)}{2} - \arccos\frac{q(x,z_2)}{2})d(\arccos\frac{x}{2}).$$

Corollary 2.5.

$$\frac{\pi^2}{p} = \int_{2\cos\alpha_h}^{-2} \frac{1}{2}(\arccos\frac{q(x,z_1)}{2} - \arccos\frac{q(x,z_2)}{2})d(\arccos\frac{x}{2}).$$

Proof. $2pV(S^3,p/q,\pi) = V(S^3) = 2\pi^2$. $\square$

Next we can apply Theorem 1.6 to find the Chern-Simons invariant and the volume of the hyperbolic manifolds obtained as cyclic coverings of S^3 branched over 2-bridge knots.

Theorem 2.6. *Let $M_n(p/q)$ be the n-cyclic cover of S^3 branched over the 2-bridge knot (p/q). Let r be the integer $0 < r < p$, such that $qr \equiv 1 \pmod{p}$). The Chern-Simons invariant of the hyperbolic manifold $M_n(p/q)$, $n > 3$ can be obtained by the* *following formulas. When r is odd*

$$CS(M_n(p/q)) \equiv nI_{\frac{1}{2}}(S^3,(p/q)_{2\pi/n}) \equiv nI_{\frac{1}{2}}(S^3,(p/q)_{2\pi/n}) \equiv \frac{n(r+q)}{4p}$$

$$-\int_{-2}^{2\cos\alpha} \frac{n}{4\pi^2}(\arccos\frac{q(x,z_1)}{2} + \arccos\frac{q(x,z_2)}{2})d(\arccos\frac{x}{2}) \quad \begin{cases} (\mathrm{mod}\ 1) & n\ \textit{even} \\ (\mathrm{mod}\ \frac{1}{2}) & n\ \textit{odd} \end{cases}$$

and if r is even

$$CS(M_n(p/q)) \equiv nI_{\frac{1}{2}}(S^3,(p/q)_{2\pi/n}) \equiv nI_{\frac{1}{2}}(S^3,(p/q)_{2\pi/n}) \equiv \frac{n(r+q-p)}{4p}$$

$$-\int_{-2}^{2\cos\alpha} \frac{n}{4\pi^2}(\arccos\frac{q(x,z_1)}{2} + \arccos\frac{q(x,z_2)}{2})d(\arccos\frac{x}{2}) \quad \begin{cases} (\mathrm{mod}\ 1) & n\ \textit{even} \\ (\mathrm{mod}\ \frac{1}{2}) & n\ \textit{odd} \end{cases}$$

The volume of the hyperbolic manifold $M_n(p/q)$, $n > 3$ is

$$V(M_n(p/q)) = nV(S^3,(p/q)_{2\pi/n}) =$$
$$\int_{2\cos\alpha}^{2\cos\alpha_h} \frac{ni}{2}(\arccos\frac{q(x,z_1)}{2} - \arccos\frac{q(x,z_2)}{2})d(\arccos\frac{x}{2}).$$

3. Computations

We have made the computation for the knots 7/3, 9/5, 11/3 and 15/11 as examples, and the results are given in the following plots and tables, which were obtained using the program MATHEMATICA ([W]).

Plots I, II, III, IV show the plots of the length δ and the jump β of the cone-manifolds $(S^3,(p/q)_\alpha)$, against the parameter α ranging between 0 and π. Recall that for the value α_h the geometry is euclidean, for $0 \leq \alpha < \alpha_h$ the geometry is hyperbolic and for $\alpha_h < \alpha \leq \pi$ is spherical. Plots V, VI, VII, VIII show the plot of the volume $V(S^3,p/q,\alpha)$ against the parameter α ranging between 0 and π. Table 0 displays the values of α_h, the jump $\beta(\alpha_h)$ and the twist $tw((p/q)_{\alpha_h})$ for the euclidean cone manifold structures which are numerical invariants of the knot p/q.

knot	α_h	$\beta(\alpha_h)$	$tw((p/q)_{\alpha_h})$
7/3	2.40717	-2.70606	-1.03673
9/5	2.57414	5.30341	2.172774
11/3	2.68404	-5.20606	-2.22392
15/11	2.80821	-4.13017	-1.84593

TABLE 0

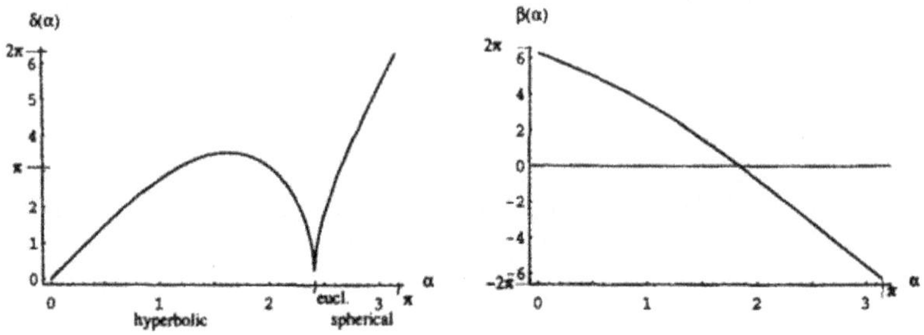

PLOT I. Length and jump of the knot 7/3

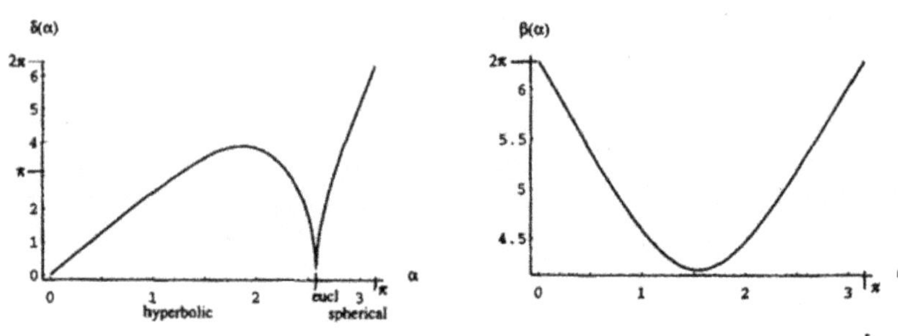

PLOT II. Length and jump of the knot 9/5

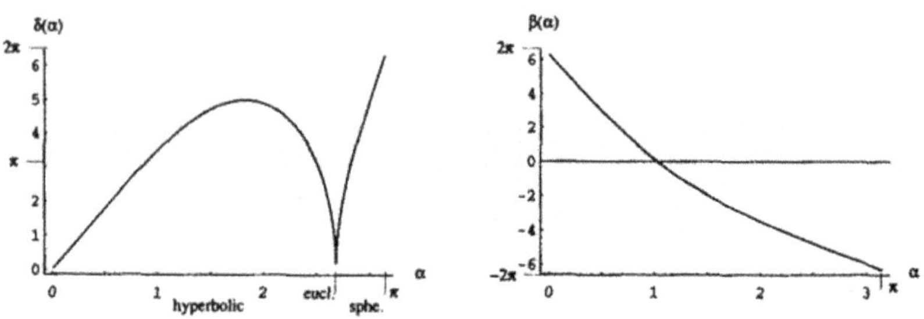

PLOT III. Length and jump of the knot 11/3

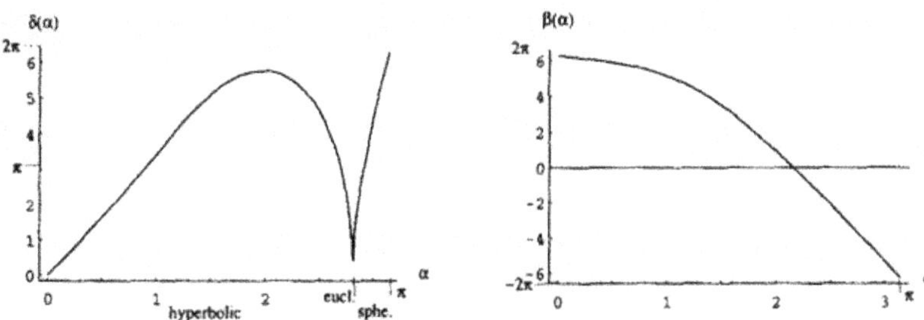

PLOT IV. Length and jump of the knot 15/11

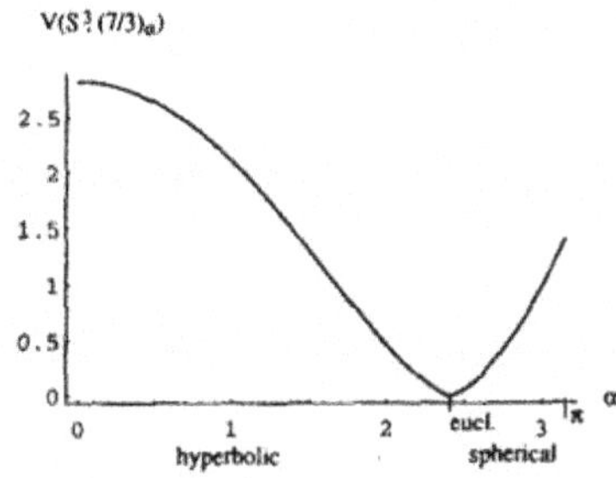

PLOT V. Volume of the cone-manifolds $(S^3, (7/3)_\alpha)$

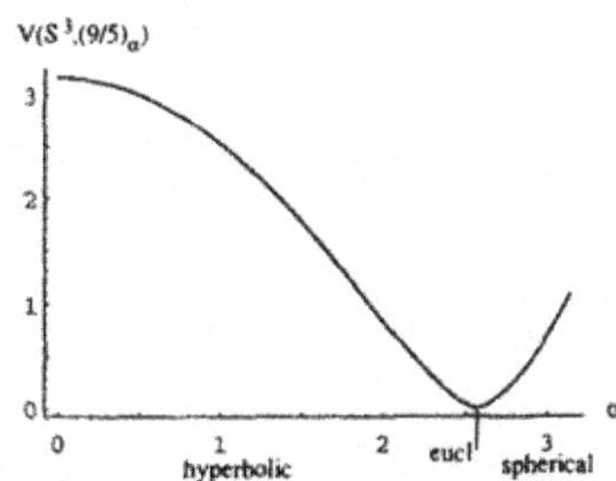

PLOT VI. Volume of the cone-manifolds $(S^3, (9/5)_\alpha)$

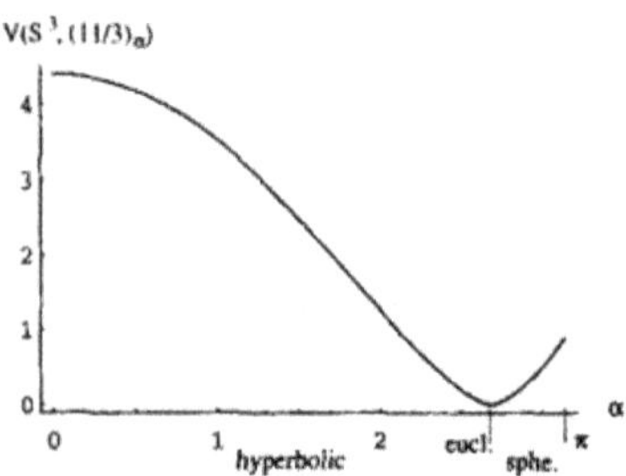

PLOT VII. Volume of the cone-manifolds $(S^3, (11/3)_\alpha)$

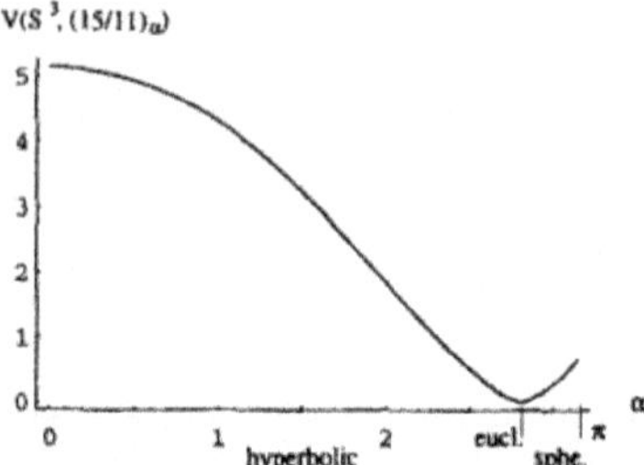

PLOT VIII. Volume of the cone-manifolds $(S^3, (15/11)_\alpha)$

TABLE I. The invariants length δ, jump β and the volume $V(S^3, 7/3, \alpha)$ for some values of α, where all the angles are measured in radians.

α	δ	β	$V(S^3, 7/3, \alpha)$
0	0	6.28319	2.82812
0.1	0.297767	6.03387	2.82068
0.2	0.594463	5.78283	2.79836
0.3	0.888989	5.52826	2.76127
0.4	1.18018	5.26823	2.70952
0.5	1.4667	5.00062	2.64333
0.6	1.74698	4.72316	2.56295
0.7	2.019	4.43345	2.46876
0.8	2.28021	4.12916	2.36123
0.9	2.52741	3.80822	2.24098
1.	2.75677	3.46912	2.10879
1.1	2.96406	3.11116	1.96567
1.2	3.14485	2.73452	1.81282
1.3	3.29483	2.34022	1.6517
1.4	3.40991	1.92987	1.48392
1.5	3.48624	1.50543	1.31135
1.6	3.51994	1.06893	1.13601
1.7	3.50676	0.622273	0.960138
1.8	3.44142	0.167207	0.786203
1.9	3.31674	-0.294782	0.616983
2.	3.12189	-0.762438	0.455695
2.1	2.83898	-1.23472	0.306253
2.2	2.43391	-1.71076	0.173811
2.3	1.82275	-2.18986	0.066203
2.40717	0	-2.70606	0
2.5	1.8273	-3.15504	0.0557996
2.6	2.72591	-3.64022	0.170633
2.7	3.47172	-4.12667	0.325959
2.8	4.15052	-4.6141	0.516711
2.9	4.79355	-5.10224	0.740422
3.	5.4166	-5.59088	0.995736
3.1	6.02948	-6.0798	1.28191
π	6.28319	-6.28319	1.40994

TABLE II. The Chern-Simons invariant $CS(S^3, 7/3, n)$ $\pmod{\frac{1}{\epsilon n}}$ of the orbifold, and the Chern-Simons invariant $CS(M_n(7/3))$ $\pmod{\frac{1}{\epsilon}}$ and the volume $V(M_n(7/3))$ for the hyperbolic manifold $M_n(7/3)$, n-cyclic overing of S^3 branched over the knot 7/3. The first table corresponds to n even, and $\epsilon = 1$. The second table corresponds to n odd and $\epsilon = 2$.

n even	$CS(S^3, 7/3, n)$ $\pmod{\frac{1}{n}}$	$CS(M_n(7/3))$ $\pmod 1$	$V(M_n(7/3))$
4	0.186811	0.747246	4.7495
6	0.0504622	0.302773	12.2552
8	0.11699	0.935921	19.022
10	0.0595432	0.595432	25.3766
12	0.02225	0.267	31.5075
14	0.06751	0.945134	37.5061
16	0.0392035	0.627256	43.4206
18	0.0173328	0.31199	49.2785
20	0.0499265	0.99853	55.0965
22	0.0311988	0.686373	60.8853
24	0.0156328	0.375187	66.652
26	0.00249013	0.064743	72.402
28	0.0269601	0.754882	78.139
30	0.0148495	0.445485	83.865

n odd	$CS(S^3, 7/3, n)$ $\pmod{\frac{1}{2n}}$	$CS(M_n(7/3))$ $\pmod{\frac{1}{2}}$	$V(M_n(7/3))$
3	0.0200144	0.0600431	0.942707
5	0.00166667	0.0083333	8.6124
7	0.0163442	0.11441	15.7081
9	0.0292902	0.263612	22.2358
11	0.039105	0.430155	28.4629
13	0.0081113	0.105447	34.5197
15	0.0190534	0.285801	40.4718
17	0.0276092	0.469356	46.3554
19	0.0081616	0.155071	52.1917
21	0.0163006	0.342313	57.994
23	0.00133367	0.0306745	63.7711
25	0.0087953	0.219883	69.529
27	0.0151759	0.409748	75.272
29	0.00345281	0.100132	81.003
31	0.0093849	0.290932	86.724

TABLE III. The invariants length δ, jump β and the volume $V(S^3, 9/5, \alpha)$ for some values of α, where all the angles are measured in radians.

α	δ	β	$V(S^3, 9/5, \alpha)$
0	0	6.28319	3.16396
0.1	0.256365	6.1006	3.15755
0.2	0.512048	5.91858	3.13834
0.3	0.766413	5.73772	3.10637
0.4	1.01891	5.5587	3.06173
0.5	1.26912	5.38233	3.00452
0.6	1.51673	5.20959	2.93486
0.7	1.76156	5.04175	2.85289
0.8	2.0035	4.88043	2.75875
0.9	2.24241	4.72771	2.65259
1.	2.47791	4.58627	2.53457
1.1	2.70914	4.45942	2.40487
1.2	2.9343	4.35112	2.26375
1.3	3.15017	4.26567	2.11159
1.4	3.35175	4.20718	1.94897
1.5	3.53214	4.17871	1.77677
1.6	3.68303	4.18154	1.59625
1.7	3.79545	4.21486	1.40911
1.8	3.86051	4.27603	1.21749
1.9	3.86952	4.36135	1.02399
2.	3.81341	4.46681	0.831627
2.1	3.68126	4.58862	0.643916
2.2	3.45757	4.72349	0.465021
2.3	3.11656	4.8687	0.300101
2.4	2.60634	5.02204	0.156147
2.5	1.77899	5.18174	0.0445329
2.57414	0	5.30341	0
2.6	1.09619	5.3464	0.0094085
2.7	2.51766	5.51485	0.103529
2.8	3.50481	5.68618	0.254867
2.9	4.36852	5.8596	0.452023
3.	5.17704	6.03442	0.690816
3.1	5.96029	6.21004	0.969312
π	6.28319	6.28319	1.09662

   H. M. HILDEN, M. T. LOZANO, AND J. M. MONTESINOS

TABLE IV. The Chern-Simons invariant $CS(S^3, 9/5, n)$ $(\mathrm{mod}\ \frac{1}{\epsilon n})$ of the orbifold, and the Chern-Simons invariant $CS(M_n(9/5))$ $(\mathrm{mod}\ \frac{1}{\epsilon})$ and the volume $V(M_n(9/5))$ for the hyperbolic manifold $M_n(9/5)$, n-cyclic covering of S^3 branched over the knot 79/5. The first table corresponds to n even, and $\epsilon = 1$. The second table corresponds to n odd and $\epsilon = 2$.

n even	$CS(S^3, 9/5, n)$ $(\mathrm{mod}\ \frac{1}{n})$	$CS(M_n(9/5))$ $(\mathrm{mod}\ 1)$	$V(M_n(9/5))$
4	0.444444	0.888889	6.59895
6	0.144925	0.579699	14.8488
8	0.0351571	0.210943	22.186
10	0.108039	0.864313	29.1289
12	0.0530574	0.530574	35.8703
14	0.0169859	0.203831	42.4952
16	0.062944	0.881216	49.0466
18	0.0350765	0.561224	55.5488
20	0.0134998	0.242996	62.0163
22	0.0463006	0.926012	68.459
24	0.0277243	0.609936	74.882
26	0.0122726	0.294542	81.291
28	0.0376798	0.979675	87.688
30	0.0237579	0.665222	94.076

n odd	$CS(S^3, 9/5, n)$ $(\mathrm{mod}\ \frac{1}{2n})$	$CS(M_n(9/5))$ $(\mathrm{mod}\ \frac{1}{2})$	$V(M_n(9/5))$
3	0.08753	0.26259	1.96274
5	0.078458	0.392288	10.8945
7	0.00506505	0.0354553	18.587
9	0.0218112	0.196301	25.6908
11	0.0333229	0.366552	32.5181
13	0.00324019	0.0421224	39.194
15	0.0147304	0.220956	45.7783
17	0.0236428	0.401928	52.3028
19	0.00444069	0.084373	58.7862
21	0.012756	0.267876	65.24
23	0.0196593	0.452164	71.672
25	0.00548202	0.13705	78.088
27	0.0119408	0.322402	84.491
29	0.000280151	0.0081244	90.883
31	0.00626275	0.194145	97.267

TABLE V. The invariants length δ, jump β and the volume $V(S^3, 11/3, \alpha)$ for some values of α, where all the angles are measured in radians.

α	δ	β	$V(S^3, 11/3, \alpha)$
0	0	6.28319	4.40083
0.1	0.349958	5.60934	4.39208
0.2	0.70048	4.93906	4.36583
0.3	1.05197	4.27595	4.32202
0.4	1.40453	3.6237	4.26061
0.5	1.75776	2.98612	4.18156
0.6	2.11061	2.36704	4.08484
0.7	2.46124	1.7702	3.97053
0.8	2.80691	1.19912	3.8388
0.9	3.14394	0.656771	3.68998
1.	3.46776	0.145353	3.52463
1.1	3.7732	-0.333982	3.34351
1.2	4.05473	-0.781281	3.1477
1.3	4.30688	-1.19777	2.93853
1.4	4.52455	-1.58566	2.71759
1.5	4.70321	-1.94782	2.48673
1.6	4.8388	-2.28744	2.24799
1.7	4.92765	-2.60772	2.00362
1.8	4.96604	-2.91162	1.75606
1.9	4.94982	-3.20182	1.50793
2.	4.87379	-3.48064	1.26208
2.1	4.73088	-3.75002	1.02166
2.2	4.51073	-4.01162	0.790274
2.3	4.19716	-4.26682	0.572142
2.4	3.76235	-4.51677	0.372566
2.5	3.15081	-4.76245	0.198814
2.6	2.21181	-5.00467	0.062738
2.68404	0	-5.20606	0
2.7	1.00031	-5.24414	0.00531085
2.8	2.79514	-5.48147	0.106262
2.9	3.95314	-5.71722	0.275978
3.	4.95586	-5.95189	0.499087
3.1	5.89791	-6.18592	0.770574
π	6.28319	-6.28319	0.897237

TABLE VI. Table VI. The Chern-Simons invariant $CS(S^3, 11/3, n)$ (mod $\frac{1}{\epsilon n}$) of the orbifold, and the Chern-Simons invariant $CS(M_n(11/3))$ (mod $\frac{1}{\epsilon}$) and the volume $V(M_n(11/3))$ for the hyperbolic manifold $M_n(11/3)$, n-cyclic covering of S^3 branched over the knot 11/3. The first table corresponds to n even, and $\epsilon = 1$. The second table corresponds to n odd and $\epsilon = 2$.

n even	$CS(S^3, 11/3, n)$ (mod $\frac{1}{n}$)	$CS(M_n(11/3))$ (mod 1)	$V(M_n(11/3))$
4	0.23583	0.943322	9.2736
6	0.0532272	0.319363	20.6463
8	0.098741	0.78993	30.8729
10	0.030636	0.30636	40.5425
12	0.070641	0.847686	49.9239
14	0.0288449	0.403828	59.1394
16	0.0605904	0.969446	68.251
18	0.0300817	0.541471	77.294
20	0.00590128	0.118026	86.288
22	0.0317225	0.697895	95.247
24	0.0116776	0.280263	104.18
26	0.0332523	0.864561	113.093
28	0.0160849	0.450378	121.989
30	0.00124713	0.0374139	130.873

n odd	$CS(S^3, 11/3, n)$ (mod $\frac{1}{2n}$)	$CS(M_n(11/3))$ (mod $\frac{1}{2}$)	$V(M_n(11/3))$
3	0.108856	0.326569	3.10481
5	0.022888	0.11444	15.1539
7	0.00669311	0.0468517	25.8563
9	0.00489938	0.0440944	35.7555
11	0.00679065	0.074697	45.2596
13	0.0095624	0.124311	54.5476
15	0.0123787	0.185681	63.7056
17	0.0149879	0.254795	72.779
19	0.01733	0.32927	81.796
21	0.0194097	0.407604	90.771
23	0.0212525	0.488807	99.716
25	0.00288798	0.0722	108.638
27	0.00582594	0.1573	117.543
29	0.0084055	0.243759	126.433
31	0.0106876	0.331316	135.311

TABLE VII. The invariants length δ, jump β and the volume $V(S^3, 15/11, \alpha)$ for some values of α, where all the angles are measured in radians.

α	δ	β	$V(S^3, 15/11, \alpha)$
0	0	6.28319	5.13794
0.1	0.321184	6.21461	5.12992
0.2	0.644055	6.14378	5.10579
0.3	0.97026	6.06832	5.06545
0.4	1.30135	5.98559	5.00869
0.5	1.63869	5.89259	4.93521
0.6	1.98334	5.78581	4.8447
0.7	2.33579	5.66112	4.73675
0.8	2.69572	5.51381	4.61099
0.9	3.06162	5.33865	4.46708
1.	3.43037	5.13021	4.30478
1.1	3.79706	4.88346	4.12407
1.2	4.15497	4.59444	3.92522
1.3	4.49602	4.26098	3.70886
1.4	4.81154	3.88307	3.47604
1.5	5.09314	3.46279	3.22827
1.6	5.33328	3.00381	2.96742
1.7	5.52556	2.5107	2.69574
1.8	5.66452	1.98826	2.41575
1.9	5.74518	1.44111	2.13026
2.	5.76245	0.873396	1.84229
2.1	5.71038	0.288717	1.55517
2.2	5.58114	-0.309887	1.27254
2.3	5.3635	-0.919873	0.998525
2.4	5.04018	-1.53913	0.737944
2.5	4.58209	-2.16588	0.49674
2.6	3.93367	-2.79865	0.282872
2.7	2.9585	-3.43619	0.108668
2.8	0.849615	-4.0774	0.00232827
2.80821	0	-4.13017	0
2.9	2.96191	-4.72133	0.089248
3.	4.46781	-5.36713	0.276506
3.1	5.76329	-6.014	0.532706
π	6.28319	-6.28319	0.657974

TABLE VIII. The Chern-Simons invariant $CS(S^3, 15/11, n)$ $(\mathrm{mod}\ \frac{1}{\epsilon n})$ of the orbifold, and the Chern-Simons invariant $CS(M_n(15/11))$ $(\mathrm{mod}\ \frac{1}{\epsilon})$ and the volume $V(M_n(15/11))$ for the hyperbolic manifold $M_n(15/11)$, n-cyclic covering of S^3 branched over the knot 15/11. The first table corresponds to n even, and $\epsilon = 1$. The second table corresponds to n odd and $\epsilon = 2$.

n even	$CS(S^3, 15/11, n)$ $(\mathrm{mod}\ \frac{1}{n})$	$CS(M_n(15/11))$ $(\mathrm{mod}\ 1)$	$V(M_n(15/11))$
4	0.366667	0.733333	12.1793
6	0.0625	0.25	25.3307
8	0.0345888	0.207533	37.0438
10	0.0280513	0.22441	48.1591
12	0.0255297	0.255297	58.9848
14	0.0242803	0.291364	69.649
16	0.0235654	0.329915	80.214
18	0.0231162	0.36986	90.714
20	0.0228149	0.410668	101.168
22	0.0226025	0.452051	111.59
24	0.0224471	0.493837	121.987
26	0.0223299	0.535918	132.365
28	0.0222393	0.578221	142.729
30	0.0221677	0.620695	153.08

n odd	$CS(S^3, 15/11, n)$ $(\mathrm{mod}\ \frac{1}{2n})$	$CS(M_n(15/11))$ $(\mathrm{mod}\ \frac{1}{2})$	$V(M_n(15/11))$
3	0.12215	0.366451	4.71354
5	0.0428241	0.21412	19.0239
7	0.0304384	0.213069	31.2952
9	0.0265452	0.238907	42.6504
11	0.0248101	0.272911	53.5979
13	0.0238782	0.310417	64.332
15	0.023317	0.349754	74.942
17	0.0229516	0.390177	85.471
19	0.0227	0.4313	95.946
21	0.0225191	0.472901	106.383
23	0.00064548	0.014846	116.791
25	0.00228181	0.0570453	127.178
27	0.00368293	0.099439	137.549
29	0.00489604	0.141985	147.906
31	0.00595652	0.184652	158.253

References

[C]   H.S.M.Coxeter, *Non-Euclidean Geometry*, University of Toronto Press, 1968.

[C-S]   S.S. Chern and J. Simons, *Characteristic forms and geometric invariants*, Annals of Math **99** (1974), 48–69.

[HLM$_1$] H.M. Hilden, M.T. Lozano and J.M.Montesinos-Amilibia, *A Characterization of Arithmetic Subgroups of SL(2,$\mathbb{R}$) and SL(2,$\mathbb{C}$)*, Math. Nach. **159** (1992), 245–270.

[HLM$_2$] H.M. Hilden, M.T. Lozano and J.M.Montesinos-Amilibia, *On the arithmetic 2-bridge knots and link orbifolds and a new knot invariant*, Journal of Knots Theory ands Its Ramifications **4** (1995), 81–114.

[HLM$_3$] H.M. Hilden, M.T. Lozano and J.M.Montesinos-Amilibia, *On a remarkable polyhedron geometrizing the figure eight knot cone manifolds*, Journal of Math. Sciences, Tokyo, (to appear).

[HLM$_4$] H.M. Hilden, M.T. Lozano and J.M.Montesinos-Amilibia, *Euclidean representations of 2-bridge knots*, in preparation.

[HLM$_5$] H.M. Hilden, M.T. Lozano and J.M.Montesinos-Amilibia, *On volumes and Chern-Simons invariants of geometric 3-manifolds*, preprint.

[H]   C. Hodgson, *Degeneration and regeneration of geometric structures on three-manifolds*, Ph.D. Thesis, Princeton University (1986).

[M]   J.Milnor, *The Schläffli differential equality*, Collected papers. Volume 1, Geometry, Publish or Perish, Inc., 1994, pp. 281–295.

[Me]   R. Meyerhoff, *Density of the Chern-Simons invariant for hyperbolic 3-manifolds*, Low-dimensional topology and Kleinian groups,London, (D.B.A.Epstein, eds), London Math. Soc. Lect. Notes 112, Cambridge University Press. 1987. pp. 217–240

[M-R]   R. Meyerhoff and D. Ruberman, *Mutation and the η-invariant*, J. Differential Geometry **31** (1990), 101–130.

[Mon]   J.M.Montesinos, *Classical Tessellations and Three-Manifolds*, Universitext. Springer-Verlag, 1987.

[Mos]   G.D. Mostow, *Quasi-conformal mappings in n-space and the rigidity of hyperbolic space forms*, Publ.IHS **34** (1968), 53–104.

[Ri]   R. Riley, *Algebra for Heckoid Groups*, Trans of AMS, **334** (1992), 389–409.

[Ro]   D. Rolfsen, *Knots and links*, Publish or Perish, Inc., 1976.

[S]   E. Suárez-Peiró, *On the uniformization of 2-bridge knot and link cone-manifolds*, in preparation.

[Y]   T. Yoshida, *The η-invariant of hyperbolic 3-manifolds*, Invent. Math. **8** (1985), 473–514.

[T]   W. Thurston, *The Geometry and Topology of 3-Manifolds*, Notes 1976–1978., Princeton University Press., (to appear).

[Vi]   E.B.Vinberg, *Geometry II*, Encyclopaedia of Mathematical Sciences. Vol♯29, Springer-Verlag, 1992.

[W]   Wolfram Research,Inc., *MATHEMATICA*, Wolfram Research,Inc.,Champaign,Illinois, 1994.

DEPARTMENT OF MATHEMATICS, UNIVERSITY OF HAWAII, HONOLULU, HI 96822, U.S.A.
E-mail address: mike@kahuna.math.hawaii.edu

DEPARTAMENTO DE MATEMÁTICAS, UNIVERSIDAD DE ZARAGOZA, 50009 ZARAGOZA, SPAIN
E-mail address: tlozano@posta.unizar.es

DEPARTAMENTO DE GEOMETRÍA Y TOPOLOGÍA, FACULTAD DE MATEMÁTICAS, UNIVERSIDAD COMPLUTENSE, 28040 MADRID, SPAIN
E-mail address: montesin@eucmax.sim.ucm.es

Proceedings of
THE 37TH TANIGUCHI SYMPOSIUM ON
TOPOLOGY AND TEICHMÜLLER SPACES
held in Finland, July 1995
ed. by Sadayoshi KOJIMA *et al.*
©1996 World Scientific Publishing Co.
pp. 57–78

AN ESTIMATE OF THE NUMBER OF NON-CONSTANT HOLOMORPHIC MAPS BETWEEN RIEMANN SURFACES

YOICH IMAYOSHI

(Received December 20, 1995)

Introduction

Let $\mathrm{Hol}(R, S)$ be the set of all non-constant holomorphic maps of a closed Riemann surface R of genus g onto another closed Riemann surface S of genus g' with $2 \le g' \le g$. Denote by $\#\mathrm{Hol}(R, S)$ the number of elements of $\mathrm{Hol}(R, S)$.

In this paper, using complex analysis and hyperbolic geometry, we will give an upper bound for $\#\mathrm{Hol}(R, S)$, which depends only on g and g'.

First, assume that $g = g' \ge 2$. In this case, Riemann-Hurwitz relation implies that every non-constant holomorphic map of R onto S is biholomorphic. Thus $\mathrm{Hol}(R, S)$ is either empty or identified with the analytic automorphism group $\mathrm{Aut}(R)$ of R. Since $\#\mathrm{Aut}(R) \le 84(g - 1)$, we have $\#\mathrm{Hol}(R, S) \le 84(g - 1)$.

Now, let us recall a proof of the assertion $\#\mathrm{Aut}(R) \le 84(g-1)$. Take a fixed-point-free Fuchsian group Γ, acting on the unit disk Δ, such that the quotient space Δ/Γ is biholomorphic to R. Denote by $N(\Gamma)$ the normalizer of Γ in the group of conformal self-maps of Δ. Let h-area(R) be the hyperbolic area of R, and h-area$(\Delta/N(\Gamma))$ the hyperbolic area of $\Delta/N(\Gamma)$. It is well-known that h-area$(R) = 4\pi(g - 1)$ and h-area$(\Delta/N(\Gamma)) \ge \pi/21$. Accordingly, we see that $\#\mathrm{Aut}(R) = \#[N(\Gamma) : \Gamma] = $ h-area$(R)/$ h-area$(\Delta/N(\Gamma)) \le 84(g - 1)$.

In order to estimate $\#\mathrm{Hol}(R, S)$ with $2 \le g' < g$, we will use similar area estimate. Our main results are as follows. First, considering in the homotopy category, we have the following theorem.

Theorem 1. *Let R be a closed Riemann surface of genus g, and S another closed Riemann surface of genus g' with $2 \le g' \le g$. Then there exists a positive constant M_1 such that*

$$\#\mathrm{Hol}(R, S) \le e^{M_1 g^3}.$$

Next, considering in the homology category, we have the following theorem, which is better.

Theorem 2. *Let R be a closed Riemann surface of genus g, and S another closed Riemann surface of genus g' with $2 \leq g' \leq g$. Then there exists a positive constant M_2 such that*

$$\# \operatorname{Hol}(R, S) \leq e^{M_2 g^2}.$$

Here, we note that the following results are known:

(1) (Howard and Sommese [3])   There exists a positive constant M_3 such that

$$\# \operatorname{Hol}(R, S) \leq e^{M_3 (\log g) g^2}.$$

(2) (Kani [7])   There exists a positive constant M_4 such that

$$\# \operatorname{Hol}(R, S) \leq e^{M_4 g^2}.$$

(3) (Tanabe [11])   There exists a positive constant M_5 such that

$$\# \operatorname{Hol}(R, S) \leq e^{M_5 (\log g) g}.$$

The results are not the main achievment of this paper but rather the methods which are geometric and probably subject to improvements. Our method is applicable to the case of analytically finite Riemann surfaces of (g, n) with $2g - 2 + n > 0$. Furthermore, by this idea, we will be able to treat problems on estimating numbers of objects for Severi's theorem (cf. Imayoshi [5]), and Mordell conjecture and Shafarevich conjecture in the function field case (cf. Imayoshi and Shiga [6]).

Unless otherwise stated, all Riemann surfaces considered in this paper will be assumed to be closed, of genus ≥ 2 and endowed with hyperbolic metric.

The author is grateful to Professor Masahiko Taniguchi for his helpful comments and valuable conversations with him. The author would like to express his hearty gratitude to the Tanuguchi Foundation for financial support.

1. Preliminaries

Let us recall rigidity theorems for non-constant holomorphic maps. Let Γ be a Fuchsian group acting on the unit disc $\Delta = \{ z \in \mathbf{C} \mid |z| < 1 \}$ in the complex plane. A Fuchsian group Γ is said to be of divergence type if

$$\sum_{\gamma \in \Gamma} (1 - |\gamma(0)|) = +\infty.$$

Myrberg's Theorem asserts that Γ is of divergence type if and only if the Riemann surface Δ/Γ has no Green's functions. Thus every finitely generated Fuchsian group of the first kind is of divergence type.

Let G be a fixed-point-free discrete subgroup on the analytic automorphism group $\operatorname{Aut}(D)$ of a bounded domain D in the complex plane. For any holomorphic map $\varphi \colon \Delta/\Gamma \to D/G$, the Monodromy Theorem implies that there exists a holomorphic map $\tilde{\varphi} \colon \Delta \to D$ with $\varphi \circ \pi = p \circ \tilde{\varphi}$, where $\pi \colon \Delta \to \Delta/\Gamma$, $p \colon D \to D/G$ are canonical projections. This holomorphic map $\tilde{\varphi}$ is called a *lift* of φ. Note that $\tilde{\varphi}$ is not unique, for we may replace $\tilde{\varphi}$ by $\tilde{\varphi}_1 = g_0 \circ \varphi \circ \gamma_0$, where $\gamma_0 \in \Gamma$ and $g_0 \in G$. A lift $\tilde{\varphi}$ induces a group homomorphism $\tilde{\varphi}_* \colon \Gamma \to G$ satisfying the following relation:

$$\tilde{\varphi} \circ \gamma = \tilde{\varphi}_*(\gamma) \circ \tilde{\varphi}, \quad \gamma \in \Gamma.$$

We call $\tilde{\varphi}_*$ the *monodromy* induced by $\tilde{\varphi}$.

Our method to estimate $\# \operatorname{Hol}(R, S)$ is to determine maps of $\operatorname{Hol}(R, S)$ by using the following rigidity theorem. (See Imayoshi [4], Theorem 1.)

Rigidity Theorem. *Let Γ be a Fuchsian group of divergence type acting on the unit disc Δ, and let G be a fixed-point-free discrete subgroup of $\operatorname{Aut}(D)$ of a bounded domain D in the complex plane. If non-constant holomorphic maps $\varphi, \psi \colon \Delta/\Gamma \to D/G$ induce the same homomorphism $\tilde{\varphi}_* = \tilde{\psi}_* \colon \Gamma \to G$, then $\varphi = \psi$.*

Instead of this rigidity theorem, we may use the following rigidity theorem due to Martens [8, 9].

Martens' Rigidity Theorem. *Let R, S_1, S_2 be closed Riemann surfaces of genera ≥ 2 and assume that holomorphic maps $f_1 \colon R \to S_1$ and $f_2 \colon R \to S_2$ are given. Then a necessary and sufficient condition for the existence of a holomorphic map $h \colon S_1 \to S_2$ making the diagram*

$$
\begin{array}{ccc}
R & \xrightarrow{f_1} & S_1 \\
{\scriptstyle id}\downarrow & & \downarrow{\scriptstyle h} \\
R & \xrightarrow{f_2} & S_2
\end{array}
$$

commutative is the existence of a homomorphism $h_* \colon H_1(S_1, \mathbf{Z}) \to H_1(S_2, \mathbf{Z})$ of the first homology groups making the diagram

$$
\begin{array}{ccc}
H_1(R, \mathbf{Z}) & \xrightarrow{(f_1)_*} & H_1(S_1, \mathbf{Z}) \\
{\scriptstyle id}\downarrow & & \downarrow{\scriptstyle h_*} \\
H_1(R, \mathbf{Z}) & \xrightarrow{(f_2)_*} & H_1(S_2, \mathbf{Z})
\end{array}
$$

commutative, and the holomorphic map h is uniquely determined by the condition that it induces the homomorphism h_*.

In order to use the rigidity theorem, we recall some results of hyperbolic geometry on surfaces. As a background reference we cite the textbook Buser [2], to which we refer throughout this paper.

We always assume that a closed Riemann surface R of genus ≥ 2 is endowed with the hyperbolic metric. Denote by $d_R(p, q)$ the hyperbolic distance between $p, q \in R$. Denote by $\ell_R(C)$ the hyperbolic length of a curve C on R. Let h-area(D) be the hyperbolic area of a set D in R.

For any point $p \in R$, the supremum of all r for which $\{q \in R \,|\, d_R(p, q) < r\}$ is isometric to a disk in the hyperbolic plane is called the *injectivity radius of R at p*, and is denoted by $r_p(R)$. The *injectivity radius of R* is defined as

$$
r_{\mathrm{inj}}(R) = \inf\{r_p(R) \,|\, p \in R\}.
$$

It is known that $r_p(R) = \ell_R(L_p)/2$, where L_p is the shortest geodesic loop at p, and $r_{\mathrm{inj}}(R) = \ell_R(L)/2$, where L is the shortest closed geodesic on R. Moreover, the geodesics L_p and L are simple. (See Buser [2], Lemma 4.1.5.)

Let $\mathcal{L} = \{L_j\}_{j=1}^{3g-3}$ be a set of pairwise disjoint $3g-3$ simple closed geodesics on R such that $R \setminus \cup_{j=1}^{3g-3} L_j$ has $2g-2$ connected components P_k of type $(0,3)$, i.e., surfaces homeomorphic to a sphere with three holes. We call $\mathcal{L}$ a *system of decomposing curves* of R, and $\mathcal{P} = \{P_j\}_{j=1}^{2g-2}$ the *pants decomposition* on R corresponding to $\mathcal{L}$. Each P_j is called a *pair of pants*, and a connected component of the boundary of P_j is called a *boundary geodesic* of P_j.

Now, we recall Collar theorems. (See Buser [2], Chapter 4.) Let R be a closed Riemann surface of genus $g \geq 2$. Let L be a boundary geodesic of a pants decomposition of R. The *collar* $\mathcal{C}_R(L)$ around L with width $w_R(L) = \log \coth(\ell_R(L)/4)$ is defined by

$$\mathcal{C}_R(L) = \{p \in R \,|\, d_R(p, L) \leq w_R(L)\}.$$

It is known that $\mathcal{C}_R(L)$ is isometric to the cylinder $[-w_R(L), w_R(L)] \times S^1$ with Riemannian metric $ds^2 = d\rho^2 + \ell_R(L)^2 \cosh^2 \rho \, dt^2$. We also define $\mathcal{C}_R^*(L)$ by

$$\mathcal{C}_R^*(L) = \{p \in R \,|\, d_R(p, L) \leq w_R^*(L)\},$$

where $w_R^*(L) = \log \coth(\ell_R(L))$.

Collar theorem I. *Let R be a closed Riemann surface of genus $g \geq 2$. Let $L_1, \cdots, L_m$ be pairwise disjoint simple closed geodesics on R. Then the following hold.*

    (1) *$m \leq 3g-3$.*

    (2) *There exist simple closed geodesics $L_{m+1}, \cdots, L_{3g-3}$ such that $L_1, \cdots, L_{3g-3}$ decompose R into pairs of pants.*

    (3) *Collars $\mathcal{C}_R(L_j) = \{p \in R \,|\, d_R(p, L_j) \leq w_R(L_j)\}, j = 1, \cdots, 3g-3$, are pairwise disjoint.*

    (4) *Each $\mathcal{C}_R(L_j)$ is isomorphic to the cylinder $[-w_R(L_j), w_R(L_j)] \times S^1$ with Riemannian metric $ds^2 = d\rho^2 + \ell_R(L_j)^2 \cosh^2 \rho \, dt^2$.*

Collar theorem II. *Let R be a closed Riemann surface of genus $g \geq 2$. Let $L_1, \cdots, L_{j_0}$ be the set of all simple closed geodesics of length $\leq 2 \operatorname{arcsinh} 1 = 2 \log(1 + \sqrt{2})$ on R. The following hold:*

    (1) *$j_0 \leq 3g-3$.*

    (2) *$L_1, \cdots, L_{j_0}$ are pairwise disjoint.*

    (3) *The injectivity radius $r_p(R)$ satisfies $r_p(R) > \operatorname{arcsinh} 1$ for $p \in R \setminus \{\mathcal{C}_R(L_1) \cup \cdots \cup \mathcal{C}_R(L_{j_0})\}$.*

    (4) *If $p \in \mathcal{C}_R(L_j), 1 \leq j \leq j_0$, and $d(p) = d_R(p, \partial \mathcal{C}_R(L_j))$, then*

$$\sinh r_p(R) = \cosh \frac{\ell_R(L_j)}{2} \cosh d(p) - \sinh d(p).$$

Next, recall Bers' theorem on pants decompositions of a closed hyperbolic Riemann surface R of genus $g \geq 2$. He proves that R has a system of decomposing curves $\mathcal{L} = \{L_j\}_{j=1}^{3g-3}$ satisfying

$$\ell_R(L_j) \leq \ell_g, \quad j = 1, \ldots, 3g-3,$$

where ℓ_g is Bers' constant, which depends only on g. Using erea estimate, Buser obtains

$$\ell_R(L_j) \leq 4j \log \frac{8\pi(g-1)}{j}, \quad j = 1, \ldots, 3g-3,$$

which yields $\ell_g \leq 26(g-1)$. (See Buser [2], Theorem 5.1.2.) It is also known that $\ell_g \geq \sqrt{6g} - 2$.

By induction with respect to genus g, for a given system of decomposing curves $\mathcal{L} = \{L_j\}_{j=1}^{3g-3}$ of R, it is shown that we can find $2g$ simple closed curves $A_1, \ldots, A_g$, $B_1, \ldots, B_g$ on R satisfying the following conditions:

(1) Each $A_j \in \mathcal{L}$.
(2) Each B_j is a union of simple paths with endpoints on decomposing curves.
(3) $A_j \cap B_j = \{p_j\}$, (one point), for $1 \leq j \leq g$, and $A_j \cap A_k = B_j \cap B_k = A_j \cap B_k = \emptyset$ for $1 \leq j, k \leq g$ with $j \neq k$.
(4) These A_j, B_j induce a canonical basis for the first homology group $H_1(R, \mathbf{Z})$, and every $L_j \in \mathcal{L}$ is homologous to $\sum_{i=1}^{g} n_i A_i, n_i \in \mathbf{Z}$.
(5) For a given point $p_0 \in R$, there exist a simple curve C_j from p_0 to p_j such that C_j is a union of simple paths with endpoints in $\{p_0\} \cup \{\cup_{j=1}^{3g-3} L_j\}$ and such that these $C_j \cdot A_j \cdot C_j^{-1}$, $C_j \cdot B_j \cdot C_j^{-1}$, $j = 1, \cdots, g$, induce a system of generators for the fundamental group $\pi_1(R, p_0)$ of R.

Throughout this paper we fix these symbols A_j, B_j, C_j, p_0, and $r_0 = \operatorname{arcsinh} 0.2 = 0.19869 \cdots$. We always assume that $\ell_g \geq 2$.

2. Estimate of $\#\mathrm{Hol}(R, S)$ with injectivity radius $r_{\mathrm{inj}}(R) \geq r_0$

In this section, we will give an upper bound for $\#\mathrm{Hol}(R, S)$ with $r_{\mathrm{inj}}(R) \geq r_0$. Let us begin with the following lemma:

Lemma 1. *If a closed Riemann surface R of genus ≥ 2 has the injectivity radius $r_{\mathrm{inj}}(R) \geq r_0$, then the diameter $\mathrm{diam}(R)$ of R satisfies*

$$\mathrm{diam}(R) \leq 3r_0 + \frac{2r_0}{\sinh^2 \frac{r_0}{2}}(g-1) \leq 41(g-1).$$

Proof. Take any two points $p, q \in R$. Let L be a geodesic on R with endpoints p, q and $d_R(p, q) = \ell_R(L)$. Let $\pi\colon \mathbf{H} \to R$ be the universal covering of R, where $\mathbf{H}$ is the upper half-plane. We may assume that a connected component of $\pi^{-1}(L)$ is a straight line segment between ia and ib with $0 < a < b$. Take disks $\{\tilde{D}_j\}_{j=1}^n$ on $\mathbf{H}$ satisfying the following conditions;

(1) each $\tilde{D}_j$ is a hyperbolic disk with center iy_j and radius r_0,
(2) $a = y_1 < y_2 < \cdots < y_n$, and $ib \in \tilde{D}_n$,
(3) $\{\tilde{D}_j\}_{j=1}^n$ are pairwise disjoint, and $\tilde{D}_j$ is tangent to $\tilde{D}_{j+1}$ for $j = 1, \cdots, n-1$.

We set $D_j = \pi(\tilde{D}_j)$. Since $d_R(p, q) = \ell_R(L)$, we see that $D_j \cap D_k = \emptyset$ for all j, k with $1 \leq j < k \leq n - 2$. Thus we find

$$\sum_{j=1}^{n-2} \text{h-area}(D_j) \leq \text{h-area}(R) = 4\pi(g-1).$$

Noting $\text{h-area}(D_j) = \text{h-area}(\tilde{D}_j) = 4\pi \sinh^2(r_0/2)$, we get

$$n - 2 \leq \frac{g-1}{\sinh^2 \frac{r_0}{2}}.$$

Therefore, we obtain

$$d_R(p, q) \leq 3r_0 + 2r_0(n-1) \leq 3r_0 + \frac{2r_0}{\sinh^2 \frac{r_0}{2}} (g-1) \leq 41(g-1).$$

$\square$

Next, we estimate hyperbolic lengths of B_j and C_j.

Lemma 2. *Let R be a closed Riemann surface of genus $g \geq 2$ with injectivity radius $r_{\text{inj}}(R) \geq r_0$. If R has a system of decomposing curves $\mathcal{L} = \{L_j\}_{j=1}^{3g-3}$ with $\ell_R(L_j) \leq \ell_g$, then B_j and C_j are taken so that*

$$\ell_R(B_j) \leq 2 \left[\log 2 + \ell_g + \log \left(\frac{\cosh \frac{\ell_g}{2}}{\sinh^2 r_0} + \coth^2 r_0 \right) \right] (g-1)$$

$$\leq (8 + 3\ell_g)(g-1),$$

$$\ell_R(C_j) \leq 2 \left[\log 2 + \ell_g + \log \left(\frac{\cosh \frac{\ell_g}{2}}{\sinh^2 r_0} + \coth^2 r_0 \right) \right] g - \ell_g$$

$$\leq (8 + 3\ell_g)(g-1) + 2\ell_g + 8.$$

Proof. For any pair of pants with boundary geodesics $L_{j_1}, L_{j_2}, L_{j_3} \in \mathcal{L}$, drop the common perpendicular H between L_{j_1}, L_{j_2}. By hyperbolic trigonometry, we have

$$\cosh \ell_R(H) = \frac{\cosh c}{\sinh a \sinh b} + \coth a \coth b,$$

where $a = \ell_R(L_{j_1})/2, b = \ell_R(L_{j_2})/2$ and $c = \ell_R(L_{j_3})/2$. (See Buser [2], Theorem 2.4.1 (i).) Thus we find

$$\cosh \ell_R(H) \leq \frac{\cosh \frac{\ell_g}{2}}{\sinh^2 r_0} + \coth^2 r_0.$$

Hence we obtain

$$\ell_R(H) \leq \log(2 \cosh \ell_R(H))$$

$$\leq \log 2 + \log \left(\frac{\cosh \frac{\ell_g}{2}}{\sinh^2 r_0} + \coth^2 r_0 \right)$$

$$= \log 2 + \log \left(25 \cosh \frac{\ell_g}{2} + 26 \right)$$

$$\leq \log 2 + 2\log 5 + \frac{\ell_g}{2}$$

$$\leq 4 + \frac{\ell_g}{2}.$$

Since B_j passes at most $2(g-1)$ pairs of pants, counting these $\ell_R(H)$ and twist parameters, we see that

$$\ell_R(B_j) \leq \left[\log 2 + \log\left(\frac{\cosh\frac{\ell_g}{2}}{\sinh^2 r_0} + \coth^2 r_0\right)\right] \times 2(g-1) + \ell_g \times 2(g-1)$$
$$\leq (3\ell_g + 8)(g-1).$$

By the same argument, and counting also the diameter of a pair of pants, we obtain the estimate for $\ell_R(C_j)$. $\square$

Now, we assert the following proposition:

Proposition 1. *Let R be closed Riemann surface of genus g with injectivity radius $r_{\mathrm{inj}}(R) \geq r_0$, and S another closed Riemann surface of genus g' with $2 \leq g' \leq g$. Set*

$$r_1(g) = 2(3\ell_g + 49)(g-1) + 5\ell_g + 16,$$
$$r_2(g) = (9\ell_g + 106)(g-1) + 4\ell_g + 16,$$
$$n_1(g) = \left[\frac{\cosh\left(r_1(g) + \log(1+\sqrt{2})\right) - 1}{\sqrt{2}-1}\right],$$
$$n_2(g) = \left[\frac{\cosh\left(r_2(g) + \log(1+\sqrt{2})\right) - 1}{\sqrt{2}-1}\right],$$

where $[\,\cdot\,]$ means the Gaussian symbol. Then

$$\# \operatorname{Hol}(R,S) \leq n_1(g)^g \, n_2(g)^g.$$

Here note that

$$n_j(g) \leq \left[\frac{(1+\sqrt{2})^2}{2}\, e^{r_j(g)}\right], \quad j = 1, 2.$$

Proof. We construct concretely the universal covering surface $\tilde{R}$ of R as follows. Fix a base point p_0 of R. Then $\tilde{R}$ is the set of all homotopy equivalence classes $[C, p]$ of pairs (C, p), where $p \in R$ and C is a path on R from p_0 to p. The fundamental group $\pi_1(R, p_0)$ is canonically identified with the covering transformation group of $\tilde{R}$. Similarly, we construct the universal covering surface $\tilde{S}$ of S with base point q_0. We may assume that the injectivity radius $r_{q_0}(S)$ of S at q_0 satisfies

$$r_{q_0}(S) \geq \operatorname{arcsinh} 1 = \log(1+\sqrt{2}).$$

(See Buser [2], Theorem 4.1.6.)

For any $\varphi \in \operatorname{Hol}(R, S)$, take a point $p_\varphi \in R$ with $\varphi(p_\varphi) = q_0$. By Lemma 1, we can find a path C_φ on R from p_0 to p_φ so that

$$\ell_R(C_\varphi) \leq 3r_0 + \frac{2r_0}{\sinh^2\frac{r_0}{2}}(g-1).$$

A lift $\tilde{\varphi} \colon \tilde{R} \to \tilde{S}$ of $\varphi \colon R \to S$ is defined by

$$\tilde{\varphi}([C, p]) = [\varphi(C_\varphi^{-1} \cdot C), \varphi(p)].$$

Then the monodromy $\tilde{\varphi}_*$ of $\tilde{\varphi}$ is given by

$$\tilde{\varphi}_*([C_0]) = \varphi_*([C_\varphi^{-1} \cdot C_0 \cdot C_\varphi]), \quad [C_0] \in \pi_1(R, p_0).$$

We set $\tilde{p}_0 = [I_{p_0}, p_0]$, $\tilde{p}_\varphi = [C_\varphi, p_\varphi] \in \tilde{R}$, and $\tilde{q}_0 = [I_{q_0}, q_0] \in \tilde{S}$, where I_{p_0}, I_{q_0} are paths on R and S, respectively such that $I_{p_0}(t) = p_0$ and $I_{q_0}(t) = q_0$ for any t in the unit interval $[0, 1]$. Note that $\tilde{\varphi}(\tilde{p}_\varphi) = \tilde{q}_0$.

Denote by $\gamma_j = C_j \cdot A_j \cdot C_j^{-1}$, $\gamma_{g+j} = C_j \cdot B_j \cdot C_j^{-1}$ for $j = 1, 2, \cdots, g$. Let $d_{\tilde{R}}$ and $d_{\tilde{S}}$ be the hyperbolic distances on $\tilde{R}$ and $\tilde{S}$, respectively. Then, by the distance decreasing property of the hyperbolic distances for holomorphic maps, we have

$$\begin{aligned}
d_{\tilde{S}}(\tilde{q}_0, \tilde{\varphi}_*(\gamma_j)(\tilde{q}_0)) &= d_{\tilde{S}}(\tilde{\varphi}(\tilde{p}_\varphi), \tilde{\varphi}_*(\gamma_j) \circ \tilde{\varphi}(\tilde{p}_\varphi)) \\
&= d_{\tilde{S}}(\tilde{\varphi}(\tilde{p}_\varphi), \tilde{\varphi} \circ \gamma_j(\tilde{p}_\varphi)) \\
&\leq d_{\tilde{R}}(\tilde{p}_\varphi, \gamma_j(\tilde{p}_\varphi)) \\
&\leq 2\ell_R(C_\varphi) + \ell_R(C_0) \\
&\leq r_1(g),
\end{aligned}$$

where $C_0 = C_j \cdot A_j \cdot C_j^{-1}$. Thus we obtain

$$(2.1) \qquad\qquad d_{\tilde{S}}(\tilde{q}_0, \tilde{\varphi}_*(\gamma_j)(\tilde{q}_0)) \leq r_1(g), \quad j = 1, \cdots, g.$$

Similarly, we get

$$d_{\tilde{S}}(\tilde{q}_0, \tilde{\varphi}_*(\gamma_j)(\tilde{q}_0)) \leq r_2(g), \quad j = g + 1, \cdots, 2g.$$

Denote by $\Delta(\tilde{q}_0, r)$ the hyperbolic disk in $\tilde{S}$ with center $\tilde{q}_0$ and radius $r > 0$. From (2.1), we have

$$\tilde{\varphi}_*(\gamma_j)(\Delta(\tilde{q}_0, r_1)) \subset \Delta(\tilde{q}_0, r_1 + r_1(g)), \quad j = 1, \cdots, g,$$

where $r_1 = \operatorname{arcsinh} 1$. Thus we get

$$(2.2) \qquad \#\{\tilde{\varphi}_*(\gamma_j) \,|\, \varphi \in \operatorname{Hol}(R, S)\} \leq \left\lceil \frac{\text{h-area}(\Delta(\tilde{q}_0, r_1 + r_1(g)))}{\text{h-area}(\Delta(\tilde{q}_0, r_1))} \right\rceil \leq n_1(g)$$

for all $j = 1, \cdots, g$. Similarly, we obtain

$$(2.3) \qquad \#\{\tilde{\varphi}_*(\gamma_j) \,|\, \varphi \in \operatorname{Hol}(R, S)\} \leq \left\lceil \frac{\text{h-area}(\Delta(\tilde{q}_0, r_1 + r_2(g)))}{\text{h-area}(\Delta(\tilde{q}_0, r_1))} \right\rceil \leq n_2(g)$$

for all $j = g + 1, \cdots, 2g$.

Now, suppose that $\# \operatorname{Hol}(R, S) > n_1(g)^g n_2(g)^g$. Then, by (2.2) and (2.3), we see that $\operatorname{Hol}(R, S)$ has a subset $\mathcal{F}_1$ consisting of at least $n_1(g)^{g-1} n_2(g)^g + 1$ distinct elements so that

$$\tilde{\varphi}_*(\gamma_1) = \tilde{\psi}_*(\gamma_1), \quad \varphi, \psi \in \mathcal{F}_1.$$

Applying this argument to $\mathcal{F}_1$, we find a subset $\mathcal{F}_2$ consisting of at least $n_1(g)^{g-2} n_2(g)^g + 1$ distinct elements in $\mathcal{F}_1$ such that

$$\tilde{\varphi}_*(\gamma_j) = \tilde{\psi}_*(\gamma_j), \quad \varphi, \psi \in \mathcal{F}_2, \text{ and } j = 1, 2.$$

Continuing these processes $2g$ times, we conclude that $\mathrm{Hol}(R, S)$ has two distinct elements φ, ψ such that

$$\tilde{\varphi}_*(\gamma_j) = \tilde{\psi}_*(\gamma_j), \quad j = 1, 2, \cdots, 2g.$$

Therefore, the Rigidity Theorem shows that $\varphi = \psi$, which is a contradiction. This completes the proof of Proposition 1. $\square$

Instead of the rigidity theorem stated above, using Martens' rigidity theorem, we shall get a better estimate for $\mathrm{Hol}(R, S)$. For this purpose, we need the following lemma.

Lemma 3. *Let R and S be closed Riemann surfaces of genus g and g', respectively, with $2 \leq g' \leq g$. Let $\{[A_j], [B_j]\}_{j=1}^{g}$ and $\{[A_j'], [B_j']\}_{j=1}^{g'}$ be canonical homology bases for $H_1(R, \mathbf{Z})$ and $H_1(S, \mathbf{Z})$, respectively. If $\varphi, \psi \in \mathrm{Hol}(R, S)$ satisfy $\varphi_*([A_j]) = \psi_*([A_j])$ in $H_1(S, \mathbf{Z})$ for all $j = 1, 2, \cdots, g$, then $\varphi = \psi$.*

Proof. Denote by $M(\varphi)$ the matrix representation of $\varphi \in \mathrm{Hol}(R, S)$ with respect to homology bases $\{[A_j], [B_j]\}_{j=1}^{g}$ and $\{[A_j'], [B_j']\}_{j=1}^{g'}$, that is,

$$(\varphi_*([A_1]), \cdots, \varphi_*([A_g]), \varphi_*([B_1]), \cdots, \varphi_*([B_g]))$$
$$= \left([A_1'], \cdots, [A_{g'}'], [B_1'], \cdots, [B_{g'}']\right) \times M(\varphi),$$

where

$$M(\varphi) = \begin{pmatrix} M_1(\varphi) & M_2(\varphi) \\ M_3(\varphi) & M_4(\varphi) \end{pmatrix}.$$

We put

$$J = \begin{pmatrix} 0 & E \\ -E & 0 \end{pmatrix}, \quad J' = \begin{pmatrix} 0 & E' \\ -E' & 0 \end{pmatrix},$$

where E is the identity matrix of size $g \times g$, and E' is the identity matrix of size $g' \times g'$. We set

$$M = M(\varphi) - M(\psi) = \begin{pmatrix} M_1 & M_2 \\ M_3 & M_4 \end{pmatrix}.$$

By a direct calculation, we see that

$$J\,{}^tM(J')^{-1}M = \begin{pmatrix} -{}^tM_2M_3 + {}^tM_4M_1 & * \\ * & {}^tM_1M_4 - {}^tM_3M_2 \end{pmatrix}.$$

Since $\varphi_*([A_j]) = \psi_*([A_j])$ in $H_1(S, \mathbf{Z})$ for each j, it follows that $M_1(\varphi) = M_1(\psi)$, $M_3(\varphi) = M_3(\psi)$, and so $M_1 = 0, M_3 = 0$. This asserts that

$$\mathrm{tr}\left(J\,{}^tM(J')^{-1}M\right) = 0.$$

Thus, from a lemma due to Martens, we have $M = 0$, i.e., $M(\varphi) = M(\psi)$. (See Martens [9], Lemma in section 2.) Therefore, Martens' Rigidity Theorem implies that $\varphi = \psi$. $\square$

Now, we have the following proposition:

Proposition 2. *Let R be closed Riemann surface of genus g with injectivity radius $r_{\mathrm{inj}}(R) \geq r_0$, and S another closed Riemann surface of genus g' with $2 \leq g' \leq g$. Set*

$$r_1'(g) = 2\left\{3r_0 + \frac{2r_0}{\sinh^2 \frac{r_0}{2}}\,(g-1)\right\} + \ell_g,$$

$$n_1'(g) = \left[\frac{\cosh\left(r_1'(g) + \log(1+\sqrt{2})\right) - 1}{\sqrt{2}-1}\right],$$

where $[\,\cdot\,]$ means the Gaussian symbol. Then

$$\#\,\mathrm{Hol}(R, S) \leq n_1'(g)^g.$$

Here note that

$$r_1'(g) \leq 82(g-1) + \ell_g,$$

$$n_1(g) \leq \left[\frac{(1+\sqrt{2})^2}{2}\, e^{r_1'(g)}\right].$$

Proof. We use the notation in the proof of Proposition 1. Put $\{p_j\} = A_j \cap B_j$. By Lemma 1, we can take a path X_j on R from p_φ to p_j so that

$$\ell_R(X_j) \leq 3r_0 + \frac{2r_0}{\sinh^2 \frac{r_0}{2}}\,(g-1).$$

By the same argument as the proof of Proposition 1, we have

$$d_{\tilde{S}}(\tilde{q}_0, \tilde{\varphi}_*([X_j \cdot A_j \cdot X_j^{-1}])(\tilde{q}_0)) \leq 2\ell_R(X_j) + \ell_R(A_j) \leq r_1'(g).$$

Thus, we obtain

$$\#\{\tilde{\varphi}_*([X_j \cdot A_j \cdot X_j^{-1}]) \in \pi_1(S, q_0)\,|\,\varphi \in \mathrm{Hol}(R, S)\} \leq n_1'(g).$$

Since $C_j \cdot A_j \cdot C_j^{-1}$ is homologous to $X_j \cdot A_j \cdot X_j^{-1}$, we get

$$\#\{\tilde{\varphi}_*([C_j \cdot A_j \cdot C_j^{-1}]) \in H_1(S, \mathbf{Z})\,|\,\varphi \in \mathrm{Hol}(R, S)\} \leq n_1'(g).$$

By the same reasoning as the proof of Proposition 1, Lemma 3 implies that

$$\#\,\mathrm{Hol}(R, S) \leq n_1'(g)^g.$$

$\square$

3. Estimate of $\#\mathrm{Hol}(R, S)$ with injectivity radius $r_{\mathrm{inj}}(R) < r_0$

In this section, we will give an upper bound for $\#\,\mathrm{Hol}(R, S)$ with $r_{\mathrm{inj}}(R) < r_0$. First, we show the following lemma:

Lemma 4. *Let R be a closed Riemann surface of genus $g \geq 2$ with injectivity radius $r_{\mathrm{inj}}(R) < r_0$. Let $L_1, \cdots, L_{j_0}$ be the set of all simple closed geodesics of length $< 2r_0$ on R. Take a connected component R_0 of $R \setminus \cup_{j=1}^{j_0} C_R(L_j)$. Let R_0' be the connected component of $R \setminus \cup_{j=1}^{j_0} L_j$ with $R_0 \subset R_0'$.*

Then for any $p_1, p_2 \in R_0$, there exists a geodesic L on R with endpoints p_1, p_2 such that $L \subset R'_0$ and

$$\ell_R(L) \leq 3r_0 + 2r_0 \, \frac{\text{h-area}(R'_0)}{4\pi \sinh^2(\frac{r_0}{2})}.$$

Proof. For each $j = 1, \cdots, j_0$, we set

$$\ell_j = \ell_R(L_j),$$

$$w_j = \log \coth \frac{\ell_j}{4},$$

$$w_j^* = \log \coth \ell_j,$$

$$C_j = \{p \in R \mid d_R(p, L_j) \leq w_j\},$$

$$C_j^* = \{p \in R \mid d_R(p, L_j) \leq w_j^*\},$$

$$\mathcal{A}_j = \{(\rho, t) \mid -w_j \leq \rho \leq w_j, \ t \in \mathbf{R}/\mathbf{Z}\},$$

$$\mathcal{A}_j^* = \{(\rho, t) \mid -w_j^* \leq \rho \leq w_j^*, \ t \in \mathbf{R}/\mathbf{Z}\}.$$

Then C_j and C_j^* are isometric to $\mathcal{A}_j$ and $\mathcal{A}_j^*$ with Riemannian metric $ds^2 = d\rho^2 + \ell_j^2 \cosh^2 \rho \, dt^2$, respectively. Now, we will prove the following estimates:

$$(3.1) \qquad \log 2 < w_j - w_j^* < 2 \log 2,$$

$$(3.2) \qquad \frac{1}{2} \ell_j \coth \frac{\ell_j}{2} < 2(w_j - w_j^*),$$

$$(3.3) \qquad r_0 < r_p(R), \quad p \in C_j \setminus C_j^*,$$

$$(3.4) \qquad r_p(R) < d_R(p, L_j), \quad p \in C_j \setminus C_j^*.$$

Setting $x = \ell_j/4$, we have

$$w_j - w_j^* = \log \coth x - \log \coth(4x)$$

$$= \log \frac{8 \cosh^4 x - 4 \cosh^2 x}{8 \cosh^4 x - 8 \cosh^2 x + 1}.$$

Since $\ell_j \leq 2r_0 = 2 \operatorname{arcsinh} 0.2$, we get $\cosh^2 x \leq \cosh^2(2r_0) = 1.1644 \cdots$. Thus we find

$$2 < \frac{8 \cosh^4 x - 4 \cosh^2 x}{8 \cosh^4 x - 8 \cosh^2 x + 1} < 4,$$

which implies (3.1).

Since $2 \log 2 = 1.38692 \cdots$, $r_0 \coth r_0 = 1.0131 \cdots$, and $\ell_j < 2r_0$, it follows that

$$2(w_j - w_j^*) > 2 \log 2 > r_0 \coth r_0 > \frac{1}{2} \ell_j \coth \frac{\ell_j}{2},$$

which yields (3.2).

If $p \in C_j \setminus C_j^*$, then p corresponds to $(-\log \coth(y/2), t)$ or $(\log \coth(y/2), t)$ in $\mathcal{A}_j \setminus \mathcal{A}_j^*$ with $\ell_j/2 \leq y \leq 2\ell_j$. Noting

$$d = d_R(p, \partial C_j) = \log \frac{\coth \frac{\ell_j}{4}}{\coth \frac{y}{2}},$$

we have

$$r_p(R) = \operatorname{arcsinh}\left\{\cosh\frac{\ell_j}{2}\cosh d - \sinh d\right\}$$

$$= \operatorname{arcsinh}\left\{\sinh\frac{\ell_j}{2}\coth y\right\}$$

$$\geq \operatorname{arcsinh}\left\{\sinh\frac{\ell_j}{2}\coth(2\ell_j)\right\}.$$

Putting $z = \ell_j/2$, we see

$$\sinh\frac{\ell_j}{2}\coth(2\ell_j) = \frac{8\cosh^4 z - 8\cosh^2 z + 1}{8\cosh^3 z - 4\cosh z} \geq \frac{1}{4}, \quad 0 < z < r_0.$$

Hence we have

$$r_p(R) \geq \operatorname{arcsinh}\frac{1}{4} > r_0, \quad p \in C_j \setminus C_j^*,$$

which shows (3.3).

For the above $p \in C_j \setminus C_j^*$, we find

$$d_R(p, L_j) - r_p(R) = \log\coth\frac{y}{2} - \operatorname{arcsinh}\left\{\sinh\frac{\ell_j}{2}\coth y\right\}$$

$$\geq \log\coth\ell_j - \operatorname{arcsinh}\left\{\sinh\frac{\ell_j}{2}\coth\frac{\ell_j}{2}\right\}$$

$$= \log\coth\ell_j - \operatorname{arcsinh}\left\{\cosh\frac{\ell_j}{2}\right\}$$

$$\geq \log\coth(2r_0) - \operatorname{arcsinh}\left\{\cosh\frac{r_0}{2}\right\}$$

$$= 0.078335\cdots,$$

which asserts (3.4).

For $p_1, p_2 \in R_0$, we set

$$\ell = \inf\{\ell_R(C') \,|\, C' \text{ is a curve on } R \text{ with endpoints } p_1, p_2$$
$$\text{such that } C' \text{ is homotopic on } R \text{ to a curve}$$
$$C_0 \,(\subset R_0) \text{ with the same endpoints}\}.$$

Then there exists a geodesic L on R with endpoints p_1, p_2 so that L is homotopic on R to a curve $C_0 (\subset R_0)$ with the same endpoints, and $\ell = d_R(L)$. Since the length of each boundary curve of C_j is $\ell_j \coth(\ell_j/2)$, by using (3.2), we see that $L \subset R_0^*$, where R_0^* is the connected component of $R \setminus \cup_{j=1}^{j_0} C_j^*$ with $R_0 \subset R_0^*$. Thus by the same reasoning as the proof of Lemma 1, using (3.3) and (3.4), we find a set of hyperbolic disks $\{D_k\}_{k=1}^n$ satisfying the following conditions;

(1) each D_k is a hyperbolic disk with center $q_k \in L_j$ and radius r_0 such that $D_k \subset R_0'$,

(2) $q_1 = p_1$, and $p_2 \in D_n$,

(3) $\{D_k\}_{k=1}^n$ are pairwise disjoint, and D_k is tangent to D_{k+1} for $k = 1, \cdots, n-2$.

Hence we have

$$\sum_{k=1}^{n-2} \text{h-area}(D_k) \le \text{h-area}(R_0').$$

Noting $\text{h-area}(D_k) = 4\pi \sinh^2(r_0/2)$, we obtain

$$n - 2 \le \frac{\text{h-area}(R_0')}{4\pi \sinh^2 \frac{r_0}{2}}.$$

Therefore, it follows that

$$d_R(L) \le 3r_0 + 2r_0(n - 2) \le 3r_0 + 2r_0 \frac{\text{h-area}(R_0')}{4\pi \sinh^2 \frac{r_0}{2}}.$$

This completes the proof of Lemma 4. $\square$

Next, we show the following lemma:

Lemma 5. *Let L be a simple closed geodesic of a closed Riemann surface R of genus ≥ 2 with $\ell_R(L) < 2r_0$. If for $\varphi \in \text{Hol}(R, S)$, the curve $\varphi(L)$ is freely homotopically trivial in S, then*

$$d_S(\varphi(p_1), \varphi(p_2)) \le 2\log 2 + 2r_0 \coth r_0, \quad p_1, p_2 \in C_R(L).$$

In particular, for any $p_1, p_2 \in \partial C_R(L)$ corresponding to $(-w_R(L), t)$, $(w_R(L), t)$ in $[-w_R(L), w_R(L)] \times \mathbf{R}/\mathbf{Z}$, the image $\varphi(C)$ of the geodesic C in $C_R(L)$ from p_1 to p_2 is homotopic to a geodesic C' $(\subset S)$ from $\varphi(p_1)$ to $\varphi(p_2)$ such that

$$\ell_S(C') \le 2\log 2 + 2r_0 \coth r_0.$$

Proof. Let $\pi \colon \mathbf{H} \to R$ be the universal covering with covering transformation Γ. We may assume that L corresponds to an element $\gamma \in \Gamma$ such that $\gamma(\zeta) = \lambda\zeta$ with $\lambda = \exp(\ell_R(L))$. We set

$$\theta = \operatorname{arccos} \tanh \frac{\ell_R(L)}{2}, \quad 0 < \theta < \frac{\pi}{2},$$

$$r = \exp\left\{-\frac{\pi^2}{\ell_R(L)}\right\},$$

$$r_1 = \exp\left\{-\frac{\pi(\pi + 2\theta)}{\ell_R(L)}\right\},$$

$$r_2 = \exp\left\{-\frac{\pi(\pi - 2\theta)}{\ell_R(L)}\right\}.$$

Define a universal covering map $\pi_0 \colon \mathbf{H} \to \mathcal{A} = \{z \in \mathbf{C} \mid r^2 < |z| < 1\}$ by

$$\pi_0(\zeta) = \exp\left\{2\pi i \frac{\log \zeta}{\log \lambda}\right\}.$$

Then $\pi_1 = \pi \circ \pi_0^{-1}$ induces a covering map of $\mathcal{A}$ onto R. From Collar theorem II, we see that π_1 induces a conformal map of $\mathcal{A}_0 = \{z \in \mathbf{C} \mid r_1 < |z| < r_2\}$ onto $C_R(L)$. In this way, we identify $C_R(L)$ with $\mathcal{A}_0$.

Let $\rho\colon \Delta \to S$ be the universal covering of S, where Δ is the unit disk $\{w \in \mathbf{C} \,|\, |w| < 1\}$. Since $\varphi(L)$ is freely homotopically trivial, $\varphi\colon R \to S$ induces a holomorphic map $\Phi\colon \mathcal{A} \to \Delta$ with $\rho \circ \Phi = \varphi \circ \pi_1$.

Without loss of generality, we may assume that $p_1 \in C_1 = \{z \in \mathbf{C} \,|\, |z| = r_1\}$ and $p_2 \in C_2 = \{z \in \mathbf{C} \,|\, |z| = r_2\}$. We may also assume that $\Phi(p_1) = 0$. For any $s \in (r^2, 1)$, we put

$$M(s) = \max\{|\Phi(z)| \,|\, |z| = s\}.$$

Then Hadamard's three-circle theorem asserts that

$$\log M(r) \leq \frac{\log r_2 - \log r}{\log r_2 - \log r_1} \log M(r_1) + \frac{\log r - \log r_1}{\log r_2 - \log r_1} \log M(r_2).$$

Since $0 < r_1 < r < r_2 < 1$, $r^2 = r_1 r_2$ and $M(r_2) < 1$, we have

$$\log M(r) \leq \frac{1}{2} \log M(r_1),$$

which implies that

$$(3.5) \qquad\qquad M(r) \leq \sqrt{M(r_1)}.$$

Since $\Phi(p_1) = 0$ and $\ell_R(C_1) = \ell_R(L) \coth(\ell_R(L)/2) \leq 2r_0 \coth(r_0)$, it follows that

$$\Phi(C_1) \subset \left\{w \in \mathbf{C} \,|\, |w| < \tanh \frac{\ell_1}{4}\right\},$$

where $\ell_1 = 2r_0 \coth r_0$. Thus (3.5) implies that

$$\Phi(L) \subset D' = \left\{w \in \mathbf{C} \,|\, |w| < \sqrt{\tanh \frac{\ell_1}{4}}\right\},$$

where L is identified with $\{z \in \mathbf{C} \,|\, |z| = r\}$. Note that for any $w \in D'$, we obtain

$$d_\Delta(0, w) \leq d_\Delta\left(0, \sqrt{\tanh \frac{\ell_1}{4}}\right)$$

$$= \log\left\{\exp\left(\frac{\ell_1}{2}\right) + \sqrt{\exp \ell_1 - 1}\right\}$$

$$\leq \log\left\{2\exp\left(\frac{\ell_1}{2}\right)\right\} = \log 2 + \frac{\ell_1}{2}.$$

This shows that

$$d_S(\varphi(p_1), q) \leq \log 2 + \frac{\ell_1}{2}, \quad q \in \varphi(L).$$

Similarly, it follows that

$$d_S(\varphi(p_2), q) \leq \log 2 + \frac{\ell_1}{2}, \quad q \in \varphi(L).$$

Therefore, we obtain

$$d_S(\varphi(p_1), \varphi(p_2)) \leq 2\log 2 + \ell_1 = 2\log 2 + 2r_0 \coth r_0.$$

This completes the proof of Lemma 4. $\qquad \square$

Now, assume that R is a closed Riemann surface of genus $g \geq 2$ with injectivity radius $r_{\mathrm{inj}}(R) < r_0 = \mathrm{arcsinh}\, 0.2 = 0.19869 \cdots$. Denote by $L_1, \cdots, L_{j_0}$ the set of all simple closed geodesics of length $< 2r_0$ on R. Note that $1 \leq j_0 \leq 3g - 3$. Take simple closed geodesics $L_{j_0+1}, \cdots, L_{3g-3}$ on R so that $\mathcal{L} = \{L_j\}_{j=1}^{3g-3}$ is a system of decomposing curves on R and $\ell_R(L_j) \leq 26(g-1)$ for $j = 1, \cdots, 3g-3$. Let S be a closed Riemann surface of genus g' with $2 \leq g' \leq g$. Let $L_1', \cdots, L_{k_0}'$ be the set of all simple closed geodesics of length $\leq 2\,\mathrm{arcsinh}\, 1$.

Take an element $\varphi \in \mathrm{Hol}(R, S)$. Since $r_q(S) > \mathrm{arcsinh}\, 1 = 0.88137 \cdots$ for $q \in S \setminus \cup_{k=1}^{k_0} C_S(L_k)$ and $\ell_R(L_j) \coth(2\ell_R(L_j)) \leq 2r_0 \coth(4r_0) = 0.60109 \cdots$ for $j = 1, \cdots, j_0$, the distance decreasing property of the hyperbolic metrics for holomorphic maps implies that for each $j = 1, \cdots, j_0$, one of the following holds;

(3.6) $\varphi(L_j)$ is freely homotopically trivial in S,

(3.7) $\varphi(L_j)$ is not freely homotopically trivial in S and $\varphi(C_R^*(L_j)) \subset C_S(L_k')$ for some $k \in \{1, \cdots, k_0\}$.

Let $\{j_1, \cdots, j_{\nu_0}\}$ be the set of all elements $j \in \{1, \cdots, j_0\}$ such that $\varphi(L_j)$ is not freely homotopically trivial in S. Denote by $R_1, \cdots, R_{m_0}$ the set of connected components of $R \setminus \cup_{\nu=1}^{\nu_0} C_R(L_{j_\nu})$. For $m = 1, \cdots, m_0$, let R_m^* and R_m' be the connected components of $R \setminus \cup_{\nu=1}^{\nu_0} C_R^*(L_{j_\nu})$ and $R \setminus \cup_{\nu=1}^{\nu_0} L_{j_\nu}$, respectively such that $R_m \subset R_m^* \subset R_m'$.

Let $L_{k(j_\nu)}'$ be the simple closed geodesic on S such that $\varphi(L_{j_\nu})$ is freely homotopic to $L_{k(j_\nu)}'$ in S. Denote by $S_1, \cdots, S_{n_0}$ the set of connected components of $S \setminus \cup_{\nu=1}^{\nu_0} C_S(L_{k(j_\nu)}')$.

Since φ is an open map of R onto S, from (3.7) we conclude that for any $m = 1, \cdots, m_0$, one of the following holds;

(3.8) $\varphi(R_m^*) \subset C_S(L_k')$ for some $k \in \{1, \cdots, k_0\}$,

(3.9) $\varphi(R_m') \supset S_n$ for some $n \in \{1, \cdots, n_0\}$.

Now, we assert the following lemma:

Lemma 6. *In the situation stated above, the following hold.*

(1) *For any $p_1, p_2 \in R_m^*$, there exists a curve $C\, (\subset R_m')$ with endpoints p_1, p_2 such that $\varphi(C)$ is homotopic to a curve $C'\, (\subset S)$ satisfying*

$$\ell_S(C') \leq \left\{ 2r_0 \left(1 + \frac{1}{\sinh^2 \frac{r_0}{2}} \frac{\text{h-area}(R_m')}{4\pi(g-1)} \right) + 8 \right\} (g-1).$$

(2) *For every $j = 1, \cdots, g$, each component $\varphi(B_j \cap R_m^*), m = 1, \cdots, m_0$, is homotopic to a curve $B_{j,m}'\, (\subset S)$, and*

$$\sum_{m=1}^{m_0} \ell_S(B_{j,m}') \leq (3\ell_g + 16)(g-1).$$

(3) *For every $j = 1, \cdots, g$, each component $\varphi(C_j \cap R_m^*), m = 1, \cdots, m_0$, is homotopic to a curve $C_{j,m}'\, (\subset S)$, and*

$$\sum_{m=1}^{m_0} \ell_S(C_{j,m}') \leq (3\ell_g + 16)(g-1) + 2\ell_g + 8.$$

Proof. Denote by $\{\mathcal{R}_\mu\}_{\mu=1}^{\mu_0}$ the set of all connected components of $R \setminus \cup_{j=1}^{j_0} L_j$ with $\mathcal{R}_\mu \subset R'_m$.

Now, assume that $p_1, p_2 \in R_m^*$. By Lemmas 4 and 5, we can find geodesics $X_1, \cdots, X_{\alpha_0}$ such that $C = X_1 \cdot \cdots \cdot X_{\alpha_0}$ is a curve in R'_m from p_1 to p_2, and such that each X_α satisfies one of the following conditions:

(3.10) $X_\alpha \subset \mathcal{R}_{\mu_\alpha}$ for some $\mu_\alpha \in \{1, \cdots, \mu_0\}$, and

$$\ell_R(X_\alpha) \le 3r_0 + 2r_0 \frac{\text{h-area}(\mathcal{R}_{\mu_\alpha})}{4\pi \sinh^2 \frac{r_0}{2}}.$$

(3.11) $X_\alpha \subset \mathcal{C}_R(L_{j_\alpha})$ for some $j_\alpha \in \{1, \cdots, j_0\}$, $\varphi(L_{j_\alpha})$ is freely homotopically trivial, and $\varphi(X_\alpha)$ is homotopic to a geodesic Y_α in S with

$$\ell_S(Y_\alpha) \le 2\log 2 + 2r_0 \coth r_0.$$

When (3.10) holds, then we set $X'_\alpha = \varphi(X_\alpha)$. When (3.11) holds, then we set $X'_\alpha = Y_\alpha$. Then $C' = X'_1 \cdot \cdots \cdot X'_{\alpha_0}$ is homotopic to $\varphi(C)$ and satisfies

$$
\begin{aligned}
\ell_S(C') &= \sum_{\alpha=1}^{\alpha_0} \ell_S(X'_\alpha) \\
&\le \sum_{\mu=1}^{\mu_0} \left\{ 3r_0 + 2r_0 \frac{\text{h-area}(\mathcal{R}_\mu)}{4\pi \sinh^2 \frac{r_0}{2}} \right\} + 4(\log 2 + r_0 \coth r_0)(g-1) \\
&\le \left\{ 2r_0 \left(1 + \frac{1}{\sinh^2 \frac{r_0}{2}} \frac{\text{h-area}(R'_m)}{4\pi(g-1)} \right) + 8 \right\} (g-1).
\end{aligned}
$$

Hence we have the first assertion of Lemma 6.

To show the second assertion, take an arbitrary pair of pants with boundary geodesics $L_{j_1}, L_{j_2}, L_{j_3} \in \mathcal{L}$. As in the proof of Lemma 2, dropping the common perpendicular H between L_{j_1}, L_{j_2}, we know that

$$\cosh \ell_R(H) = \frac{\cosh c}{\sinh a \sinh b} + \coth a \coth b,$$

where $a = \ell_R(L_{j_1})/2, b = \ell_R(L_{j_2})/2$, and $c = \ell_R(L_{j_3})/2$. Setting $t = \cosh c + \cosh a \cosh b$, we have

$$
\begin{aligned}
\ell_R(H) &= \log \frac{t + \sqrt{t^2 - \sinh^2 a \sinh^2 b}}{\sinh a \sinh b} \\
&\le \log \frac{2t}{\sinh a \sinh b} \\
&= \log 2 + \log \left(\frac{\cosh c}{\sinh a \sinh b} + \coth a \coth b \right).
\end{aligned}
$$

If $a, b \geq r_0$, then we know that

$$\ell_R(H) \leq \log 2 + \log\left(\frac{\cosh \frac{\ell_g}{2}}{\sinh^2 r_0} + \coth^2 r_0\right)$$

$$= \log 2 + \log\left(25 \coth \frac{\ell_g}{2} + 26\right)$$

$$\leq 4 + \frac{\ell_g}{2}.$$

If $a < r_0$ and $b \geq r_0$, then we obtain

$$\ell_R(H \setminus C_R^*(L_{j_1})) = \ell_R(H) - \log\coth(2a)$$

$$= \ell_R(H) - \log\coth a + \log\frac{\coth a}{\coth(2a)}$$

$$\leq 2\log 2 + \log\left(\frac{\cosh c}{\cosh a \sinh b} + \coth b\right)$$

$$\leq 2\log 2 + \log\left(\frac{\cosh \frac{\ell_g}{2}}{\sinh r_0} + \coth r_0\right)$$

$$= 2\log 2 + \log\left(5\cosh \frac{\ell_g}{2} + \sqrt{26}\right) \leq 4 + \frac{\ell_g}{2}.$$

If $a, b < r_0$, we get

$$\ell_R(H \setminus \{C_R^*(L_{j_1}) \cup C_R^*(L_{j_2})\})$$

$$= \ell_R(H) - \log\coth(2a) - \log\coth(2b)$$

$$= \ell_R(H) - \log\coth a - \log\coth b + \log\frac{\coth a \coth b}{\coth(2a)\coth(2b)}$$

$$\leq 3\log 2 + \log\left(\cosh \frac{\ell_g}{2} + 1\right) \leq 4 + \frac{\ell_g}{2}.$$

Since B_j passes at most $2(g-1)$ pairs of pants, counting these $\ell_R(H), \ell_R(H \setminus C_R^*(L_{j_1})), \ell_R(H \setminus \{C_R^*(L_{j_1}) \cup C_R^*(L_{j_2})\})$, twist papameters, and using Lemma 5, we conclude that each component $\varphi(B_j \cap R_m^*), m = 1, \cdots, m_0$, is homotopic to a curve $B_{j,m}'$ $(\subset S)$, and

$$\sum_{m=1}^{m_0} \ell_S(B_{j,m}') \leq \{(4 + \frac{\ell_g}{2}) + \ell_g + 4\} \times 2(g-1) = (3\ell_g + 16)(g-1).$$

This implies the second assertion.

Similarly, counting also the diameter of a pair of pants, we find that each component $\varphi(C_j \cap R_m^*), m = 1, \cdots, m_0$, is homotopic to a curve $C_{j,m}'$ $(\subset S)$, and

$$\sum_{m=1}^{m_0} \ell_S(C_{j,m}') \leq \{(4 + \frac{\ell_g}{2}) + \ell_g + 4\} \times 2(g-1) + (4 + \ell_g) \times 2$$

$$= (3\ell_g + 16)(g-1) + 2\ell_g + 8.$$

This completes the proof of the third assertion of Lemma 6. $\square$

Now, we have the following proposition:

Proposition 3. *Let R be a closed Riemann surface of genus $g \geq 2$ with injectivity radius $r_{\mathrm{inj}}(R) < r_0$, and S another closed Riemann surface of genus g' with $2 \leq g' \leq g$. Set*

$$r_3(g) = 108(g-1)^2 + 4(3\ell_g + 57)(g-1) + 10\ell_g + 32,$$
$$r_4(g) = 108(g-1)^2 + 2(9\ell_g + 130)(g-1) + 8\ell_g + 32,$$
$$n_3(g) = \left[3^{2(g-1)} e^{r_3(g)}\right] 4^{2(g-1)}$$
$$n_4(g) = \left[3^{2(g-1)} e^{r_4(g)}\right] 4^{2(g-1)},$$

where $[\,\cdot\,]$ means the Gaussian symbol. Then

$$\# \operatorname{Hol}(R,S) \leq \{3(g'-1)+1\}^{3(g-1)} \{5(g'-1)\}^{2(g-1)} n_3(g)^g n_4(g)^g.$$

Proof. As in the proof of Proposition 1, let $\pi\colon \tilde{R} \to R$ be the universal covering of R which is constructed canonically by homotopical equivalence classes $[C,p]$ with base point p_0.

We take a Kleinian group G_0 satisfying the following conditions:

(1) There exists a bounded connected component D of the region of discontinuity of G_0 such that the invariant subgroup $G = \{g \in G_0 \,|\, g(D) = D\}$ has no fixed-points in D and the quotient space D/G is biholomorphic to S.

(2) Every simple closed geodesic L'_k on S with $\ell_S(L'_k) < 2\operatorname{arcsinh} 1$ corresponds to an elliptic element in G of order 4.

In fact, such a group G_0 is constructed by using the limit circle theorem of Klein and Poincaré, Klein's combination theorem, Maskit's second combination theorem, and quasiconformal deformation. (See Bers [1], §3.) Let $\rho\colon D \to D/G = S$ be the canonical projection.

Fix $\varphi \in \operatorname{Hol}(R,S)$, and assume that φ satisfies conditions (3.7), (3.8), and (3.9). We may also assume that $p_0 \in R_1$ and $S_1 \subset \varphi(R_1^*)$. Fix $q_0 \in S_1$ with $r_{\mathrm{inj}}(S) \geq \operatorname{arcsinh} 1 = \log(1 + \sqrt{2})$. We also fix $\tilde{q}_0 \in D$ with $\rho(\tilde{q}_0) = q_0$.

Take a point $p_\varphi \in R_1^*$ such that $\varphi(p_\varphi) = q_0$. By Lemma 6, there exists a curve $Y_\varphi (\subset R_1')$ from p_φ to p_0 so that $\varphi(Y_\varphi)$ is homotopic to a curve $Y'_\varphi (\subset S)$ satisfying

$$\ell_S(Y'_\varphi) \leq K(R_1),$$

where

$$K(R_1) = \left\{ 2r_0 \left(1 + \frac{1}{\sinh^2 \frac{r_0}{2}} \frac{\text{h-area}(R_1')}{4\pi(g-1)} \right) + 8 \right\} (g-1).$$

For any curve $C (\subset R)$ from p_0 to p, we put $\tilde{\varphi}([C,p]) = \tilde{q}$, where $\tilde{q} \in D$ is the terminal point of the lift of the curve $\varphi(Y_\varphi \cdot C)$ to D by ρ with initial point $\tilde{q}_0$. In this way, we obtain a lift $\tilde{\varphi}\colon \tilde{R} \to D$ of φ.

Denote by $\mathcal{F}_\varphi$ the set of all $\psi \in \operatorname{Hol}(R,S)$ such that ψ satisfies the same conditions (3.7), (3.8), and (3.9) as φ. Let $p_\psi \in R_1^*$ with $\psi(p_\psi) = q_0$, and $Y_\psi (\subset R_1')$ a curve from

p_ψ to p_0 such that $\psi(Y_\psi)$ is homotopic to a curve Y'_ψ with $\ell_S(Y'_\psi) \le K(R_1)$. Using this Y_ψ, we construct similarly a lift $\tilde{\psi} \colon \tilde{R} \to D$ of ψ.

Now, we will estimate the number of monodromies $\tilde{\psi}_*$ of $\psi \in \mathcal{F}_\varphi$. If $\ell_R(A_j) < 2r_0$, then we replace with a boundary curve of $\mathcal{C}^*_R(A_j)$, which is denoted by the same symbol. We set $R^* = \cup_{m=1}^{m_0} R^*_m$.

Let X_1 be the component of $(C_j \cdot A_j \cdot C_j^{-1}) \cap R^*_1$ with initial point p_0 and terminal point $p_1 \in \partial \mathcal{C}^*_R(L_{j_\nu})$ for some $j_\nu \in \{j_1, \cdots, j_{\nu_0}\}$. By Lemma 6, we find curves $Z_\varphi (\subset R'_1)$ from p_φ to p_1 and $Z_\psi (\subset R'_1)$ from p_ψ to p_1 so that $\varphi(Z_\varphi), \psi(Z_\psi)$ are homotopic to curves Z'_φ, Z'_ψ, respectively, and $\ell_S(Z'_\varphi) \le K(R_1), \ell_S(Z'_\psi) \le K(R_1)$. By condition (3.7), it follows that $\varphi(p_1), \psi(p_1) \in \mathcal{C}_S(L'_{k_1})$ for some $k_1 \in \{1, \cdots, k_0\}$. Thus we can find a geodesic $H_1 (\subset \mathcal{C}_S(L'_{k_1}))$ from $\varphi(p_1)$ to $\psi(p_1)$ with $\ell_S(H_1) \le 2K(R_1)$. Then the closed curve $g_{1,\psi} = \psi(Y_\psi \cdot X_1) \cdot H_1^{-1} \cdot \varphi(Y_\varphi \cdot X_1)^{-1}$ with base point q_0 satisfies

$$\ell_S(g_{1,\psi}) \le 4K(R_1) + 2\ell_R(X_1).$$

Here we identify $g_{1,\psi}$ with an element of G, which is denoted by the same symbol. By the same argument as the proof of Proposition 1, the number of elements of $G_1 = \{g_{1,\psi} \in G \mid \psi \in \mathcal{F}_\varphi\}$ is estimated as

$$\begin{aligned}
\#G_1 &\le \left\lceil \frac{\text{h-area}(\Delta(\tilde{q}_0, r_1 + 4K(R_1) + 2\ell_R(X_1)))}{\text{h-area}(\Delta(\tilde{q}_0, r_1))} \right\rceil \\[2mm]
&\le \left\lceil \frac{(1+\sqrt{2})^2}{2} e^{4K(R_1)+2\ell_R(X_1)} \right\rceil \\[2mm]
&\le \left\lceil 3 e^{4K(R_1)+2\ell_R(X_1)} \right\rceil.
\end{aligned}$$

Next, we set $X_1^* = (C_j \cdot A_j \cdot C_j^{-1}) \cap \mathcal{C}^*_R(L_{j_\nu})$. Note that both $\varphi(X_1^*)$ and $\psi(X_1^*)$ are included in the same cylinder $\mathcal{C}_S(L'_{k_1})$. Let p_2 be the endpoint of X_1^* with $p_1 \ne p_2$. Let X_2 be the component of $(C_j \cdot A_j \cdot C_j^{-1}) \cap R^*$ with initial point p_2 and terminal point p_3. Here, assume that $\varphi(p_3) \in \mathcal{C}_S(L_{k_2})$ for some $k_2 \in \{1, \cdots, k_0\}$. Denote by R_2^* the component of R^* with $X_2 \subset R_2^*$. If $S_2 \subset \varphi(R_2^*)$, then by the same reasoning as above, we can find a curve $Y_{2,\varphi}$ with terminal point p_2 and geodesics $H_2 (\subset \mathcal{C}_S(L_{k_1}))$, $H_3 (\subset \mathcal{C}_S(L_{k_2}))$ so that

$$\begin{aligned}
\psi(Y_\psi \cdot X_1 \cdot X_1^* \cdot X_2) =& g_{1,\psi} \cdot \varphi(Y_\varphi \cdot X_1 \cdot X_1^*) \cdot \{\varphi(X_1^*)^{-1} \cdot H_1 \cdot \psi(X_1^*) \cdot H_2^{-1}\} \cdot \\
& g_{2,\psi} \cdot \varphi(Y_{2,\varphi} \cdot X_2) \cdot H_3.
\end{aligned}$$

Note that $\varphi(X_1^*)^{-1} \cdot H_1 \cdot \psi(X_1^*) \cdot H_2^{-1}$ corresponds to a power of an elliptic element of order 4 in G. By the same argument as above, the set G_2 of all these elements $g_{2,\psi}$ is estimated as

$$\#G_2 \le \left\lceil 3 e^{4K(R_2)+2\ell_R(X_2)} \right\rceil.$$

If $\varphi(R_2^*) \subset \mathcal{C}_S(L'_{k_1})$, then it follows that

$$\psi(Y_\psi \cdot X_1 \cdot X_1^* \cdot X_2) = g_{1,\psi} \cdot \varphi(Y_\varphi \cdot X_1 \cdot X_1^*) \cdot \{\varphi(X_1^*)^{-1} \cdot H_1 \cdot \psi(X_1^* \cdot X_2)\},$$

$$\varphi(X_1^*)^{-1} \cdot H_1 \cdot \psi(X_1^* \cdot X_2) \subset \mathcal{C}_S(L'_{k_1}).$$

Then we go to the next step. Continuing these procedures, we conclude that

$$\#\{\tilde{\psi}_*([C_j \cdot A_j \cdot C_j^{-1}]) \mid \psi \in \mathcal{F}_\varphi\} \leq \left[3^{2(g-1)} e^{4K_1 + 2K_2}\right] 4^{2(g-1)},$$

where

$$K_1 = \sum_{m=1}^{m_0} \left\{ 2r_0 \left(1 + \frac{1}{\sinh^2 \frac{r_0}{2}} \frac{\text{h-area}(R'_m)}{4\pi(g-1)}\right) + 8 \right\} (g-1),$$
$$K_2 = 2\ell_R(C_j) + \ell_R(A_j).$$

Note that

$$K_1 \leq 4(r_0 + 4)(g-1)^2 + \frac{2r_0}{\sinh^2 \frac{r_0}{2}}(g-1)$$
$$\leq 17(g-1)^2 + 41(g-1).$$

By Lemma 6, we get

$$K_2 \leq 2(3\ell_g + 16)(g-1) + 5\ell_g + 16.$$

Hence we have

$$\#\{\tilde{\psi}_*([C_j \cdot A_j \cdot C_j^{-1}]) \mid \psi \in \mathcal{F}_\varphi\} \leq \left[3^{2(g-1)} e^{r_3(g)}\right] 4^{2(g-1)},$$

where

$$r_3(g) = 108(g-1)^2 + 4(3\ell_g + 57)(g-1) + 10\ell_g + 32.$$

Similarly, we obtain

$$\#\{\tilde{\psi}_*([C_j \cdot B_j \cdot C_j^{-1}]) \mid \psi \in \mathcal{F}_\varphi\} \leq \left[3^{2(g-1)} e^{r_4(g)}\right] 4^{2(g-1)},$$

where

$$r_4(g) = 108(g-1)^2 + 2(9\ell_g + 130)(g-1) + 8\ell_g + 32.$$

We put

$$n_3(g) = \left[3^{2(g-1)} e^{r_3(g)}\right] 4^{2(g-1)}$$
$$n_4(g) = \left[3^{2(g-1)} e^{r_4(g)}\right] 4^{2(g-1)}.$$

Then, by the same reasoning as the proof of Proposition 1, we show that

$$\#\mathcal{F}_\varphi \leq n_3(g)^g n_4(g)^g.$$

The number of cases where $\varphi \in \text{Hol}(R, S)$ satisfies conditions (3.7), (3.8), and (3.9) is at most

$$\sum_{\nu_0}^{j_0} {}_{j_0}C_{\nu_0} k_0^{\nu_0} \left(\sum_{m=1}^{m_0} {}_{m_0}C_m n_0^m k_0^{m_0 - m}\right)$$
$$\leq (k_0 + 1)^{j_0} (n_0 + k_0)^{m_0}$$
$$\leq \{3(g'-1) + 1\}^{3(g-1)} \{5(g'-1)\}^{2(g-1)}.$$

Therefore, we conclude that

$$\# \text{Hol}(R, S) \leq \{3(g'-1) + 1\}^{3(g-1)} \{5(g'-1)\}^{2(g-1)} n_3(g)^g n_4(g)^g.$$

This complete the proof of Proposition 3. $\square$

Proof of Theorem 1. By Propositions 1 and 3, it is easy to get Theorem 1.   □

To obtain a better estimate for $\# \operatorname{Hol}(R, S)$ with $r_{\mathrm{inj}}(R) < r_0$, we use the following rigidity theorem due to Tanabe. (See Tanabe [10], Theorem 3.)

Tanabe's Rigidity Theorem. *Let R and S be closed Riemann surfaces of genus g and g', respectively, with $2 \leq g' \leq g$. Let $\{[A_j], [B_j]\}_{j=1}^{g}$ and $\{[A'_j], [B'_j]\}_{j=1}^{g'}$ be canonical homology bases for $H_1(R, \mathbf{Z})$ and $H_1(S, \mathbf{Z})$, respectively. Let N_g be an integer with $N_g > 8(g - 1)$. If $\varphi, \psi \in \operatorname{Hol}(R, S)$ satisfy $\varphi_*([A_j]) \equiv \psi_*([A_j])$ $(\bmod N_g)$ in $H_1(S, \mathbf{Z})$ for all $j = 1, 2, \cdots, g$, then $\varphi = \psi$.*

Now, we have the following proposition.

Proposition 4. *Let R be a closed Riemann surface of genus $g \geq 2$ with injectivity radius $r_{\mathrm{inj}}(R) < r_0$, and S another closed Riemann surface of genus g' with $2 \leq g' \leq g$. Set*

$$n'_2(g) = \left[3e^{98(g-1)+\ell_g} \right],$$

where $[\cdot]$ means the Gaussian symbol. Then

$$\# \operatorname{Hol}(R, S) \leq \left\{ 1 + 3(g' - 1)N_g \right\}^{3(g-1)} \left\{ 2(g' - 1)n'_2(g) + 3(g' - 1)N_g \right\}^{3(g-1)},$$

where N_g is an integer with $N_g > 8(g - 1)$.

Proof. Let $L_1, \cdots, L_{j_0}$ the set of all simple closed geodesics of length $< 2r_0$ on R. Note that $1 \leq j_0 \leq 3g - 3$. Take simple closed geodesics $L_{j_0+1}, \cdots, L_{3g-3}$ on R so that $\mathcal{L} = \{L_j\}_{j=1}^{3g-3}$ is a system of decomposing curves on R and $\ell_R(L_j) \leq 26(g - 1)$ for $j = 1, \cdots, 3g - 3$. By virtue of Tanabe's rigidity theorem, in order to determine $\varphi \in \operatorname{Hol}(R, S)$, it is sufficient to know $\varphi_*([L_j]) \in H_1(S, \mathbf{Z}), j = 1, \cdots, 3g - 3$.

Denote by $L'_1, \cdots, L'_{k_0}$ the set of all simple closed geodesics on S of length $\leq 2\operatorname{arcsinh} 1$. Let $S_1, \cdots, S_{n_0}$ be connected components of $S \setminus \cup_{k=1}^{k_0} \mathcal{C}_S(L'_k)$. Note that $k_0 \leq 3(g' - 1)$ and $n_0 \leq 2(g' - 1)$.

For every $\mu = j_0 + 1, \cdots, 3g - 3$ and $n = 1, \cdots, n_0$, we set

$$\mathcal{F}_{\mu,n} = \{\varphi \in \operatorname{Hol}(R, S) \,|\, \varphi(L_\mu) \subset S_n\}.$$

Fix $p_0 \in L_\mu$ and $q_0 \in S_n$ with $r_S(q_0) > \operatorname{arcsinh} 1 = \log(1 + \sqrt{2})$. Let $\pi \colon \tilde{R} \to R$ be the universal covering of R which is constructed canonically by homotopical equivalence classes $[C, p]$ with base point p_0. By the same argument of the proof of Proposition 3, using Lemma 6, we can show that for any $\varphi \in \mathcal{F}_{\mu,n}$, there are a point $p_\varphi \in R$ and a curve $C_\varphi (\subset R)$ from p_φ to p_0 such that $\varphi(C_\varphi)$ is homotopic to a curve $C'_\varphi (\subset S)$ satisfying

$$\ell_S(C'_\varphi) \leq 49(g - 1).$$

Hence, in the same way of the proof of Proposition 2, we see that

$$\#\{\varphi_*([L_\mu]) \in H_1(S, \mathbf{Z}) \,|\, \varphi \in \mathcal{F}_{\mu,n}\} \leq n'_2(g),$$

where

$$n'_2(g) = \left[3e^{98(g-1)+\ell_g} \right].$$

Therefore, by a similar reasoning to the proof of Proposition 3, we conclude that

$$\# \operatorname{Hol}(R, S) \leq \sum_{\nu_0=0}^{j_0} {}_{j_0}C_{\nu_0}\, k_0^{\nu_0} \left\{ \sum_{m=1}^{3g-3-j_0} {}_{3g-3-j_0}C_m\, n_0^m\, k_0^{3g-3-j_0-m} \right\} \times$$

$$N_g^{\nu_0}\, N_g^{3g-3-j_0-m} n_2'(g)^m$$

$$\leq \sum_{\nu_0=0}^{j_0} {}_{j_0}C_{\nu_0}\, (k_0 N_g)^{\nu_0} \times$$

$$\left\{ \sum_{m=0}^{3g-3-j_0} {}_{3g-3-j_0}C_m\, (n_0 n_2'(g))^m\, (k_0 N_g)^{3g-3-j_0-m} \right\}$$

$$= \left\{ 1 + k_0 N_g \right\}^{j_0} \left\{ n_0 n_2'(g) + k_0 N_g \right\}^{3g-3-j_0}$$

$$\leq \left\{ 1 + 3(g'-1)N_g \right\}^{3(g-1)} \left\{ 2(g'-1)n_2'(g) + 3(g'-1)N_g \right\}^{3(g-1)}.$$

This completes the proof of Proposition 4. $\square$

Proof of Theorem 2. By Propositions 2 and 4, it is easy to get Theorem 2. $\square$

References

1. L. Bers, *Spaces of degenerating Riemann surfaces*, in " Discontinuous Groups and Riemann Surfaces, 1973 Maryland Conference", edited by L. Greenberg, Ann. Math. Studies No. 79, Princeton Univ. Press, 1974, pp. 43-55.
2. P. Buser, *Geometry and Spectra of Compact Riemann Surfaces*, Birkhäuser, Boston, 1992.
3. A. Howard and A. J. Sommese, *On the theorem of de Franchis*, Ann. Scoula. Norm. Sup. Pisa **10** (1983), 429-436.
4. Y. Imayoshi, *Generalizations of de Franchis theorem*, Duke Math. J. **50** (1983), 393-408.
5. Y. Imayoshi, *An analytic proof of Severi's theorem*, Complex Variables **2** (1983), 151-155.
6. Y. Imayoshi and H. Shiga, *A finiteness theorem of holomorphic families of Riemann surfaces*, Holomorphic Functions and Moduli, 1986 MSRI Conference, Vol. II, pp. 207-219, Drasin, D. et al: (eds.), Springer-Verlag, Berlin and New York, 1988.
7. E. Kani, *Bounds on the number of non-rational subfields of a function field*, Invent. math. **85** (1986), 185-198.
8. H. H. Martens, *Observations on morphisms of closed Riemann surfaces*, Bull. London Math. Soc. **10** (1978), 209-212.
9. H. H. Martens, *Mappings of closed Riemann surfaces*, Proceedings. of Symposia in Pure Math. **49** (1989), Part 1, 531-539.
10. M. Tanabe, *On rigidity of holomorphic maps of Riemann surfaces*, to appear in Osaka Math. J.
11. M. Tanabe, *An explicit bound for the theorem of de Franchis*, to appear.

DEPARTMENT OF MATHEMATICS, OSAKA CITY UNIVERSITY, SUGIMOTO, SUMIYOSHI-KU, OS-AKA 558, JAPAN

E-mail address: h1667@ocugw.cc.osaka-cu.ac.jp

Proceedings of
THE 37TH TANIGUCHI SYMPOSIUM ON
TOPOLOGY AND TEICHMÜLLER SPACES
held in Finland, July 1995
ed. by Sadayoshi KOJIMA *et al.*
©1996 World Scientific Publishing Co.
pp. 79–100

AN INFINITESIMAL APPROACH TO THE STABLE COHOMOLOGY OF THE MODULI OF RIEMANN SURFACES

NARIYA KAWAZUMI

(Received December 25, 1995)

ABSTRACT. In this note we review an infinitesimal approach to the stable cohomology of the moduli of compact Riemann surfaces by means of complex analytic Gel'fand-Fuks cohomology of open Riemann surfaces. Under a hypothesis (a certain kind of the Frobenius Reciprocity Laws) we prove the (p, q)-equivariant cohomology of the dressed moduli of compact Riemann surfaces coincides with the polynomial algebra generated by the Morita-Mumford classes $e_n = \kappa_n$ $(n \geq 1)$ [Mo] [Mu] for $p \geq q$. This suggests it is reasonable to conjecture that the stable cohomology algebra of the moduli of compact Riemann surfaces would be generated by the Morita-Mumford classes e_n's. For a more detailed description the reader is referred to [Ka1, Ka2, Ka4, Ka5].

CONTENTS

1991 *Mathematics Subject Classification.* 1991 Mathematical Subject Classification. Primary 14H15. Secondary 57R32.

Introduction

The Riemann moduli space M_g of compact Riemann surfaces of genus $g \geq 2$, i.e., the complex analytic space of all isomorphism classes of (almost) complex structures on the 2-dimensional closed oriented C^∞-manifold Σ_g of genus g, is one of the most important objects not only in complex analytic geometry but in low-dimensional topology. In fact, in view of the contractibility of the Teichmüller spaces, we have a natural isomorphism of rational cohomology algebras [EE]:

$$H^*(M_g; \mathbb{Q}) \cong H^*(B\operatorname{Diff}^+ \Sigma_g; \mathbb{Q})$$

$$= \{\text{rational characteristic classes of oriented } \Sigma_g\text{-bundles}\}$$

The group $\operatorname{Diff}^+ \Sigma_g$ means the topological group consisting of all orientation-preserving diffeomorphisms on the manifold Σ_g endowed with C^∞-topology. The Harer stability theorem [H] [Iv] states that the q-th rational cohomology group $H^q(M_g; \mathbb{Q})$ does not depend upon the genus g if $q < g/2$. This enables us to consider the stable cohomology group of the moduli of Riemann surfaces

$$\lim_{g \to \infty} H^*(M_g; \mathbb{Q}).$$

As was established by Miller [Mi] and Morita [Mo] independently, a polynomial algebra in countable many generators e_n's ($n \in \mathbb{N}_{\geq 1}$) is imbedded in the stable cohomology group

$$(0.1) \qquad\qquad \mathbb{Q}[e_n; n \geq 1] \subset \lim_{g \to \infty} H^*(M_g; \mathbb{Q}).$$

Here $e_n = \kappa_n \in H^{2n}(M_g; \mathbb{Q})$ is the n-th Morita-Mumford characteristic class defined as follows [Mo] [Mu]. Let $C_g \to M_g$ denote the universal family of compact Riemann surfaces of genus g and let $e \in H^2(C_g)$ be the Euler class (the first Chern class) of the relative tangent bundle T_{C_g/M_g}. The n-th Morita-Mumford class e_n is defined as the fiber integral of the $(n+1)$-th power of the class e:

$$(0.2) \qquad e_n := \int_{\text{fiber}} e^{n+1} \in H^{2n}(M_g; \mathbb{Q}).$$

Though it has been studied energetically by many mathematicians, the entirety of the stable cohomology algebra remains to be clarified. The purpose of this note is to review an infinitesimal approach to the stable cohomology of the moduli of compact Riemann surfaces, which gives an affirmative evidence to the conjecture that the stable cohomology algebra of the moduli of compact Riemann surfaces would be generated by the Morita-Mumford classes $e_n = \kappa_n$'s, $n \geq 1$.

Our approach is based on an observation of Kontsevich [Kon] and Beilinson-Manin-Schechtmann [BMS]. We follow the formulation given by Arbarello-DeConcini-Kac-Procesi [ADKP]. Let $\rho > 0$ be a fixed positive real number. Instead of the moduli M_g consider the moduli space $M_{g,\rho}$ of triples (C, p, z), where C is a compact Riemann surface of genus g, $p \in C$ and z is a suitable complex analytic coordinate centered at p (§1.2). The space $M_{g,\rho}$ is (homotopy equivalent to) the classifying space of the topological group $\text{Diff}^+(\Sigma_{g,1})$ of all orientation-preserving diffeomorphisms of the surface $\Sigma_{g,1} := \Sigma_g - $ (interior of an imbedded disk) fixing the boundary <u>pointwise</u> (endowed with C^∞-topology):

$$M_{g,\rho} = B\,\text{Diff}^+(\Sigma_{g,1}).$$

As a consequence of the Harer stability theorem quoted above we have

$$H^*(M_{g,\rho}; \mathbb{Q}) = H^*(M_g; \mathbb{Q}) \quad \text{for } * < g/2.$$

Beilinson-Manin-Schechtmann and Kontsevich observed that the Lie algebra $\mathfrak{d}_\rho$ of all complex analytic vector fields defined near $\{z \in \mathbb{C}; 0 < |z| \leq \rho\}$ acts on the dressed moduli $M_{g,\rho}$ infinitesimal homogeneously (§1.4). We study the stable cohomology of $M_{g,\rho}$ through this action.

Let $C^\times$ be a once punctured compact Riemann surface and $L(C^\times)$ the Lie algebra of all complex analytic vectors on $C^\times$. We compute the q-th cohomology group of $L(C^\times)$ with coefficients in the complex analytic quadratic differentials on the p-fold product space $(C^\times)^p$ for the case $p \geq q$ (Theorem 4.4.1). The cohomology group vanishes for $p > q$, and, for $p = q$, it forms a trivial constant sheaf on the dressed moduli $M_{g,\rho}$ of compact Riemann surfaces of genus g. (The stalk does not depend on the genus g.) Through the infinitesimal action of $\mathfrak{d}_\rho$ this induces a natural map of the cohomology group for $p = q$ into the (p, p)-cohomology of the moduli $M_{g,\rho}$ (§2.3). We proved the map is a stable isomorphism onto the subalgebra generated by the Morita-Mumford classes $e_n = \kappa_n$'s (Theorem 2.4.1). This gives an affirmative evidence for the conjecture stated above (Theorem 3.4.1).

In §5 and §6 we study the (dressed) relative Picard variety $F_{g,\rho}$ on the universal curve over the dressed moduli $M_{g,\rho}$ following Arbarello et.al.[ADKP]. Similarly we have

$$F_{g,\rho} = B(H_1(\Sigma; \mathbb{Z}) \rtimes \mathrm{Diff}^+ \Sigma_{g,1}).$$

Instead of $\mathfrak{d}_\rho$ the Lie algebra $\mathcal{D}_\rho$ of all complex analytic differential operators of order ≤ 1 defined near $\{z \in \mathbb{C}; 0 < |z| \leq \rho\}$ acts on the Picard variety $F_{g,\rho}$. We obtain the vanishing of the q-th cohomology of the Lie algebra of all complex analytic differential operators of order ≤ 1 on $C^\times$ with coefficients in the complex analytic alternating 1-jets of 1-forms on the p-fold product space $(C^\times)^p$ for the case $p > q$ (Theorem 5.0.1). Our computation for the diagonal part $(p = q)$ is incomplete. In §6 we compute the stable rational cohomology of $F_{g,\rho}$ modulo that of $M_{g,\rho}$ (Theorem 6.0.1). Then we introduce *the generalized Morita-Mumford class* $\widetilde{m_{i,j}} \in H^{2i+2j-2}(F_{g,\rho}; \mathbb{Z})$, $(i, j \geq 0,$ $i + j \geq 2)$. When $j = 0$, the class $\widetilde{m_{i+1,0}}$ is equal to the i-th Morita-Mumford class e_i. The class $\widetilde{m_{i,j}}$'s are contained in the image of the diagonal part $(p = q)$. It is unknown whether the diagonal part is stably isomorphic to the polynomial algebra generated by $\widetilde{m_{i,j}}$'s.

1. Homogeneous structure on the dressed moduli

1.1. Infinitesimal action

First of all we review the notion of an *infinitesimal action* of a Lie algebra on a manifold. Let M be a (possibly infinite dimensional) complex analytic manifold on which a Lie algebra $\mathfrak{g}$ acts complex analytically. This means a homomorphism of Lie algebras

$$\mu : \mathfrak{g} \to \mathrm{Vect}(M),$$

called the *action*, is given, where $\mathrm{Vect}(M)$ denotes the complex Lie algebra of complex analytic vector fields on M. The kernel of the composite of the evaluation map ev_x at a point $x \in M$ and the action μ

$$\mathrm{ev}_x \circ \mu : \mathfrak{g} \to \mathrm{Vect}(M) \to T_x M$$

is denoted by $\mathfrak{g}_x$ and called the *isotropy subalgebra* of $\mathfrak{g}$ at the point $x \in M$. We call the action μ *homogeneous* when the composite map $\mathrm{ev}_x \circ \mu$ is surjective at each point $x \in M$.

1.2. Dressed moduli $M_{g,\rho}$

Now we define the dressed moduli $M_{g,\rho}$. For a <u>non-negative</u> real number $\rho \geq 0$ we denote

$$D_\rho := \{z \in \mathbb{C}; |z| < \rho\}, \quad \overline{D_\rho} := \{z \in \mathbb{C}; |z| \leq \rho\},$$
$$D_\rho^\times := \{z \in \mathbb{C}; 0 < |z| < \rho\} \quad \text{and} \quad \overline{D_\rho^\times} := \{z \in \mathbb{C}; 0 < |z| \leq \rho\}.$$

Fix an integer $g \geq 0$. *The dressed moduli $M_{g,\rho}$ of compact Riemann surfaces of genus g is the space consisting of all triples (C, p, z), where C is a compact Riemann surface of genus g, p is a point of C, and z is a complex coordinate of a neighborhood U of p satisfying the conditions*

$$(1.2.1) \qquad\qquad z(p) = 0 \quad \text{and} \quad z(U) \supset \overline{D_\rho}.$$

For a point $(C, p, z) \in M_{g,\rho}$, we denote

$$C^\times := C - \{p\}, \quad C_\rho := C - z^{-1}(\overline{D_\rho}) \quad \text{and}$$

$$C_{g,\rho} := \bigcup_{(C,p,z) \in M_{g,\rho}} C_\rho, \quad C_{g,\rho}^\times := \bigcup_{(C,p,z) \in M_{g,\rho}} C^\times$$

$$\overline{D_{g,\rho}^\times} := \bigcup_{(C,p,z) \in M_{g,\rho}} \overline{D_\rho^\times} = M_{g,\rho} \times \overline{D_\rho^\times} \subset C_{g,\rho}^\times.$$

We have inclusions $C_{g,\rho} \subset C_{g,\rho}^\times$ and $\overline{D_{g,\rho}^\times} \subset C_{g,\rho}^\times$ and a natural projection $\pi_{g,\rho} : C_{g,\rho}^\times \to M_{g,\rho}$, $(C, p, z, p_1) \mapsto (C, p, z)$. Each of these spaces has a natural fiber structure over $M_{g,\rho}$ through the projection $\pi_{g,\rho}$.

1.3. Lie algebra $\mathfrak{d}_\rho$

Let $L(U)$ denote the topological Lie algebra consisting of all complex analytic vector fields on an open Riemann surface U with the Fréchet topology of uniform convergence on compact sets. The closed subalgebra consisting of all vector fields which have a zero at a (fixed) point $p_1 \in U$ is denoted by $L(U, p_1)$. The topological Lie algebra $\mathfrak{d}_\rho$ is defined by

$$\mathfrak{d}_\rho := \varinjlim_{\rho_1 \downarrow \rho} L(D_{\rho_1}^\times),$$

which is endowed with the inductive limit locally convex topology [G,TVS] [Ko] [Ko1].

1.4. $\mathfrak{d}_\rho$-action on the dressed moduli

Now we introduce the infinitesimal action [BMS] [Kon] of the topological Lie algebra $\mathfrak{d}_\rho$ on the spaces $M_{g,\rho}$ and $C_{g,\rho}^\times$ following the formulation in [ADKP](3.19).

Fix $X \in \mathfrak{d}_\rho$ and $(C, p, z) \in M_{g,\rho}$ (resp. $(C, p, z, p_1) \in C_{g,\rho}^\times$, i.e., $(C, p, z) \in M_{g,\rho}$ and $p_1 \in C^\times = C - \{p\}$). For some $\rho_1 > \rho_2 > \rho$, $z(U)$ includes $\overline{D_{\rho_1}}$, and the integral $\exp \epsilon X : D_{\rho_2}^\times \to D_{\rho_1}^\times$ of the vector field X exists for sufficiently small complex number ϵ. A complex analytic path

$$\epsilon \mapsto (C(\epsilon), p(\epsilon), z(\epsilon)) \in M_{g,\rho}, \quad (\text{resp. } \mapsto (C(\epsilon), p(\epsilon), z(\epsilon), p_1(\epsilon)) \in C_{g,\rho}^\times)$$

is defined by

$$C(\epsilon) := (C^\times \amalg D_{\rho_2}) / \sim_\epsilon, \quad z \sim_\epsilon (\exp \epsilon X)(z), \quad z \in D_{\rho_2}$$

$$p(\epsilon) := \text{the image of } 0 \in D_{\rho_2} \text{ on } C(\epsilon)$$

$$z(\epsilon) := \text{the coordinate of } C(\epsilon) \text{ induced by the identity } D_{\rho_2} \hookrightarrow D_{\rho_2}$$

$$p_1(\epsilon) := \text{the image of } p_1 \in C^\times \text{ on } C(\epsilon).$$

Differentiating the path $(C(\epsilon), p(\epsilon), z(\epsilon))$ and $(C(\epsilon), p(\epsilon), z(\epsilon), p_1(\epsilon))$, we obtain the maps $P = P_{(C,p,z)}$ and $P = P_{(C,p,z,p_1)}$:

$$P : \mathfrak{d}_\rho \to T_{(C,p,z)} M_{g,\rho}, \quad X \mapsto (d/d\epsilon)\big|_{\epsilon=0}(C(\epsilon), p(\epsilon), z(\epsilon))$$

$$P : \mathfrak{d}_\rho \to T_{(C,p,z,p_1)} C_{g,\rho}^\times, \quad X \mapsto (d/d\epsilon)\big|_{\epsilon=0}(C(\epsilon), p(\epsilon), z(\epsilon), p_1(\epsilon)),$$

which define an infinitesimal action of $\mathfrak{d}_\rho$ on the dressed moduli $M_{g,\rho}$ and the universal family $C_{g,\rho}^\times$. These actions are compatible to the fiber structure on the space $C_{g,\rho}^\times$ over $M_{g,\rho}$.

1.5. Homogeneity

Lemma 1.5.1 ([ADKP] **Proposition 3.19**). *The sequences*

$$(1.5.1) \qquad\qquad 0 \to L(C^\times) \hookrightarrow \mathfrak{d}_\rho \xrightarrow{P} T_{(C,p,z)}M_{g,\rho} \to 0$$

$$(1.5.2) \qquad\qquad 0 \to L(C^\times, p_1) \hookrightarrow \mathfrak{d}_\rho \xrightarrow{P} T_{(C,p,z,p_1)}C_{g,\rho} \to 0$$

are exact. Here the algebras $L(C^\times)$ and $L(C^\times, p_1)$ are regarded as subalgebras of $\mathfrak{d}_\rho$ through the coordinate z.

This gives us a presentation of the cotangent spaces of $M_{g,\rho}$ and $C_{g,\rho}$.

Corollary 1.5.2.

$$T^*_{(C,p,z)}M_{g,\rho} = (\mathfrak{d}_\rho / L(C^\times))^*$$
$$T^*_{(C,p,z,p_1)}C_{g,\rho} = (\mathfrak{d}_\rho / L(C^\times, p_1))^*$$

Here and throughout this note the asterisk $*$ means the strong dual.

2. Results

2.1. Standard cochain complex

We begin this section by fixing our notation on the cohomology theory for Lie algebras. Let $\mathfrak{g}$ be a complex topological Lie algebra. By a $\mathfrak{g}$-*module* we mean a complex topological vector space on which $\mathfrak{g}$ acts continuously. The standard continuous cochain complex of the topological Lie algebra $\mathfrak{g}$ with coefficients in a $\mathfrak{g}$-module N is denoted by

$$C^*(\mathfrak{g}; N) = \bigoplus_{p \geq 0} C^p(\mathfrak{g}; N),$$

where $C^p(\mathfrak{g}; N)$ is the linear space of continuous alternating multilinear mappings $c : \mathfrak{g}^{\otimes p} \to N$. The cohomology group of the complex $C^*(\mathfrak{g}; N)$ is called the (continuous) cohomology group of $\mathfrak{g}$ with coefficients in N and denoted by $H^*(\mathfrak{g}; N)$. When N is the trivial $\mathfrak{g}$-module $\mathbb{C}$, we abbreviate them to $C^*(\mathfrak{g})$ and $H^*(\mathfrak{g})$ respectively. (For details, see for example [HS].)

2.2. Main Theorem (1/2)

Theorem 2.2.1. *For any $\overline{x} \in M_{g,\rho}$ and $x \in C_{g,\rho}$ we have*

$$(2.2.1)$$
$$H^q((\mathfrak{d}_\rho)_{\overline{x}}; \textstyle\bigwedge^p T^*_{\overline{x}}M_{g,\rho}) = H^q((\mathfrak{d}_\rho)_x; \textstyle\bigwedge^p T^*_x C_{g,\rho}) = 0 \quad (\forall p > \forall q)$$

$$(2.2.2)$$
$$\bigoplus_{p \geq 0} H^p((\mathfrak{d}_\rho)_{\overline{x}}; \textstyle\bigwedge^p T^*_{\overline{x}}M_{g,\rho}) = \mathbb{C}[\kappa_n; n \geq 1]/relations,$$

$$(2.2.3)$$
$$\bigoplus_{p \geq 0} H^p((\mathfrak{d}_\rho)_x; \textstyle\bigwedge^p T^*_x C_{g,\rho}) = \mathbb{C}[\epsilon, \kappa_n; n \geq 1]/relations.$$

where $\kappa_n \in H^n((\mathfrak{d}_\rho)_{\overline{x}}; \bigwedge^n T^*_{\overline{x}}M_{g,\rho})$ is defined in (4.6.1) and $\epsilon \in H^1((\mathfrak{d}_\rho)_x; T^*_x C_{g,\rho})$ is defined in (4.5.1).

Here and throughout this paper we mean by $\wedge^n$ the completed n-fold alternating tensor product. Furthermore we have

Proposition 2.2.2. *The cohomology groups $H^p((\mathfrak{d}_\rho)_{\bar{x}}; \wedge^p T^*_{\bar{x}} M_{g,\rho})$, $\bar{x} \in M_{g,\rho}$, (resp. $H^p((\mathfrak{d}_\rho)_x; \wedge^p T^*_x C_{g,\rho})$, $x \in C_{g,\rho}$) form a constant sheaf on the moduli $M_{g,\rho}$ (resp. the universal curve $C_{g,\rho}$).*

2.3. Framework for construction of cohomology classes

Let M be a complex manifold on which a Lie algebra $\mathfrak{g}$ acts infinitesimally as in §1.1, and $E \to M$ a complex analytic vector bundle on which the algebra $\mathfrak{g}$ acts complex analytically and compatibly with the action on M. This means $\mathfrak{g}$ acts on each $\mathcal{O}_M(E)(O)$ $(O \overset{\text{open}}{\subset} M)$ such that

(1) each restriction map is $\mathfrak{g}$-equivariant, and
(2) the formula

$$X(f\sigma) = (Xf)\sigma + f(X\sigma), \quad X \in \mathfrak{g},\ f \in \mathcal{O}_M(O),\ \sigma \in \mathcal{O}_M(E)(O)$$

holds for any open subset $O \subset M$.

In the sequal we call such a vector bundle a $\mathfrak{g}$-*vector bundle over M* in short. The fiber E_x at $x \in M$ is a $\mathfrak{g}_x$-module in an obvious manner.

Let $n \in \mathbf{N}_{\geq 0}$ be a fixed non-negative integer. We assume

$$(\mathrm{A}(n)) \qquad \forall x \in M \quad \forall n' < n \quad H^{n'}(\mathfrak{g}_x; E_x) = 0.$$

Under the assumption $(\mathrm{A}(n))$ we have an exact sequence of complex analytic 'vector bundles' over M

$$0 \to E \to \coprod_{x \in M} C^1(\mathfrak{g}_x; E_x) \to \coprod_{x \in M} C^2(\mathfrak{g}_x; E_x) \to \cdots$$

$$\cdots \to \coprod_{x \in M} C^{n-1}(\mathfrak{g}_x; E_x) \to \coprod_{x \in M} Z^n(\mathfrak{g}_x; E_x) \to \coprod_{x \in M} H^n(\mathfrak{g}_x; E_x) \to 0,$$

where Z^n means the n cocycles. This exact sequence induces the n-fold composite of the connecting homomorphisms

$$D : H^0(M; \mathcal{O}_M(\coprod_{x \in M} H^n(\mathfrak{g}_x; E_x))) \to H^n(M; \mathcal{O}_M(E)).$$

The map D has a multiplicative property and a functorial property in a natural manner.

2.4. Main Theorem (2/2)

Theorem 2.4.1. *For $g \geq 0$ and $\rho > 0$ we have*

$$(\sqrt{-1}/2\pi)^n D(\kappa_n) = e_n \in H^{n,n}(M_{g,\rho}), \quad n \geq 1,$$

where e_n is the n-th Morita-Mumford class (0.2).

As corollaries we obtain

Corollary 2.4.2. *If $\rho > 0$, the map*

$$D : \bigoplus_{n \geq 0} H^n((\eth_\rho)_{\overline{x}}; \textstyle\bigwedge^n T_{\overline{x}}^* M_{g,\rho}) \to \bigoplus_{n \geq 0} H^{n,n}(M_{g,\rho})$$

$$(resp. \quad D : \bigoplus_{n \geq 0} H^n((\eth_\rho)_x; \textstyle\bigwedge^n T_x^* C_{g,\rho}) \to \bigoplus_{n \geq 0} H^{n,n}(C_{g,\rho}))$$

is a stable isomorphism onto the subalgebra generated by the Morita-Mumford classes e_n's (resp. the Euler class e and the Morita-Mumford classes e_n's).

Corollary 2.4.3. *There exist no algebraic relations among the classes κ_n's, i.e., we have isomorphisms of $\mathbb{C}$-algebras*

$$\bigoplus_{n \geq 0} H^n((\eth_\rho)_{\overline{x}}; \textstyle\bigwedge^n T_{\overline{x}}^* M_{g,\rho}) = \mathbb{C}[\kappa_n; n \geq 1]$$

$$\bigoplus_{n \geq 0} H^n((\eth_\rho)_x; \textstyle\bigwedge^n T_x^* C_{g,\rho}) = \mathbb{C}[\epsilon, \kappa_n; n \geq 1].$$

These follow from the theorem of Miller [Mi] and Morita [Mo] quoted in (0.1).

3. Equivariant cohomology

Now we introduce an equivariant cohomology theory for Lie algebras, which gives us a reformulation of the results stated in the preceding section. It plays an important role also in the proof of the results.

3.1. Definition of equivariant cohomology

As in §1.1 and §2.3, let M be a (possibly infinite dimensional) complex analytic manifold on which a complex Lie algebra $\mathfrak{g}$ acts complex analytically and let E be a $\mathfrak{g}$-vector bundle over M. Here we assume that the action of $\mathfrak{g}$ on M is homogeneous, i.e., that the composite $\mathrm{ev}_x \circ \mu : \mathfrak{g} \to \mathrm{Vect}(M) \to T_x M$ is surjective for each $x \in M$.

Since $\mathcal{O}_M(E)$ is a sheaf of $\mathfrak{g}$-modules, the cochain complex of sheaves over M

$$C^*(\mathfrak{g}; \mathcal{O}_M(E)) : M \overset{\mathrm{open}}{\supset} O \mapsto C^*(\mathfrak{g}; \mathcal{O}_M(E)(O))$$

is defined, where $C^*(\mathfrak{g}; \cdot)$ is the standard cochain complex of the Lie algebra $\mathfrak{g}$ with values in a $\mathfrak{g}$-module $\cdot$ introduced in §2.1. We denote by $H_\mathfrak{g}^*(M; \mathcal{O}_M(E))$ the hypercohomology group of the cochain complex of sheaves over M with respect to the functor $\Gamma(M; \cdot)$ (= the sections of $\cdot$ over M) ([G,E] ch.0, §11.4, pp.32-) and call it *the $\mathfrak{g}$-equivariant cohomology group of M with values in the $\mathfrak{g}$-vector bundle E.* Namely we define

$$H_\mathfrak{g}^*(M; \mathcal{O}_M(E)) := H^*(\mathrm{Total}(\Gamma(M; C^{*,*})))$$

for an injective right Cartan-Eilenberg resolution $C^{*,*} = (C^{i,j})_{i,j \geq 0}$ of the complex $C^*(\mathfrak{g}; \mathcal{O}_M(E))$ (cf. ibid. loc. cit.). Especially, if E is the n-cotangent bundle $\bigwedge^n T^* M$, we denote

$$H_\mathfrak{g}^{n,*}(M) := H_\mathfrak{g}^*(M; \mathcal{O}_M(\textstyle\bigwedge^n T^* M))$$

and call it *the $\mathfrak{g}$-equivariant $(n, *)$ cohomology of M.*

There exist two spectral sequences converging to $H_{\mathfrak{g}}^{\bullet}(M; \mathcal{O}_M(E))$

$$(3.1.1) \qquad {}^{\backprime}E_2^{p,q} = H^p(H^q(M; C^{\bullet}(\mathfrak{g}; \mathcal{O}_M(E))))$$

$$(3.1.2) \qquad {}^{\backprime\backprime}E_2^{p,q} = H^p(M; H^q(\mathfrak{g}; \mathcal{O}_M(E))),$$

where we denote $H^{\bullet}(\mathfrak{g}; \mathcal{O}_M(E))$ is the sheaf over M defined as the cohomology of the cochain complex of sheaves $C^{\bullet}(\mathfrak{g}; \mathcal{O}_M(E))$.

3.2. Finite dimensional case

The equivariant cohomology theory on finite dimensional manifolds is essential to our computation of the cohomology groups of the isotropy subalgebras $(\mathfrak{d}_\rho)_{\overline{x}}$ and $(\mathfrak{d}_\rho)_x$. In this subsection we assume the $\mathfrak{g}$-manifold M and $\mathfrak{g}$-vector bundle E are *finite dimensional.* Then the stalk of the sheaf $H^{\bullet}(\mathfrak{g}; \mathcal{O}_M(E))$ at a point $x \in M$ is given by

Lemma 3.2.1. *If $\mathfrak{g}$ is a Fréchet nuclear space, the stalk at a point $x \in M$ is naturally isomorphic to the cohomology group of the stalk $\mathcal{O}_M(E)_x$:*

$$H^{\bullet}(\mathfrak{g}; \mathcal{O}_M(E))_x \cong H^{\bullet}(\mathfrak{g}; \mathcal{O}_M(E)_x).$$

This follows from Lemma 4.3 in [Ka] and the exactness of the injective limit functor. Next we look at the natural map

$$\varphi^{p,q} : {}^{\backprime}E_2^{p,q} = H^p(H^q(M; C^{\bullet}(\mathfrak{g}; \mathcal{O}_M(E)))) \to H^p(\mathfrak{g}; H^q(M; \mathcal{O}_M(E))).$$

Especially we have the natural map

$$H_{\mathfrak{g}}^n(M; \mathcal{O}_M(E)) \to H^0(\mathfrak{g}; H^n(M; \mathcal{O}_M(E))) = H^n(M; \mathcal{O}_M(E))^{\mathfrak{g}}.$$

Proposition 3.2.2. *Let M and E be as above. We assume that the Lie algebra $\mathfrak{g}$ is topologically isomorphic to the topological vector space consisting of all complex analytic sections on a finite dimensional vector bundle over a finite dimensional paracompact complex analytic manifold, (hence $\mathfrak{g}$ is a Fréchet nuclear space), and that M, E and $H^q(M; \mathcal{O}_M(E))$ for $q \neq 0$ are all finite dimensional. Then the natural map $\varphi^{p,q}$ is an isomorphism*

$$\varphi^{p,q} : {}^{\backprime}E_2^{p,q} \cong H^p(\mathfrak{g}; H^q(M; \mathcal{O}_M(E))).$$

Especially if M is a finite dimensional *Stein* manifold, we have

$$H_{\mathfrak{g}}^{\bullet}(M; \mathcal{O}_M(E)) = H^{\bullet}(\mathfrak{g}; \mathcal{O}_M(E)(M)).$$

Hence we obtain a spectral sequence

$$(3.2.1) \qquad {}^{\backprime\backprime}E_2^{p,q} = H^p(M; \mathcal{H}^q) \Rightarrow H^{p+q}(\mathfrak{g}; \mathcal{O}_M(E)(M)),$$

where $\mathcal{H}^q$ is a sheaf over M whose stalk at $x \in M$ is given by

$$\mathcal{H}_x^q = H^q(\mathfrak{g}; \mathcal{O}_M(E)_x).$$

We call this sequence *the Rešetnikov spectral sequence* (see [R] and [Ka] §9).

Let E be a finite dimensional complex analytic vector bundle over $\mathbb{C}^n$ and let $M = \mathbb{C}^n - \{0\}$. Scheja [Sc] has computed the cohomology group $H^{\bullet}(\mathbb{C}^n - \{0\}; \mathcal{O}_M(E))$.

So, though the cohomology group $H^{n-1}(\mathbb{C}^n - \{0\}; \mathcal{O}_M(E))$ is *not* finite dimensional, we obtain a similar result to Proposition 3.2.2 ([Ka1] Proposition 1.9):

Proposition 3.2.3. *If $\mathfrak{g}$ satisfies the condition given in Proposition 3.2.2 then the 'E_2 term (3.1.1) converging to the equivariant cohomology $H^{\bullet}_{\mathfrak{g}}(\mathbb{C}^n - \{0\}; \mathcal{O}_{\mathbb{C}^n}(E))$ is given by*

$$'E_2^{p,q} = \begin{cases} H^p(\mathfrak{g}; \mathcal{O}_{\mathbb{C}^n}(E)(\mathbb{C}^n)), & \text{if } q = 0, \\ H^p(\mathfrak{g}; H^{n-1}(\mathbb{C}^n - \{0\}; \mathcal{O}_{\mathbb{C}^n}(E))), & \text{if } q = n - 1, \\ 0, & \text{otherwise.} \end{cases}$$

3.3. Frobenius Reciprocity Law

In a *general situation* we relate the map D in §2.3 to the second spectral sequence "$E_2^{p,q}$ (3.1.2). For any $x \in M$ the evaluation map $\mathrm{ev}_x : \mathcal{O}_M(E)(O) \to E_x$, ($x \in O \subset M$) induces a homomorphism

$$(3.3.1) \qquad\qquad (\mathrm{ev}_x)_{\bullet} : H^n(\mathfrak{g}; \mathcal{O}_M(E))_x \to H^n(\mathfrak{g}_x; E_x),$$

which we call *the evaluation homomorphism.*

Here we assume the action of $\mathfrak{g}$ on M is *homogeneous*. Then, taking into consideration the finite dimensional case studied by Bott [B] (although the $\eth_\rho$-manifolds $M_{\mathfrak{g},\rho}$ and $C_{\mathfrak{g},\rho}$ are infinite dimensional in our case), we may regard the $\mathfrak{g}$-module $(\mathcal{O}_M(E))_x$ as the (co-)induced module of the $\mathfrak{g}_x$-module E_x ($x \in M$). Hence we put a general hypothesis:

Hypothesis 3.3.1 (Frobenius Reciprocity Law). *The evaluation homomorphism (3.3.1) is an isomorphism for any $x \in M$.*

It could be regarded as a certain kind of the Frobenius reciprocity laws, i.e., the Shapiro isomorphisms. Through the evaluation homomorphism (3.3.1) the vector bundle $\coprod_{x \in M} H^{\bullet}(\mathfrak{g}_x; E_x)$ possesses the natural structure of a sheaf over M and we have an isomorphism

$$(3.3.2) \qquad\qquad "E_2^{p,q} \cong H^p(M; H^q(\mathfrak{g}_x; E_x)),$$

where the RHS means the cohomology of M with values in the sheaf $\coprod_{x \in M} H^{\bullet}(\mathfrak{g}_x; E_x)$. Thus, under the Hypothesis 3.3.1, the assumption (A(n)) implies that the term '$E_2^{p,q}$ vanishes for $q < n$, so that

$$H^q_{\mathfrak{g}}(M; \mathcal{O}_M(E)) = \begin{cases} 0, & \text{if } q < n, \\ H^0(M; H^n(\mathfrak{g}_x; E_x)), & \text{if } q = n. \end{cases}$$

The map D introduced in §2.3

$$D : H^0(M; H^n(\mathfrak{g}_x; E_x)) \to H^n(M; \mathcal{O}_M(E))$$

is nothing but the composite of the above isomorphism and the natural map $H^n_{\mathfrak{g}}(M; \mathcal{O}_M(E)) \to H^n(M; \mathcal{O}_M(E))$.

3.4. Equivariant cohomology of the dressed moduli

Consider the $\mathfrak{d}_\rho$-manifolds $M = M_{g,\rho}$, $C_{g,\rho}$ and the $\mathfrak{d}_\rho$-vector bundles $\bigwedge^n T^* M$. We assume $\rho > 0$. Then the Hypothesis 3.3.1 is

$$H^*(\mathfrak{d}_\rho; \mathcal{O}_M(\textstyle\bigwedge^n T^* M))_x \cong H^*((\mathfrak{d}_\rho)_x; \textstyle\bigwedge^n T_x^* M)$$

$$\text{for all } n \geq 0 \text{ and all } x \in M = M_{g,\rho} \text{ and } C_{g,\rho},$$

which seems to be true. But at present the author has no proof for this assertion.

Theorem 3.4.1. *If the Frobenius Reciprocity Law (Hypothesis 3.3.1) holds good for the $\mathfrak{d}_\rho$-manifolds $M = M_{g,\rho}$ and $C_{g,\rho}$ and the $\mathfrak{d}_\rho$-vector bundles $\bigwedge^* T^* M$, we have*

$$(3.4.1) \qquad \bigoplus_{p \geq q} H_{\mathfrak{d}_\rho}^{p,q}(M_{g,\rho}) = \mathbb{C}[e_n; n \geq 1]$$

$$(3.4.2) \qquad \bigoplus_{p \geq q} H_{\mathfrak{d}_\rho}^{p,q}(C_{g,\rho}) = \mathbb{C}[e, e_n; n \geq 1]$$

for all $g \geq 0$ and $\rho > 0$, where $e = c_1(T_{C_{g,\rho}/M_{g,\rho}}) \in H^{1,1}(C_{g,\rho})$ and $e_n \in H^{n,n}(M_{g,\rho})$ is the n-th Morita-Mumford class ($n \in \mathbf{N}_{\geq 1}$).

This gives an affirmative evidence for the conjecture: the stable cohomology algebra of the moduli of compact Riemann surfaces would be generated by the Morita-Mumford classes e_n's.

4. Outline of the proof - Complex analytic Gel'fand-Fuks cohomology

4.1. Köthe duality

First of all we identify the cotangent space of the dressed moduli $M_{g,\rho}$ with the space of quadratic differential on $C_\rho = C - z^{-1}(\overline{D_\rho})$ by a generalized Köthe duality [Koe] [Ko2]. Fix a point $(C, p, z) \in M_{g,\rho}$ and $(C, p, z, p_1) \in C_{g,\rho}^\times$. We regard the disk $\overline{D_\rho}$ as a subset of C through the coordinate z. Let E be a complex analytic line bundle over C. For an open set U in C, we abbreviate $E(U) := \mathcal{O}_C(E)(U)$, which is a Fréchet space with respect to the topology of uniform convergence on compact sets. For a subset F in C, we denote by $E(F)$ the inductive limit locally convex space

$$E(F) = \varinjlim_{F \subset U} E(U),$$

where U runs over all the open neighborhoods of F in C. The canonical bundle of C, namely, the cotangent bundle of C is denoted by K. We have $(K^{-1})(\overline{D_\rho^\times}) = \mathfrak{d}_\rho$.

Theorem 4.1.1 (generalized Köthe duality). *The map*

$$\eta : (E^{-1} \otimes K)(C_\rho) \to (E(\overline{D_\rho^\times})/E(C^\times))^*, \quad s \mapsto \left(f \mapsto \frac{1}{2\pi\sqrt{-1}} \oint f \cdot s \right)$$

is a topological isomorphism. Here $f \cdot s$ intends a 1-form defined on an annulus $\{\rho < |z| < \rho + \varepsilon\}(0 < \varepsilon \ll 1)$, and the line integral $\oint$ is carried out on the circle $\{|z| = \rho + \varepsilon/2\}$.

Denote by $Q(U)$ the Fréchet space of complex analytic quadratic differentials on a Riemann surface U:

$$Q(U) = H^0(U; \mathcal{O}_U((T^*U)^{\otimes 2})) = H^0(U; \mathcal{O}_U(K^{\otimes 2})).$$

For $\lambda \in \mathbb{Z}$ and $p_1 \in U$ we denote by $Q^\lambda(U, p_1)$ the Fréchet space of meromorphic quadratic differentials on U with a pole only at p_1 of order $\leq \lambda$.

$$Q^\lambda(U, p_1) = H^0(U; \mathcal{O}_U(K^{\otimes 2} \otimes [p_1]^{\otimes \lambda})).$$

Corollary 4.1.2. *We have topological isomorphisms*

$$T^*_{(C,p,z)}M_{g,\rho} = Q(C_\rho) \quad and \quad T^*_{(C,p,z,p_1)}C_{g,\rho} = Q^1(C_\rho, p_1).$$

These isomorphisms are respectively $(\eth_\rho)_{(C,p,z)} = L(C^\times)$ and $(\eth_\rho)_{(C,p,z,p_1)} = L(C^\times, p_1)$-equivariant by Stokes' Theorem.

4.2. Fundamental Exact Sequence

What we have to compute are the cohomology groups of the Lie algebras $(\eth_\rho)_{(C,p,z)} = L(C^\times)$ and $(\eth_\rho)_{(C,p,z,p_1)} = L(C^\times, p_1)$ (Lemma 1.5.1). These are *the complex analytic Gel'fand-Fuks cohomology groups of an open Riemann surface*. So, as is expected, it is necessary to compute the cohomology groups of the Lie algebras

$$W_1 := L(\mathbb{C}) \quad and \quad L_0 := L(\mathbb{C}, 0)$$

with coefficients in the n-fold completed tensor products $\bigwedge^n$ of

$$Q := Q(\mathbb{C}) \quad and \quad Q^1 := Q^1(\mathbb{C}, 0).$$

We should remark the cohomology group $H^*(W_1; \bigwedge^n Q)$ is embedded into the group $H^*(L_0; 1_2 \otimes \bigwedge^{n-1} Q)$:

$$(4.2.1) \qquad\qquad H^*(W_1; \bigwedge^n Q) \subset H^*(L_0; 1_2 \otimes \bigwedge^{n-1} Q)$$

by the Shapiro Lemma. Here we denote

$$1_\nu := (T_0^*\mathbb{C})^{\otimes \nu}, \quad (\nu \in \mathbb{Z}),$$

on which the Lie algebra L_0 acts by the Lie derivative.

In general we denote by τ_ν^λ, $\lambda, \nu \in \mathbb{Z}$, the line bundle $(T^*\mathbb{C})^{\otimes \nu} \otimes [0]^{\otimes \lambda}$ over the complex line $\mathbb{C}$, where $[0]$ is the line bundle induced by the divisor $0 \in \mathbb{C}$. If $\lambda = 0$, we denote $\tau_\nu := \tau_\nu^0$. We have $T_\nu^\lambda = H^0(\mathbb{C}; \mathcal{O}_\mathbb{C}(\tau_\nu^\lambda))$, $T_\nu = H^0(\mathbb{C}; \mathcal{O}_\mathbb{C}(\tau_\nu))$ and $T_\nu^\times = H^0(\mathbb{C} - \{0\}; \mathcal{O}_\mathbb{C}(\tau_\nu))$. The line bundle over $\mathbb{C}^n$ given by

$$(\mathrm{pr}_1^*\tau_{\nu_1}^{\lambda_1}) \otimes \cdots \otimes (\mathrm{pr}_n^*\tau_{\nu_n}^{\lambda_n}) \quad (\lambda_1, \ldots, \lambda_n, \nu_1, \ldots, \nu_n \in \mathbb{Z})$$

is denoted by $\tau_{\nu_1,\ldots,\nu_n}^{\lambda_1,\ldots,\lambda_n}$, where $\mathrm{pr}_i : \mathbb{C}^n \to \mathbb{C}$ is the i-th projection. If $\lambda_1 = \cdots = \lambda_n = \lambda$ and $\nu_1 = \cdots = \nu_n = \nu$, we abbreviate $(\tau_\nu^\lambda)^n := \tau_{\nu_1,\ldots,\nu_n}^{\lambda_1,\ldots,\lambda_n}$. By the diagonal Lie derivative action, $1_{\nu_0} \otimes \mathcal{O}_{\mathbb{C}^n}(\tau_{\nu_1,\ldots,\nu_n}^{\lambda_1,\ldots,\lambda_n})$ $(\nu_0 \in \mathbb{Z})$ is a sheaf of L_0-modules. From the nuclear theorem we have L_0-isomorphisms

$$1_{\nu_0} \otimes \bigotimes_{i=1}^n T_{\nu_i}^{\lambda_i} = H^0(\mathbb{C}^n; 1_{\nu_0} \otimes \mathcal{O}_{\mathbb{C}^n}(\tau_{\nu_1,\ldots,\nu_n}^{\lambda_1,\ldots,\lambda_n})),$$

and so on. We denote

$$T_\nu^\times := H^0(\mathbb{C} - \{0\}; \mathcal{O}_\mathbb{C}((T^*\mathbb{C})^{\otimes\nu})).$$

Substituting $\mathfrak{g} = L_0$ and $E = \tau_{\nu_1,\dots,\nu_n}^{\lambda_1,\dots,\lambda_n}$ into Proposition 3.2.3, we have

$$(4.2.2) \quad \cdots \to H^{q-n}(L_0; 1_{\nu_0} \otimes \bigotimes_{i=1}^n (T_{\nu_i}^\times / T_{\nu_i}^{\lambda_i})) \xrightarrow{d_1} H^q(L_0; 1_{\nu_0} \otimes \bigotimes_{i=1}^n T_{\nu_i}^{\lambda_i})$$

$$\to H_{L_0}^q(\mathbb{C}^n - \{0\}; 1_{\nu_0} \otimes \mathcal{O}_{\mathbb{C}^n}(\tau_{\nu_1,\dots,\nu_n}^{\lambda_1,\dots,\lambda_n})) \to \cdots ,$$

which we call *the fundamental exact sequence for the L_0 module $1_{\nu_0} \otimes \bigotimes_{i=1}^n T_{\nu_i}^{\lambda_i}$*.

As an application we can compute the cohomology algebra of W_1 with coefficients in the Fréchet space $F(\mathbb{C}^n)$ consisting of all complex analytic functions on $\mathbb{C}^n$:

Theorem 4.2.1.

$$H^*(W_1; F(\mathbb{C}^n)) = H^*(P_n) \otimes \mathbb{C}[v_2,\dots,v_n] \otimes \bigwedge{}^*(\nabla_0^1,\dots,\nabla_0^n)$$

Here the (holomorphic) de Rham cohomology algebra of P_n, $H^*(P_n)$, is mapped into the algebra $H^*(L(\mathbb{C}); F(\mathbb{C}^n))$ in an obvious way ([Ka] Corollary 9.9). The 2-cocycle v_i is defined by

$$v_i(\xi_1(z)\frac{d}{dz}, \xi_2(z)\frac{d}{dz}) := \int_{z_{i-1}}^{z_i} \det \begin{pmatrix} \xi_1'(z) & \xi_2'(z) \\ \xi_1''(z) & \xi_2''(z) \end{pmatrix} dz$$

and the 1-cocycle ∇_0^j by $\nabla_0^j(\xi(z)\frac{d}{dz}) := \xi'(z_j)$ for $\xi(z)\frac{d}{dz}, \xi_1(z)\frac{d}{dz}$ and $\xi_2(z)\frac{d}{dz} \in W_1$.

Taking the $\mathfrak{S}_n$-invariant part of the exact sequence (4.2.2), we have

$$(4.2.3) \quad \cdots \to H^{q-n}(L_0; 1_{\nu_0} \otimes S^n(T_\nu^\times / T_\nu^\lambda)) \xrightarrow{d_1} H^q(L_0; 1_{\nu_0} \otimes \bigwedge{}^n T_\nu^\lambda)$$

$$\to H_{L_0}^q(\mathbb{C}^n - \{0\}; 1_{\nu_0} \otimes \mathcal{O}_{\mathbb{C}^n}((\tau_\nu^\lambda)^n))^{\mathfrak{S}_n, \text{alt}} \to \cdots ,$$

which we call *the fundamental exact sequence for the L_0-module $1_{\nu_0} \otimes \bigwedge^n T_\nu^\lambda$*. Here S^n means the completed n fold symmetric tensor product of a topological vector space.

4.3. Vanishing Theorem

Now we consider the cohomology groups $H^*(W_1; \bigwedge^n Q)$ and $H^*(L_0; \bigwedge^n Q^1)$. By a straightforward computation we have

Lemma 4.3.1.

$$H^0(L_1; S^*(T_2^\times/Q)) = \mathbb{C}[q_1] \quad \text{(the polynomial algebra)},$$

where $q_1 := z^{-1}dz^2 \in T_2^\times$ and L_1 is the subalgebra of L_0 generated by $z^{k+1}\frac{d}{dz}$ $(k \geq 1)$.

By an inductive argument on n with (4.2.1), the fundamental sequence (4.2.3) and Lemma 4.3.1 we have

Theorem 4.3.2 (Vanishing Theorem). *If $q < n$,*

$$H^q(W_1; \bigwedge{}^n Q) = 0, \quad \text{and} \quad H^q(L_0; \bigwedge{}^n Q^1) = 0.$$

4.4. Rešetnikov spectral sequence

We define a sheaf $\mathcal{Q}_n$ over $C_\rho{}^n$ by

$$(4.4.1) \qquad \mathcal{Q}_n := \mathcal{O}_{C_\rho{}^n}(\bigotimes_{i=1}^{n} \mathrm{pr}_i{}^*(T^*C_\rho)^{\otimes 2}),$$

where $\mathrm{pr}_i : C_\rho{}^n \to C_\rho$ is the i-th projection $(1 \leq i \leq n)$. The symmetric group $\mathfrak{S}_n$ acts on the sheaf $\mathcal{Q}_n$ by the permutation of the i-th components $(1 \leq i \leq n)$ twisted by the sign. In view of the nuclear theorem, we have

$$\mathcal{Q}_n(C_\rho{}^n) = Q(C_\rho)^{\otimes n} \quad \text{and} \quad \mathcal{Q}_n(C_\rho{}^n)^{\mathfrak{S}_n} = \bigwedge^{n} Q(C_\rho) = \bigwedge^{n} T^*_{(C,p,z)} M_{g,\rho}.$$

Taking the $\mathfrak{S}_n$-invariant part of the Rešetnikov spectral sequence (3.2.1) converging to $H^\bullet(L(C^\times); Q(C_\rho)^{\otimes n})$, we obtain a spectral sequence

$$(4.4.2) \qquad E_2^{p,q} = H^p(C_\rho{}^n; H^q(L(C^\times); \mathcal{Q}_n))^{\mathfrak{S}_n} \Rightarrow H^{p+q}(L(C^\times); \bigwedge^{n} Q(C_\rho)).$$

Here $H^q(L(C^\times); \mathcal{Q}_n)$ is a sheaf over $C_\rho{}^n$ whose stalk at $z \in C_\rho{}^n$ is the q-th cohomology of $L(C^\times)$ with values in the stalk $(\mathcal{Q}_n)_z$ of the sheaf $\mathcal{Q}_n$. From Decomposition Theorem ([Ka] Theorem 5.3) and Vanishing Theorem 4.3.2 we can compute each stalk of the $\mathfrak{S}_n$-invariant part of the sheaf $H^q(L(C^\times); \mathcal{Q}_n)$ for $q \leq n$. Consequently we have

Theorem 4.4.1. *For $x = (C, p, z) \in M_{g,\rho}$, we have*

$$H^q(\mathfrak{d}_{\rho x}; \bigwedge^{n} T^*_x M_{g,\rho}) = H^q(L(C^\times); \bigwedge^{n} Q(C_\rho)) = \begin{cases} 0, & \text{if } q < n, \\ H^n(W_1; \wedge^n Q), & \text{if } q = n. \end{cases}$$

Decomposition Theorem quoted above is the reason why the n-th cohomology group does *not* depend on the genus g. It is derived from the Bott-Segal Addition Theorem [BS] [Ka].

One deduces the following in a similar way to $M_{g,\rho}$.

Theorem 4.4.2. *For $x = (C, p, z, p_1) \in C_{g,\rho}$, we have*

$$H^q(\mathfrak{d}_{\rho x}; \bigwedge^{n} T^*_x C_{g,\rho})$$

$$= H^q(L(C^\times, p_1); \bigwedge^{n} Q^1(C_\rho, p_1)) = \begin{cases} 0, & \text{if } q < n, \\ H^n(L_0; \wedge^n Q^1), & \text{if } q = n. \end{cases}$$

Thus our study is reduced to that of the cohomology groups $H^n(W_1; \wedge^n Q)$ and $H^n(L_0; \wedge^n Q^1)$.

4.5. Euler class

Here we reconstruct the (relative) Euler class of the universal curve $C_{g,\rho} \to M_{g,\rho}$, i.e., the first Chern class of the relative tangent bundle with respect to the canonical trivialization induced by the coordinate z on $\overline{D^\times_{g,\rho}}$

$$e := c_1(T_{C^\times_{g,\rho}/M_{g,\rho}}, \frac{d}{dz}) \in H^{1,1}(C^\times_{g,\rho}, \overline{D^\times_{g,\rho}})$$

in our framework.

We define $\epsilon \in H^1(L_0; Q^1)$ by

(4.5.1)
$$\epsilon := d(z^{-2}dz^2) \quad \text{i.e.,} \quad \epsilon(f(z)\frac{d}{dz}) := 2(z^{-2}f'(z) - z^{-3}f(z))dz^2, \quad f(z)\frac{d}{dz} \in L_0.$$

This extends to a nowhere-vanishing section of the sheaf $\coprod_{x \in C_{g,\rho}} H^1((\mathfrak{d}_\rho)_x; T_x^* C_{g,\rho})$, which we denote also by ϵ. The cohomology class $D\epsilon \in H^{1,1}(C_{g,\rho})$ is induced by an extension of vector bundles

$$0 \to Q^1(C_\rho, p_1) \to Q^2(C_\rho, p_1) \overset{\text{Res}}{\to} \mathbb{C} \to 0, \quad (x \in C_{g,\rho})$$

over $C_{g,\rho}$. Here Res means the residue of a meromorphic quadratic differential at the point p_1. So, in view of a theory of complex analytic connections on complex analytic fiber bundles [A], we have

Theorem 4.5.1.
$$(\sqrt{-1}/2\pi)D\epsilon = e \in H^{1,1}(C_{g,\rho}).$$

If $x = (C, p, z, p_1) \in \overline{D_{g,\rho}^\times} \subset C_{g,\rho}^\times$, the coordinate z induces a canonical decomposition of $(\mathfrak{d}_\rho)_x (= L(C^\times, p_1))$-modules

(4.5.2)
$$T_x^* C_{g,\rho}^\times = T_{\pi_{g,\rho}(x)}^* M_{g,\rho} \oplus T_{p_1}^* C^\times,$$

which implies $\epsilon^{n+1} = 0 \in H^{n+1}((\mathfrak{d}_\rho)_x; \wedge^{n+1} T_x^* C_{g,\rho}^\times)$ for $\forall x \in \overline{D_{g,\rho}^\times}$ and $\forall n \geq 1$. Here it should be remarked Köthe duality (Corollary 4.1.2) does not hold for the case $x \in \overline{D_{g,\rho}^\times}$.

Thus we have

Proposition 4.5.2. *If* $n \geq 1$,
$$(\sqrt{-1}/2\pi)^{n+1} D(\epsilon^{n+1}) = e^{n+1} \in H^{n+1,n+1}(C_{g,\rho}^\times, \overline{D_{g,\rho}^\times}).$$

4.6. Algebraic fiber integrals

To construct an algebraic analogue of the Morita-Mumford classes e_n i.e., the fiber integral of the $(n+1)$-th power e^{n+1} in our cohomology group $H^n(W_1; \wedge^n Q)$ we introduce a map called the algebraic fiber integral.

We introduce a L_0-module $1_1 := T_0^* \mathbb{C}$ and W_1-modules

$$K := T_1 = \mathcal{O}_{\mathbb{C}}(T^*\mathbb{C})(\mathbb{C}), \quad Q^\times := T_2^\times = \mathcal{O}_{\mathbb{C}}((T^*\mathbb{C})^{\otimes 2})(\mathbb{C} - \{0\}) \quad \text{and}$$

$$(K \otimes \wedge^n Q)^\times := \mathcal{O}_{\mathbb{C}^{n+1}}(\text{pr}_0^* T^*\mathbb{C} \otimes \bigotimes_{i=1}^{n} \text{pr}_i^*(T^*\mathbb{C})^{\otimes 2})(O_n),$$

where $\text{pr}_i : \mathbb{C}^{n+1} \to \mathbb{C}$ is the i-th projection $(0 \leq i \leq n)$ and

$$O_n := \{(t, z_1, \ldots, z_n) \in \mathbb{C}^{n+1}; t \neq z_i \quad (1 \leq \forall i \leq n)\}.$$

Then the Shapiro Lemma gives a natural isomorphism

$$\psi_n : H^*(W_1; (K \otimes \wedge^n Q)^\times/(K \otimes \wedge^n Q)) \overset{\simeq}{\to} H^*(L_0; 1_1 \otimes (\wedge^n Q^\times/\wedge^n Q)).$$

94   NARIYA KAWAZUMI

On the other hand the map $\oint_t$ defined by

$$\oint_t : (K \otimes \bigwedge\nolimits^n Q)^\times \to \bigwedge\nolimits^n Q$$

$$\omega = f(t, z_1, \ldots, z_n) dt\, dz_1^{\ 2} \cdots dz_n^{\ 2} \mapsto \oint_t \omega$$

$$\oint_t \omega := \lim_{R \to +\infty} \left(\frac{1}{2\pi\sqrt{-1}} \int_{|t|=R} f(t, z_1, \ldots, z_n) dt \right) dz_1^{\ 2} \cdots dz_n^{\ 2}$$

induces a W_1-homomorphism $\oint_t : (K \otimes \bigwedge^n Q)^\times / (K \otimes \bigwedge^n Q) \to \bigwedge^n Q$. We call the composite map

$$\int_{\text{fiber}} := \oint_t \circ (\psi_n)^{-1} : H^*(L_0; 1_1 \otimes (\bigwedge\nolimits^n Q^\times / \bigwedge\nolimits^n Q)) \to H^*(W_1; \bigwedge\nolimits^n Q)$$

the algebraic fiber integral.

Let $n \geq 1$. From Theorem 4.3.2 we have $H^n(L_0; 1_1 \otimes \bigwedge^n Q^\times) = 0$ and so

$$0 \to H^n(L_0; 1_1 \otimes (\bigwedge\nolimits^n Q / \bigwedge\nolimits^n Q^\times)) \xrightarrow{\delta^*} H^{n+1}(L_0; 1_1 \otimes \bigwedge\nolimits^n Q)$$
$$\xrightarrow{j^*} H^{n+1}(L_0; 1_1 \otimes \bigwedge\nolimits^n Q^\times) \quad (\text{exact}).$$

The L_0-homomorphism $\bigwedge^{n+1} Q^1 \to \bigwedge^{n+1} Q^1 / \bigwedge^{n+1} Q = 1_1 \otimes \bigwedge^n Q$ maps the cohomology class ϵ^{n+1} into $\operatorname{Ker} j^* = \operatorname{Im} \delta^*$. Hence we can define

$$(4.6.1) \qquad \kappa_n := \int_{\text{fiber}} (\delta^*)^{-1} \epsilon^{n+1} \in H^n(W_1; \bigwedge\nolimits^n Q).$$

This is an algebraic analogue of the n-th Morita-Mumford class $e_n \in H^{2n}(M_g)$.

From Vanishing Theorem 4.3.2 we obtain an algebraic analogue of the Harer stability theorem:

$$\bigoplus_{n \geq 0} H^n(L_0; \bigwedge\nolimits^n Q^1) = (\bigoplus_{n \geq 0} H^n(W_1; \bigwedge\nolimits^n Q))[\epsilon], \quad \text{and}$$
$$(4.6.2)$$
$$H^n(L_0; 1_1 \otimes (\bigwedge\nolimits^n Q^\times / \bigwedge\nolimits^n Q)) = \bigoplus_{i=1}^n ((\delta^*)^{-1} \epsilon^{i+1}) H^{n-i}(W_1; \bigwedge\nolimits^{n-i} Q).$$

On the other hand, the fiber integral $\int_{\text{fiber}}$ has a natural right-inverse. Therefore we obtain

$$(4.6.3) \qquad \bigoplus_{n \geq 0} H^n(W_1; \bigwedge\nolimits^n Q) = \mathbb{C}[\kappa_n; n \geq 1]/\text{relations}.$$

Theorem 2.2.1 follows from Theorem 4.4.1, Theorem 4.4.2, (4.6.2) and (4.6.3).

Comparing our algebraic integrals with the original ones, we obtain Theorem 2.4.1:

$$(\sqrt{-1}/2\pi)^n D(\kappa_n) = e_n \in H^{n,n}(M_{g,\rho}), \quad n \geq 1.$$

5. Relative Picard variety on the universal family

Arbarello et.al.[ADKP] has studied also the (dressed) relative Picard variety $F_{g,\rho}$ on the universal curve over the dressed moduli $M_{g,\rho}$. For an integer $g \geq 0$ and a real number $\rho \geq 0$ we define the (dressed) relative Picard variety $F_{g,\rho}$ by the moduli of quintuples $(C, p, z, L, [\phi])$, where $(C, p, z) \in M_{g,\rho}$, L is a holomorphic line bundle on C of degree $g - 1$, and ϕ is a complex analytic local trivialization of L near $z^{-1}(\overline{D_\rho})$. Furthermore $[\phi]$ is the equivalent class of ϕ under the equivalent relation

$$[\phi] = [\phi'] \quad \Longleftrightarrow \quad \exists \lambda \in \mathbb{C}^\times \quad \phi' = \lambda\phi \quad \text{near } z^{-1}(\overline{D_\rho}).$$

(The space $F_{g,0}$ is equal to the space $\widehat{\mathcal{F}_{g-1}}$ in [ADKP].)

We introduce a topological Lie algebra

$$\mathcal{D}_\rho := \varinjlim\nolimits_{\rho_1 \downarrow \rho} \Delta(D_{\rho_1}^\times).$$

Here we denote by $\Delta(U)$ the topological Lie algebra consisting of all complex analytic differential operators of order ≤ 1 on an open Riemann surface U. When we denote by $F(U)$ the $L(U)$-module consisting of all complex analytic functions on U, we have a decomposition

$$\Delta(U) = F(U) \rtimes L(U).$$

So we denote an element of $\Delta(U)$ (and $\mathcal{D}_\rho$) by $f + X \in \Delta(U)$, $f \in F(U)$, $X \in L(U)$ in the sequel.

The Lie algebra $\mathcal{D}_\rho$ acts on the manifold $F_{g,\rho}$ infinitesimally. In fact, for any $f + X \in \mathcal{D}_\rho$, $(C, p, z, L, [\phi]) \in F_{g,\rho}$ and sufficient small complex numbers ϵ, a complex analytic path $(C(\epsilon), p(\epsilon), z(\epsilon), L(\epsilon), [\phi(\epsilon)]) \in F_{g,\rho}$ is defined by $(C(\epsilon), p(\epsilon), z(\epsilon))$ as in §1.4 and

$$L(\epsilon) := (L|_{C^\times} \amalg D \times \mathbb{C})/\sim_\epsilon$$
$$(z, u) \sim_\epsilon ((\exp \epsilon X)(z), u(\exp \epsilon f(z))\phi(\exp \epsilon X)) \quad (z, u) \in D \times \mathbb{C}$$
$$\phi(\epsilon) := \text{the local trivialization induced by the identity } D \times \mathbb{C} \hookrightarrow D \times \mathbb{C}.$$

Differentiating the path $(C(\epsilon), p(\epsilon), z(\epsilon), L(\epsilon), [\phi(\epsilon)])$ we obtain an infinitesimal action P of $\mathcal{D}_\rho$ on the manifold $F_{g,\rho}$. As was proved in [ADKP] Proposition 3.19, we have an exact sequence

$$0 \to \Delta(C^\times) \hookrightarrow \mathcal{D}_\rho \xrightarrow{P} T_{(C,p,z,L,[\phi])} F_{g,\rho} \to 0.$$

Especially the action P is homogeneous.

Our main result in this section is

Theorem 5.0.1. *For all $x \in F_{g,\rho}$ we have*

$$H^q((\mathcal{D}_\rho)_x; \wedge^p T_x^* F_{g,\rho}) = 0 \quad \text{if } q < p,$$

and the diagonal part $H^p((\mathcal{D}_\rho)_x; \wedge^p T_x^* F_{g,\rho})$ does not depend on the genus $g \geq 0$.

As a corollary

Corollary 5.0.2. *If the Frobenius Reciprocity Law (Hypothesis 3.3.1) holds for the* $\mathcal{D}_\rho$*-manifold* $M = F_{g,\rho}$ *and the* $\mathcal{D}_\rho$*-vector bundles* $\bigwedge^* T^* M$, *we have*

$$H^{p,q}_{\mathcal{D}_\rho}(F_{g,\rho}) = 0, \quad \text{if } q < p,$$

for all $g \geq 0$ *and* $\rho > 0$.

At present our computation for the diagonal part is incomplete. In the next section we discuss the cohomology algebra $H^*(F_{g,\rho}; \mathbf{Q})$.

Fix a point $x = (C, p, z, L, [\phi]) \in F_{g,\rho}$. The Köthe duality (Theorem 4.1.1) gives a $(\mathcal{D}_\rho)_x$-isomorphism

$$T^*_x F_{g,\rho} = (\mathcal{D}_\rho/\Delta(C^\times))^* = JQ(C_\rho),$$

where we denote by $JQ(U)$ the space consisting of all complex analytic 1-jets of 1-forms on an open Riemann surface U:

$$JQ(U) := \mathcal{O}_U(J^1 T^* U)(U) = K(U) \ltimes Q(U).$$

Thus we have

$$H^*((\mathcal{D}_\rho)_x; \bigwedge^n T^*_x F_{g,\rho}) = H^*(\Delta(C^\times); \bigwedge^n JQ(C_\rho)).$$

Similar results to those for $L(U)$ (Rešetnikov spectral sequence, Decomposition Theorem, Vanishing Theorem and so on) hold also for the Lie algebra $\Delta(U)$. So we obtain Theorem 5.0.1.

6. Stable cohomology of the extended mapping class groups

Let $g \geq 2$, $r, s \geq 0$ be integers. Let $\Sigma^s_{g,r}$ denote a 2-dimensional oriented C^∞ manifold (i.e., oriented surface) of genus g with r boundary components and (ordered) s punctures. The group of path-components $\pi_0(\mathrm{Diff}^+(\Sigma^s_{g,r}))$ is denoted by $\Gamma^s_{g,r}$ (or $\mathcal{M}^s_{g,r}$) and called the mapping class group of genus g with r boundary components and (ordered) s punctures. Here $\mathrm{Diff}^+(\Sigma^s_{g,r})$ denotes the topological group (endowed with C^∞ topology) consisting of all orientation preserving diffeomorphisms of $\Sigma^s_{g,r}$ which fix all the boundary points and the punctures <u>pointwise</u>. When $s = 0$, we drop the indices: $\Sigma_{g,r} = \Sigma^0_{g,r}$, $\Gamma_{g,r} = \Gamma^0_{g,r}$ and similarly $\Sigma_g = \Sigma^0_{g,0}$, $\Gamma_g = \Gamma^0_{g,0}$. By the contractibility of the Teichmüller space we have $B\,\mathrm{Diff}^+(\Sigma^s_{g,r}) = B\Gamma^s_{g,r}$ [EE]. We denote

$$H := H_1(\Sigma_{g,1}; \mathbf{Z}) = H_1(\Sigma_g) = H^1(\Sigma_g) = H^1(\Sigma_{g,1}),$$

on which the mapping class group Γ_g acts in an obvious way. By the *extended mapping class group* we mean the semi-direct product

$$\widetilde{\Gamma^s_{g,r}} := H \rtimes \Gamma^s_{g,r}.$$

The relative Picard variety $F_{g,\rho}$ is (homotopy equivalent to) the classifying space of $\widetilde{\Gamma_{g,1}}$. So we have $H^*(F_{g,\rho}; \mathbf{Z}) = H^*(\widetilde{\Gamma^s_{g,r}}; \mathbf{Z})$. Our main result in this section is

Theorem 6.0.1. *For $* < g/2$ we have*

$$H^*(F_{g,\rho};\mathbb{Q}) = H^*(\widetilde{\Gamma_{g,1}};\mathbb{Q}) = H^*(\Gamma_{g,1};\mathbb{Q}) \otimes \otimes_{i,j}\mathbb{Q}[\widetilde{m_{i,j}}],$$

where $\widetilde{m_{i,j}} \in H^{2i+2j-2}(\widetilde{\Gamma_{g,1}};\mathbb{Z})$ is the generalized Morita-Mumford class and the integers i and j run over the domain $\{(i,j) \in \mathbb{Z} \times \mathbb{Z};\quad i \geq 0, j \geq 1\, and\, i + j \geq 2\}$.

6.1. Cohomology of pairs of groups

Let G be a group, K a subgroup of G, and M a G-module. We denote by $H^*(G, K; M)$ the cohomology group of the kernel of the restriction map $C^*(G; M) \to C^*(K; M)$ and call it *the (relative) cohomology group of the pair of groups (G, K) with values in the G-module M*. Here we denote by $C^*(G; M)$ the <u>normalized</u> cochain complex of a group G with values in a G-module M.

Let $N \lhd G$ be a normal subgroup satisfying the condition

$$(6.1.1) \qquad\qquad\qquad KN = G.$$

Then we have the following Lyndon-Hochshild-Serre (LHS) spectral sequence [HS1]:

$$(6.1.2) \qquad E_2^{p,q} = H^p(G/N; H^q(N, N \cap K; M)) \Rightarrow H^*(G, K; M),$$

which is multiplicative in a usual manner.

6.2. Generalized Morita-Mumford classes

First we remark the surface $\Sigma_{g,1}^1$ is obtained by gluing the surfaces $\Sigma_{g,1}$ and $\Sigma_{0,2}^1$ along the boundaries. So the diffeomorphism of $\Sigma_{g,1}$ is naturally extended to that of $\Sigma_{g,1}^1$. The infinite cyclic group $\mathbb{Z}$ acts on the surface $\Sigma_{0,2}^1$ by rotating the puncture and fixing the boundaries pointwise. Similarly this action is extended to that on $\Gamma_{g,1}^1$ in a natural way. Thus we obtain a natural homomorphism $\Gamma_{g,1} \times \mathbb{Z} \to \Gamma_{g,1}^1$, which is injective ([Is] §5). In the sequel we regard the group $\Gamma_{g,1} \times \mathbb{Z}$ as a subgroup of $\Gamma_{g,1}^1$ through the injection. Especially we may consider the cohomology group $H^*(\Gamma_{g,1}^1, \Gamma_{g,1} \times \mathbb{Z}; M)$ for an arbitrary $\Gamma_{g,1}^1$-module M.

The LHS spectral sequence (6.1.2) of the forgetful extension

$$(6.2.1) \qquad\qquad 1 \to \pi_1(\Sigma_{g,1}) \to \Gamma_{g,1}^1 \xrightarrow{\pi} \Gamma_{g,1} \to 1$$

induces a Gysin exact sequence

$$\cdots \to H^{q-1}(\Gamma_{g,1}; M) \to H^{q+1}(\Gamma_{g,1}; H \otimes M)$$
$$\to H^{q+2}(\Gamma_{g,1}^1, \Gamma_{g,1} \times \mathbb{Z}; M) \xrightarrow{\pi_!} H^q(\Gamma_{g,1}; M) \to \cdots$$

for any $\Gamma_{g,1}$-module M. The map $\pi_!$ is called *the Gysin map* or *the fiber integral*. Similarly we can define *the fiber integral* $\tilde{\pi}_! : H^p(\widetilde{\Gamma_{g,1}^1}, \widetilde{\Gamma_{g,1}} \times \mathbb{Z}) \to H^{p-2}(\widetilde{\Gamma_{g,1}})$.

Choose a simple curve l inside the subsurface $\Sigma_{0,2}^1$ connecting the puncture to a point on the boundary. A 1-cocycle $\omega_l \in Z^1(\Gamma_{g,1}^1, \Gamma_{g,1} \times \mathbb{Z}; H)$ and a 2-cocycle $\tilde{\omega}_l \in Z^2(\widetilde{\Gamma_{g,1}^1}, \widetilde{\Gamma_{g,1}} \times \mathbb{Z}; \mathbb{Z})$ are given by

$$(6.2.3) \qquad \begin{aligned} &\omega_l(\gamma) = \gamma l - l \in H, \quad \gamma \in \Gamma_{g,1}^1, \\ &\widetilde{\omega}_l(u_1\gamma_1, u_2\gamma_2) := \gamma_1(\gamma_2 l - l) \cdot u_1, \quad u_1, u_2 \in H, \gamma_1, \gamma_2 \in \Gamma_{g,1}^1, \end{aligned}$$

where $\gamma l - l$ is regarded as a closed curve on $\Sigma_{g,1}$. Let $e \in H^2(\Gamma_g^1; \mathbb{Z})$ be the Euler class of the central extension $0 \to \mathbb{Z} \to \Gamma_{g,1} \to \Gamma_g^1 \to 1$.

For integers $i, j \geq 0$ satisfying $i + j \geq 2$ we define

$$(6.2.4) \qquad \begin{aligned} m_{i,j} &:= \pi_!(e^i \omega_l^{\,j}) \in H^{2i+j-2}(\Gamma_{g,1}; \textstyle\bigwedge^j H), \\ \widetilde{m_{i,j}} &:= \tilde{\pi}_!(e^i \widetilde{\omega_l}^{\,j}) \in H^{2i+2j-2}(\widetilde{\Gamma_{g,1}}). \end{aligned}$$

Clearly $m_{i+1,0}$ and $\widetilde{m_{i+1,0}}$ are equal to (the image of) the i-th Morita-Mumford (tautological) class $e_i(= \kappa_i) \in H^{2i}(\Gamma_g; \mathbb{Z})$ [Mo] [Mu]. So we call these classes $m_{i,j}$'s and $\widetilde{m_{i,j}}$'s *the generalized Morita-Mumford classes*.

6.3. Stable cohomology of the extended mapping class groups

Recently, Looijenga [L] proved the <u>rational</u> stable cohomology group of the mapping class group Γ_g (with no boundary and no punctures) with coefficients in any irreducible representation of the complex symplectic group is a free module over the rational stable cohomology algebra of the mapping class group, and described its free basis. As a consequence he determined the rational stable cohomology of $\widetilde{\Gamma_g}$ modulo that of Γ_g. His computation is involved with geometric consideration on the moduli orbifold of complex algebraic curves including a theorem in Hodge theory. Here it is remarkable that his results are based only on the Harer stability theorem [H] [Is] with trivial coefficients.

Following Looijenga [L] we have proved the following results from the Harer stability theorem with trivial coefficients, but geometric considerations including Hodge theory do not fit our situation where the surface is not closed. We use the Lyndon-Hochschild-Serre spectral sequence for a pair of groups introduced in §6.1 instead.

Theorem 6.3.1. *If $s \geq 0$, $r \geq 1$ and $n \geq 0$, we have*

$$H^*(\Gamma_{g,r}^s; H^1(\Sigma_{g,r}^s; \mathbb{Z})^{\otimes n}) = H^*(\Gamma_{g,1}; H^{\otimes n}) \otimes_{H^*(\Gamma_{g,1};\mathbb{Z})} H^*(\Gamma_{g,r}^s; \mathbb{Z})$$

for $* \leq g/2 - n$. Furthermore the cohomology group $H^*(\Gamma_{g,1}; H^{\otimes n})$ is stably a free module over the algebra $H^*(\Gamma_{g,1}; \mathbb{Z})$ and combinations of the (modified) generalized Morita-Mumford classes give its free basis.

Taking the alternating part of the second half of this theorem, we obtain

Theorem 6.3.2.

$$H^*(\Gamma_{g,1}; \textstyle\bigwedge^* H^1(\Sigma_{g,1}; \mathbb{Q})) = H^*(\Gamma_{g,1}; \mathbb{Q}) \otimes \otimes_{i,j} \mathbb{Q}[m_{i,j}]$$

if the total degree is smaller than $g/2$. Here the integers i and j run over the domain given in Theorem 6.0.1.

As a corollary we obtain Theorem 6.0.1.

Finally we mention how these classes $\widetilde{m_{i,j}}$'s are related to the equivariant cohomology of $F_{g,\rho}$. The class $\widetilde{\omega_l}$ corresponds to the Abelian differential of the third kind on a compact Riemann surface. Especially it has a position in the complex analytic Gel'fand-Fuks cohomology. This implies the classes $\widetilde{m_{i,j}}$'s are contained in the image of the equivariant (p, p)-cohomology. The author does not know whether the equivariant (p, p)-cohomology is stably isomorphic to the polynomial algebra generated by $\widetilde{m_{i,j}}$'s.

References

[A]   M. Atiyah, *Complex analytic connections in fibre bundles*, Trans. Amer. Math. Soc. **85** (1957), 181–207, (This article is *not* equal to the 5-th paper of "Collected Works" vol.1.).

[ADKP]   E. Arbarello, C. DeContini, V.G. Kac, and C. Procesi, *Moduli spaces of curves and representation theory*, Commun. Math. Phys. **117** (1988), 1–36.

[BMS]   A. A. Beilinson, Yu. I. Manin and V. V. Schechtman, *Sheaves of Virasoro and Neveu-Schwartz algebras*, Lect. Note in Math. **1289** (1987), Springer, Berlin - Heidelberg - New York.

[B]   R. Bott, *Homogeneous vector bundles*, Ann. Math. **66** (1957), 203–248.

[BS]   R. Bott and G. Segal, *The cohomology of the vector fields on a manifold*, Topology **16** (1977), 285–298.

[EE]   C.J. Earle and J. Eells, *A fiber bundle description of Teichmüller theory*, J. Diff. Geom. **3** (1969), 19–43.

[FT]   B.L. Feigin and B.L. Tsygan, *Riemann-Roch theorem and Lie algebra cohomology I*, Rend. Circ. Mat. Palermo **21** (1989), 15–52.

[G,TVS] A. Grothendieck, *Topological Vector Spaces*, Gordon and Breach, New York, London, Paris 1973.

[G,E]   ———, *Éléments de géometrie algébrique III*, Publ. I.H.E.S. **11** (1973).

[H]   J. Harer, *Stability of the homology of the mapping class group of orientable surfaces*, Ann. Math. **121** (1985), 215–249.

[H1]   ———, *The second homology group of the mapping class group of an orientable surface*, Invent. math. **72** (1983), 221–239.

[H2]   ———, *The third homology group of the moduli space of curves*, Duke Math. J. **63** (1991), 25–55.

[HS]   G. Hochschild and J.-P. Serre, *Cohomology of Lie algebras*, Ann. Math. **157** (1953), 591–603.

[HS1]   ———, *Cohomology of group extensions*, Trans. Amer. Math. Soc. **74** (1953), 110–134.

[Is]   A. Ishida, *Master Thesis*, (in Japanese) Univ. of Tokyo (1994).

[Iv]   N. Ivanov, *Complexes of curves and the Teichmüller modular group*, Russian Math. Survey **42** (1987), 55–107.

[Ka]   N. Kawazumi, *On the complex analytic Gel'fand-Fuks cohomology of open Riemann surfaces*, Ann. Inst. Fourier **43** (1993), 655–712.

[Ka1]   ———, *An application of the second Riemann continuation theorem to cohomology of the Lie algebra of vector fields on the complex line*, (preprint: UTMS 93-18) (1993).

[Ka2]   ———, *Moduli space and complex analytic Gel'fand-Fuks cohomology of Riemann surfaces*, (preprint: UTMS 93-29) (1993).

[Ka3]   ———, *Homology of hyperelliptic mapping class groups for surfaces*, preprint. Hokkaido Univ. **262** (1994).

[Ka4]   ———, *A generalization of the Morita-Mumford classes to extended mapping class groups for surfaces*, preprint. Hokkaido Univ. **292** (1995).

[Ka5]   ———, *On the stable cohomology algebra of extended mapping class groups for surfaces*, preprint. Hokkaido Univ. **311** (1995).

[Ka6]   ———, *Moduli space and complex analytic Gel'fand-Fuks cohomology of Riemann surfaces, II, III*, (in preparation).

[Koe]   G. Köthe, *Dualität in der Funktionentheorie*, J. Reine u. Angew. Math. **191** (1953) 30–49.

[Ko]   H. Komatsu, *Theory of locally convex spaces*, Dept. Math., Univ. of Tokyo, 1974

[Ko1]   ———, *Projective and injective limits of weakly compact sequences of locally convex spaces*, J. Math. Soc. Japan **19** (1955), 366–383.

[Ko2]   ______, *Choh-kansuh ron nyuhmon. (An introduction to theories of distributions and hyperfunctions.)*, (in Japanese), Iwanami, Tokyo, 1978.

[Kon]   M.L. Kontsevich, *Virasoro algebra and Teichmüller spaces*, Functional Anal. Appl. **21** (1987), 156–157.

[Koe]   G. Köthe, *Dualität in der Funktionentheorie*, J. Reine u. Angew. Math. **191** (1953), 30–49.

[L]   E. Looijenga, *Stable cohomology of the mapping class group with symplectic coefficients and of the universal Abel-Jacobi map*, Jour. Alg. Geom. (to appear).

[Mi]   E.Y. Miller, *The homology of the mapping class group*, J. Diff. Geom. **24** (1986), 1–14.

[Mo]   S. Morita, *Characteristic classes of surface bundles*, Inventiones math. **90** (1987), 551–577.

[Mo1]   ______, *Families of Jacobian manifolds and characteristic classes of surface bundles, I*, Ann. Inst. Fourier **39** (1989), 777–810.

[Mo2]   ______, *Families of Jacobian manifolds and characteristic classes of surface bundles, II*, Math. Proc. Camb. Phil. Soc. **105** (1989), 79–101.

[Mo3]   ______, *The extension of Johnson's homomorphism from the Torelli group to the mapping class group*, Invent. math. **111** (1993), 197–224.

[Mu]   D. Mumford, *Towards an enumerative geometry of the moduli space of curves*, Arithmetic and Geometry., Progr. Math. **36** (1983), 271–328.

[R]   V.N. Rešetnikov, *On the cohomology of the Lie algebra of vector fields on a manifold with non trivial coefficients*, Soviet Math. Dokl. **14** (1) (1973), 234–240.

[Sa]   K. Saito, *A generalization of Eichler integrals and certain local systems over spin Riemann surfaces*, Publ. RIMS, Kyoto Univ. **27** (1991), 431–460.

[Sc]   G. Scheja, *Riemannsche Hebbarkeitssätze für Cohomologieklassen*, Math. Ann. **144** (1961), 345–360.

DEPARTMENT OF MATHEMATICS, FACULTY OF SCIENCE, HOKKAIDO UNIVERSITY, SAPPORO 060 JAPAN

E-mail address: kawazumi@math.hokudai.ac.jp

Proceedings of
THE 37TH TANIGUCHI SYMPOSIUM ON
TOPOLOGY AND TEICHMÜLLER SPACES
held in Finland, July 1995
ed. by Sadayoshi KOJIMA *et al.*
©1996 World Scientific Publishing Co.
pp. 101–114

DEFORMATIONS OF HYPERBOLIC CONE MANIFOLDS

STEVEN P. KERCKHOFF

(Received July 1, 1996)

Variational techniques are valuable tools in studying geometric structures on manifolds. Since, by Mostow rigidity, hyperbolic structures on closed 3-manifolds are unique, one must consider manifolds with boundary or structures with singularities in order to have a deformation theory. Typically, a theory of deformations has two parts, a local theory describing small or infinitesimal variations of structure and a global theory describing the boundary of the space of structures in terms of the possible degenerations of structure.

Both of these aspects of the deformation theory of hyperbolic structures on 3-manifolds have been the object of much study on their own. They also have been important in the construction of hyperbolic structures on closed 3-manifolds. This phenomenon was exhibited by Thurston's construction of hyperbolic structures on atoroidal Haken manifolds by cutting along an incompressible surface and studying structures on the resulting manifold with boundary and by his proof of the orbifold theorem through the deformations of hyperbolic structures with conical singularities. In this note we discuss some recent extensions and refinements of these ideas.

A 3-dimensional hyperbolic *cone manifold* is a hyperbolic structure on a 3-manifold with singularities along a link. The precise definition of such a structure is that there is a hyperbolic structure on the complement of the link, which is incomplete, whose metric completion determines a singular metric with singularities along the link. The link is totally geodesic and in cylindrical co-ordinates around a component of the singular locus, the metric has the form $dr^2 + sinh^2r \ d\theta^2 + cosh^2r \ dh^2$ where r is the distance from the singular locus, h is the distance along the singular locus, θ is the angular measure around the singular locus, which measured *modulo* α for some $\alpha \in R_+$. Then α is called the cone angle at that component. Note that the metric in a disk in the (r, θ)-plane is induced from that of a wedge of angle α in H^2, with sides identified. This is the cone from which the name arises.

1991 *Mathematics Subject Classification.* Primary 57M50; Secondary 30F40.
Key words and phrases. cone manifold, hyperbolic manifold, rigidity, deformation.

This is the natural generalization of the metric structure associated to a hyperbolic orbifold, where the cone angles are of the form $\frac{2\pi}{n}$. Such structures also arise when one perturbs the representation corresponding to a non-compact, finite volume, complete structure. Although complete structures are rigid, there are always nearby representations which correspond to incomplete structures. The space of such incomplete structures (with certain control on their asymptotic behavior), called *hyperbolic Dehn surgery space*, plays a significant role in 3-dimensional topology and it is important to understand its structure. The cone manifolds are like a real subvariety of this space; nearby points off this set correspond to hyperbolic manifolds with Dehn surgery singularities. (For a definition of Dehn surgery singularities, their coefficients and hyperbolic Dehn surgery space see [16], Chapter 4 .)

In [8] rigidity properties are established for hyperbolic cone manifolds with singularities along a link, when the cone angles along the link are at most 2π. It is proved that it is impossible locally to deform such a hyperbolic structure without altering some cone angle. We will call this *rigidity rel cone angles*. This implies that the representation variety is smooth near such a structure and that it is always possible to alter the cone angles by a small amount. Furthermore, the cone manifold structures are locally parametrized by their cone angles. This gives a reasonable local description of the space of cone manifold structures on a fixed manifold with a fixed singular set, at least with this angle constraint.

To understand the deformation space of hyperbolic cone manifolds globally one must be able to describe how and predict when one reaches the boundary of this space. Various degenerations of hyperbolic structure can occur as one increases the cone angle of a hyperbolic cone manifold. The general conjecture which characterizes the boundary of the space of such hyperbolic structures is that degeneration can occur only for one of the following reasons:

DEG 1 The uniform collapsing of the structure, which, after rescaling, converges to another one of the basic 3-dimensional geometries.

DEG 2 The appearance of an incompressible Euclidean surface, possibly with cone points.

DEG 3 The singular locus intersects itself.

The first two types of degeneration provide considerable topological information about the underlying manifold. For this reason, they can be ruled out in certain situations. However, DEG 3 is much less well understood. Indeed, dealing with this type of degeneration is one of the main obstacles to improved understanding of hyperbolic Dehn surgery space. As we shall see below in Theorem 2, when this type of degeneration can be ruled out, the limiting behavior of a sequence of cone manifold structures is well-controlled. We would like to be able to predict ahead of time, starting with a complete structure (cone angle 0) or a small cone angle, how far we can increase the cone angle. In particular, we would like to know when cone angle 2π, which is a smooth structure, can be attained.

If M is a non-compact, finite volume hyperbolic 3-manifold, it is the interior of a compact 3-manifold with a finite number of torus boundary components. For each torus, there are an infinite number of distinct ways to attach a solid torus; such "Dehn fillings" are parametrized by pairs of relatively prime integers, once a basis for the fundamental group of the torus is chosen. If each torus is filled, the resulting manifold is closed. A fundamental theorem of Thurston ([16]) states that, for all but a finite number of Dehn surgeries on each boundary component, the resulting closed 3-manifold has a hyperbolic structure. However, it was unknown whether or not the number of such non-hyperbolic was bounded independent of the original non-compact hyperbolic manifold.

Recently, in joint work ([9]) with Craig Hodgson, it has been established that there is, indeed, a universal upper bound on the number of non-hyperbolic Dehn surgeries per boundary torus, independent of the manifold M. Although an explicit bound could, in theory, be given, we will not attempt to do so here because, with the current proof, such a calculation would depend on computation of Sobolev constants for (linear) elliptic operators and would be unlikely to give useful bounds.

This result should be compared with the known bounds on the number of Dehn surgeries which yield *not negatively curved* manifolds. (We use this term for manifolds which do not admit negatively curved metrics since "non-negatively curved" is generally used to mean something else.) It follows from the work of Gromov-Thurston ([7], see also [3]) that all but a universal number of surgeries on each torus yield 3-manifolds which admit negatively curved metrics. The best known estimate for this bound is currently 48 (see [3]). Having a negatively curved metric implies strong topological properties for the underlying 3-manifold; in particular, they are irreducible, atoroidal and have infinite fundamental groups. If the geometrization conjecture were true, it would imply that they actually had hyperbolic metrics. Currently, the largest number of known non-hyperbolic Dehn surgeries is 10, for the case of the complement of the figure-8 knot.

The bound on the number of not-negatively curved Dehn surgeries comes from what is usually referred to as the "2π-theorem". It can be stated as follows: Given a cusp in a complete, hyperbolic 3-manifold, remove a horoball neighborhood of the cusp, leaving a manifold with a boundary torus which has a flat metric. Let γ be an isotopy class of simple closed curve on this torus and let M_γ denote M filled in so that γ bounds a disk. Then the 2π-theorem states that, if the geodesic length of γ on the torus is greater than 2π, then M_γ can be given a metric of negative curvature which agrees with the hyperbolic metric in the region outside the horoball. The bound on the number of not-negatively curved surgeries follows from the fact that it is always possible to find an embedded horoball with boundary torus whose shortest geodesic has length at least 1. On such a torus there are a bounded number of isotopy classes of geodesics with length less than or equal to 2π.

Our bound is derived in a similar manner. We show that, if the geodesic length of γ on the torus is sufficiently long *and* its extremal length (= (geodesic length)^2 / area) is sufficiently long, then it is possible to deform the complete hyperbolic structure through cone manifold structures with γ bounding the singular meridian disk until the cone angle reaches 2π. This gives the smooth hyperbolic structure on M_γ. The important point here is that "sufficiently long" is universal, independent of M. As before, it is straightforward to show that there is an embedded horospherical torus with the property that all but a universal number of isotopy classes of simple closed curves satisfy both of these length conditions.

The bound follows from two theorems which are proved by rather different methods:

Theorem 1 [9]. *Let M be a non-compact, complete, finite volume hyperbolic 3-manifold and let N be the compact manifold with toral boundary gotten by truncating M along a horospherical torus T. Let γ be a simple closed curve on T, M_γ the Dehn filling where γ bounds a meridian disk. Let $M_\gamma(\theta)$ be a cone manifold structure on M_γ with cone angle θ along the core, Σ, of the added solid torus, gotten by increasing the angle from the complete structure. There is a universal constant K such that if the geodesic length and extremal length of γ on T is bigger than K, then there are positive lower bounds to the tube radius around Σ and the length of Σ in $M_\gamma(\theta)$ for any $2\pi \geq \theta \geq \epsilon > 0$.*

This theorem doesn't guarantee that cone angle 2π can actually be reached, just that there are lower bounds for any angle less than or equal to 2π that *is* attained. That 2π can actually be attained follows from the next theorem.

Theorem 2 [9]. *Let $M_\gamma(\theta_i)$ be a sequence of closed hyperbolic cone structures on M_γ with singular locus Σ. Suppose $\theta_i \to \theta \neq 0$ and that the volumes of the $M_\gamma(\theta_i)$ are bounded above. If there there are positive constants C, L such that the length of Σ is at least L and there is a tube of radius at least C around Σ for all i, then $M_\gamma(\theta_i)$ converges geometrically to a cone structure $M_\gamma(\theta)$.*

Remark. *The condition on the volume is satisfied if, for example, $M_\gamma(\theta_i)$ are part of a smooth family of cone manifolds with θ_i increasing. This follows from Schläfli's formula.*

An easy corollary of Theorems 1 and 2 is:

Theorem 3 [9]. *There is a bound on the number of non-hyperbolic surgeries per cusp for M, independent of M.*

The extra condition in Theorem 1, not necessary for the 2π Theorem, that the extremal length, not just the flat geodesic length, be sufficiently long is probably not necessary, but is an artifact of the proof. The construction in the 2π-theorem gives a metric which, for larger and larger lengths of γ, is closer and closer to being hyperbolic. Thus, Theorem 3 could be viewed as a kind of weak pinching theorem: If the manifold has a metric sufficiently close (in this special way) to being hyperbolic, then it actually has a hyperbolic metric.

Here is a sketch of the proof of Theorem 3.

Begin with a non-compact, finite volume hyperbolic 3-manifold M, which, for simplicity, we assume has a single end. In the general case the ends are handled independently. M is the interior of a compact manifold N which has a single torus boundary. Choose a simple closed curve γ on the torus; we wish to put a hyperbolic structure on the closed manifold M_γ gotten by Dehn filling. The metric on the open manifold M is deformed through incomplete metrics whose metric completion is a singular metric on M_γ, making it into a cone manifold. The singular set is a simple closed geodesic at the core of the added solid torus. Denote the singular set by Σ and its cone angle by θ. The complete structure can be considered as a cone manifold with angle 0. The cone angle is increased monotonically; if the angle of 2π is reached, this defines a smooth hyperbolic metric on M_γ.

As noted before, from [8] it is known that it is always possible to change the cone angle a small amount, either increase it or decrease it. Furthermore, this can be done in a unique way, at least locally; the cone angles locally parametrize the set of cone manifold structures on M_γ. Thus, to choose a 1-parameter family of cone angles is to choose a well-defined family of singular hyperbolic metrics on M_γ of this type.

Although there are always local variations of the cone manifold structure, the global structure may degenerate in various ways as a family of angles reaches a limit. In order to find a smooth hyperbolic metric on M_γ it is necessary to show that no degeneration occurs before the angle 2π is attained.

The proof has two main steps, which we have called Theorem 1 and Theorem 2, respectively, involving rather different types of arguments. The proof of Theorem 1 is fairly analytic, showing that under the length hypotheses, there is a lower bound to the tube radius and to the length of the core geodesic for any of the cone manifold structures with angle at most 2π. Given these geometric conditions, the second step consists of showing that no degeneration of the hyperbolic structure is possible. This is the content of Theorem 2. It is proved by studying possible geometric limits; the tube conditions restrict such limits to fairly tractable and well-understood types.

The argument showing that there is a lower bound to the tube radius is based on the rigidity theory for cone manifolds developed in [8]. Indeed, the key estimates are best viewed as effective versions of the local rigidity of cone manifolds. Below we choose a parametrization of the increasing family of cone angles which uniquely determines a family of cone manifold structures. We then need to control the global behavior of these metrics. The idea is to first form a model for the deformation in a neighborhood of the singular locus which changes the cone angle in the prescribed fashion and then find estimates which bound the deviation of the actual deformation from the model.

To describe the model deformation, we cut out a horospherical neighborhood of the cusp, leaving a torus boundary with an intrinsic flat metric. The normal curvatures are all $+1$ for such a torus so the geodesic curvature in the 3-manifold of a curve which is geodesic in the torus is $+1$. We deform the embedding of the torus and its intrinsic metric so it becomes the boundary of a tube of constant radius around the geodesic which forms the singular locus Σ for a cone manifold structure on M_γ.

We denote by R the distance of the torus to the singular locus Σ. Then the normal curvature of the geodesic on the torus isotopic to γ is $coth\ R$ and of a geodesic on the torus orthogonal to γ is $tanh\ R$. We denote by $\ell(t)$ the geodesic length on the torus of γ at time t and let $\ell_0 = \ell(0)$. The cone angle at time t is denoted by $\theta(t)$. Then in our model for increasing the cone angle in this torus neighborhood these quantities behave as follows:

$$\ell(t) = \ell_0 e^{-t}$$
$$\theta^2(t) = 2\ell_0^2 \left(1 - e^{-2t}\right)$$
$$\ell(t) = \theta(t)\ sinh\ R(t)$$

In this model, one reaches cone angle 2π as long as $\ell_0 > \frac{1}{\sqrt{2}} 2\pi$. Furthermore, the longer ℓ_0 is, the quicker the angle reaches 2π. For large values of this length, the time T when this is attained is $\cong (\frac{1}{\ell_0})^2$.

In [8] an infinitesimal deformation in the neighborhood of the singular locus was described which is the tangent vector for such a family of deformations. (The infinitesimal deformation there has to be suitably rescaled.) In particular, there is a precise description of an extension over the solid torus neighborhood for this deformation. In this deformation, the length of the singular locus is strictly increasing. In the actual variation of the hyperbolic structures, the deformations are completely determined by the change in the cone angle. Thus, the changes in the holonomies of the other elements of the fundamental group of the torus cannot be arbitrarily prescribed and it is necessary to alter the model to fit the actual changes. Again, an explicit description of any such necessary alteration is given in [8]. The first use of the effective rigidity theory is to give bounds on the size of this correction of the holonomy.

Infinitesimal deformations, up to infinitesimal isometry, are parametrized by a cohomology group $H^1(M; E)$ where $M = M_\gamma - \Sigma$ is the complement of the singular locus in M_γ with its incomplete metric and E is the flat bundle of infinitesimal isometries. As stated before, once the change in the cone angle is prescribed, this completely determines the variation of the hyperbolic metric on M_γ, hence on M. This, in turn, determines the cohomology class.

The first step in proving a local rigidity theorem is to represent this (deRham) class by a closed and co-closed E-valued 1-form. If M were closed this would be standard Hodge theory. In the cone manifold case, M is neither compact nor complete so more care is needed in controlling the behavior of the representative as it approaches the singular locus. First, a representative is chosen which is closed and co-closed in a neighborhood of the singular locus. This can be done so it corresponds to the modified model described in the previous paragraph. It is then possible to find a unique E-valued 1-form with certain asymptotic behavior which is closed and co-closed on all of M. It is necessary to estimate how much this process changes the representative from the modified model. This gives an estimate on how much the actual infinitesimal variation differs from the model infinitesimal variation. Bounding the infinitesimal variation, in turn, bounds the entire variation by integration.

The reason for choosing a closed and co-closed representative for the cohomology class is that it satisfies a strong partial differential equation. The real part of this representative is a 1-form with values in the tangent bundle of M. By change of type, it can be considered as a symmetric 2-tensor; this describes the infinitesimal change in the metric. The 1-form η with values in the tangent bundle satisfies a Weitzenbock-type formula:

$$D^*D\eta = -\eta$$

where D is the exterior covariant derivative on such forms and D^* is its dual. Taking the L^2 dot product of this formula with η and integrating by parts would give the formula

$$||D\eta||^2 = -||\eta||^2$$

if M were closed. Thus $\eta = 0$ and the deformation would be trivial. This is the proof of local rigidity for closed hyperbolic manifolds. (See [6] and [19] for details.)

If M is truncated to have toral boundary, there will be a boundary term after the integration by parts step. If this term is zero, the same conclusion holds: the deformation is trivial. When M_γ is a cone manifold with cone angle at most 2π, it was shown in [8] that, for a deformation which leaves the cone angle fixed, it is possible to find a representative as above for which the boundary term goes to zero on a sequence of tori which bound arbitrarily small neighborhoods of the singular locus. This is the proof of local rigidity rel cone angles. The same argument shows that, if the manifold is truncated along an embedded torus surrounding the singular locus, the boundary term is negative, with absolute value equal to the square of the L^2 norm of the representative on the part of the manifold thrown away.

Furthermore, explicit forms were found in the neighborhood of the singular locus which changed the cone angle. The boundary term is positive in this case. The model deformation described earlier has such a form as its tangent vector, normalized so that the value of the boundary term is less than or equal to 2 times the area of the boundary. In [8] it was also shown that there is a representative of such an infinitesimal deformation which differs from the model form by a harmonic form for which the boundary term on the whole manifold goes to zero. The condition that the boundary term for the actual representative (model plus this other term) be positive puts strong restrictions on the contribution to the boundary term on the truncated manifold coming from this "correction" term. This, in turn, bounds the L^2 norm of the correction term on the part of the manifold which is thrown away. There is also a bound on the L^2 norm of the entire form on the truncated manifold.

What this argument shows is that there is a bound on the "correction" term which describes the difference between the model infinitesimal deformation on the neighborhood of the singular locus and the actual infinitesimal deformation. This bound is an L^2 bound which is a constant times the area of the boundary, where the constant depends only on how the tube radius differs from the model. Up to this point everything can be made quite explicit and the constants are easily computed (and are not large). To actually bound the geometric changes introduced by the correction term requires standard but fairly ineffective estimates from elliptic PDE's. In particular, it is necessary to compute how the infinitesimal change in the tube radius is affected

by this term. The conclusion is that this contribution to the derivative of the tube
radius is bounded by a constant times this L^2 norm where the constant depends only
on the tube radius. Of course, the norm depends on the area of the boundary so, if
the boundary has large area, this is a large bound.

At this stage, we recall that the time needed to reach angle 2π is $\cong (\frac{\pi}{\ell_0})^2$. The
total correction is bounded by the total time multiplied by the infinitesimal correction
which is bounded by a constant times the area divided by ℓ_0^2. It is here that we need
the hypothesis that this ratio is bounded, hence that the extremal length not be too
small.

This completes the sketch of the proof of Theorem 1.

$\square$

The proof of Theorem 2 uses some techniques for studying geometric limits first
described in Thurston's proof of the orbifold theorem, as described in his lectures and
later outlined in some expository notes. The cases which need to be considered for
Theorem 2 are the easiest ones in that theory. The condition that there be a lower
bound on the length of Σ is unnecessary but it greatly simplifies the proof. A similar
analysis, involving more cases (but with stronger angle constraints), is carried out in
[10]. As this material has been outlined elsewhere we will not give a sketch of the
proof here. The full proof will appear in [9].

The proofs of Theorem 1 and 2 show that if there is a large enough tube around
the singular set for a cone manifold with non-zero cone angles (less than 2π), then it
is possible to increase the cone angle monotonically to 2π. Since there are many link
complements in non-hyperbolic manifolds that have complete hyperbolic structures
and these can always be perturbed to cone manifold structures with very small cone
angles, it is clear that some condition on the geometry of a cone manifold is necessary
to guarantee that angle 2π can be reached. However, it seems likely that it is always
possible to *decrease* the cone angle to zero, which corresponds to a cusp. Indeed,
there is no topological obstruction; the complement of the singular locus admits a
finite volume, complete hyperbolic structure. The issue is whether or not the cone
structure can be connected to the complete structure by a 1-parameter family of cone
structures.

Conjecture 1. *Given a hyperbolic cone manifold, it is always possible to decrease
the cone angles, through hyperbolic cone manifolds, back to a complete finite volume
structure on the complement of the singular locus.*

We divide this conjecture into two conjectures, each of which is of independent
interest.

Conjecture 1a. *As the cone angle is decreased, the distance of the singular locus
from itself is bounded below.*

Conjecture 1b. *If the volume and the distance of the singular locus from itself are bounded away from zero and the cone angles are bounded from above in a sequence of hyperbolic cone structures, then no degeneration can occur.*

Remark. *As the cone angle is being decreased, the volume increases. Thus the two conjectures above imply Conjecture 1.*

It is useful to view Conjecture 1 in the context of the 3 expected kinds of degeneration of hyperbolic structures. Since the volume is bounded below, DEG 1 cannot occur and since the cone angles are decreasing, the Gauss-Bonnet theorem implies that no Euclidean surfaces can appear (DEG 2). The third type of degeneration (DEG 3) is being ruled out by Conjecture 1a.

However, if the cone angles are increased, all of these phenomenon can occur. Dealing with the first two occupies a large part of the proof of the orbifold theorem. It is possible to show that DEG 3 can only occur when the cone angle is bigger than π; hence this phenomenon does not arise in this situation. However, if one attempts to increase the cone angle to 2π (a smooth structure), this case must be dealt with. For example, if one starts with a hyperbolic manifold, removes a knot which is not in the isotopy class of a geodesic but whose complement has a complete hyperbolic structure, then it will not be possible to deform the cone angle to 2π without some kind of degeneration. In many case the singular locus should be forced to intersect itself at some smaller angle. In some explicit examples, it is possible to show that, at least numerically, this does indeed occur.

One possible approach to Conjecture 1a is analytic. The Hodge representative guaranteed by the Hodge theorem in [8] has an explicit non-L^2 part which is expanding radially away from the singular locus. Thus, when the cone angle is decreased, the "model" deformation, as discussed in the sketch of the proof of Theorem 1, has an embedded tube which is getting monotonically larger. So it appears that this case should be easier than that considered in Theorem 1. However, to prove the conjecture, it is necessary to bound uniformly the L^2 "correction term" over the sequence, independent of the initial geometry. Attempts to find such a priori bounds have thus far proved unsuccessful. All that is currently known is that, if there is a sufficiently large (depending on the cone angle) embedded tube around the singular locus, the proofs of Theorems 1 and 2 go through, and the cone angle can be decreased back to the complete structure.

Conjecture 1b is known to be true in many situations; Theorem 2 above is one such example. In [10] further cases are considered for which the conjecture is established. All such proofs currently follow the basic outline of Thurston's arguments for the orbifold theorem.

If Conjecture 1 were true, it would imply a global, Mostow rigidity for cone manifolds (at least for those cone angles where local rigidity was known). To see this, consider two hyperbolic cone manifold structures with the same cone angles and singular locus. If the conjecture were true, they could both be deformed back to a complete structure which must be the same, by Mostow rigidity in the complete case. But local rigidity implies that there are no branch points along the paths back to (and including) the complete structure, so they must have been the same all along. This argument is carried out by S. Kojima in [10] for cone angles at most π. With this cone angle restriction, Conjectures 1a and 1b can be handled by the type of analysis needed for the orbifold theorem.

The conjectures should also hold for hyperbolic structures with Dehn surgery singularities; if arguments could be found that were independent of cone angle this more general case would likely follow. It is conjectured that hyperbolic Dehn surgery space is star-like along lines through the origin and that the map from the representation variety to hyperbolic Dehn surgery space is a local diffeomorphism at representations corresponding to such structures. In other words, both local rigidity and Conjecture 1 should be true for these singularities as well.

Finally, we discuss some recent work of Otal([15]) and of Bonahon ([5]) which connects the topics discussed above with non-singular hyperbolic structures on 3-manifolds with boundary. Let M be a compact, oriented, 3-manifold with non-empty boundary. A *compact, convex* hyperbolic structure on M is given by a discrete, faithful representation of $\pi_1 M$ in $PSL(2, C)$ whose convex core is homeomorphic to M. (There is a degenerate case, when the convex core is 2-dimensional; then a small tubular neighborhood is homeomorphic to M.) Denoting the image of $\pi_1 M$ in $PSL(2, C)$ by Γ, the convex core is defined as the hyperbolic convex hull of the limit set of Γ divided out by the action of Γ. The space of compact, convex hyperbolic structures on M (where two such structures are equivalent if there is an isometry, isotopic to the identity, between them), will be denoted by $\mathcal{C}(M)$.

The convex core sits naturally, as a convex set, inside the larger space, $M_\Gamma = H^3 \cup \Omega/\Gamma$, where Ω is the domain of discontinuity of Γ on the sphere at infinity. M_Γ is also homeomorphic to M; it inherits a complete hyperbolic structure on its interior and a conformal structure on its boundary. It follows from the work of Ahlfors, [1], Bers [2], Maskit [13], Kra [11] and others that the space $\mathcal{C}(M)$, if non-empty, is homeomorphic to the space of conformal structures (up to an appropriate equivalence relation) on the boundary components of M_Γ. It is of interest to know to what extent structures on the boundary of the convex core also parametrize $\mathcal{C}(M)$.

The boundary of the convex core consists of developable surfaces, which, therefore, inherit hyperbolic metrics. These surfaces are generally not smooth, but are bent along some geodesic lamination. They are embedded and hence are a special case of *pleated surfaces*. (See [16], Chapter 9.) The geodesic laminations come equipped with transverse measures, determined by the amount of bending along arcs transverse to the lamination. We call the k-tuple (when there are k boundary components) of these measured laminations the *bending lamination* of the compact, convex structure. It is conjectured that the map from $\mathcal{C}(M)$ to bending laminations is 1-1 when all of the components of the lamination are non-zero. This means that, except for the subset of structures where a boundary component is totally geodesic, $\mathcal{C}(M)$ is parametrized by the associated bending laminations.

An interesting special case of this conjecture is when the bending lamination consists of a system of disjoint, non-trivial, non-homotopic simple closed curves with non-zero weights. Such laminations are dense in the space of all measured geodesic laminations. In this case, the work of Otal ([15]) and of Bonahon [5] gives a fairly complete and satisfying solution to this conjecture. Otal considered only the case when M is a handlebody; however, his arguments extend readily to general 3-manifolds. Bonahon has worked out this general case as well as analyzing the situation when the bending lamination is not just a collection of simple closed curves. (See comments after the discussion of Theorem 4 below.)

To state their results, we need a few definitions. Let M be a compact, oriented, irreducible 3-manifold with non-empty boundary, consisting of surfaces of genus at least 2. Let σ be a collection of disjoint, non-trivial, non-homotopic simple closed curves on the boundary of M. We say that the pair (M, σ) is *incompressible* if the complement of σ in the boundary of M, $\partial M - \sigma$, is incompressible. Similarly, we will say that (M, σ) is (homotopy) *acylindrical* if every continuous, incompressible, proper map of an annulus a into M is homotopic, relative to ∂a, into ∂M.

For each component σ_i of σ we let θ_i denote the internal angle of bending of the boundary of the convex core along σ_i. This angle is measured so that $\theta_i \in (0, \pi]$ and $\theta_i = \pi$ means that there is no bending along σ_i. The limiting case, when $\theta_i = 0$ occurs when the group element corresponding to σ_i has become parabolic. (These values are not in our space as defined but would be if we were to allow cusps.) The corresponding weights, when considered as part of a measured geodesic lamination, are $\pi - \theta_i$.

It is not hard to see that for any compact, convex core which is 3-dimensional, the boundary of any essential, incompressible, embedded annulus must intersect a component of σ where there is non-trivial bending. Otherwise, both boundary components could be made totally geodesic relative to the boundary of M. Hence, the annulus would be homotopic, rel boundary, to a single curve. Similarly, by the Gauss-Bonnet theorem, every simple, closed curve bounding a compressing disk (i.e., a curve, non-trivial in the boundary which bounds a disk in M) must have total geodesic curvature more than 2π so the sum of the bending angles at the points of intersection of such a curve with σ must be at least 2π. To record these conditions we define (following [15]) for any such non-trivial annulus a or compressing disk m:

$$\chi_a(\Theta) \;=\; \Sigma\, i(\sigma_i, \partial a)(\pi - \theta_i)$$

$$\chi_m(\Theta) \;=\; \Sigma\, i(\sigma_i, \partial m)(\pi - \theta_i)$$

Here $i(\sigma_i, \partial a)$ and $i(\sigma_i, \partial m)$ denote the intersection number, (i.e. the minimal number of transverse intersections) of σ_i with the component(s) of ∂a and ∂m, respectively. The sum is taken over the components of σ and $\Theta = (\theta_1, \cdots, \theta_k)$ is the ordered collection of bending angles. The necessary conditions above then become: $\chi_a(\Theta) > 0$ for any such annulus a and $\chi_m(\Theta) > 2\pi$ for any such compressing disk.

The work of Otal and of Bonahon implies that these conditions are sufficient:

Theorem 4 (Otal [15], see also Bonahon [5]). *Let σ be a collection $\{\sigma_i\}$ of disjoint, non-trivial, non-homotopic simple, closed curves on ∂M so that (M,σ) has incompressible boundary and is acylindrical. Then there is a 3-dimensional, compact, convex structure on M bent along $\sigma = \{\sigma_i\}$ with angles $\theta_i \in (0,\pi]$ if and only $\chi_a(\Theta) > 0$ for any non-trivial annulus a and $\chi_m(\Theta) > 2\pi$ for any compressing disk m. Furthermore, such a structure with these angles is unique.*

We give a brief sketch of the argument, with particular emphasis on the relation to the earlier material on cone manifolds. The first step is to show that the subset $C(M,\sigma)$ of $C(M)$ with the boundary bent along σ is non-empty. This is explained below. Assuming that it is non-empty, it is necessary to show that it possible to vary the bending angles on each component σ_i independently. To do this, one doubles the convex, bent structure on M. This defines a cone manifold structure on the double of M with the image of σ as the singular locus. The cone angle along the image of σ_i is 2θ. In particular, it is at most 2π and goes to zero as θ_i approaches 0. Local rigidity implies that these structures are parametrized locally by the cone angle. Symmetry implies that nearby cone structures are also geometric doubles and, hence, that it is possible locally to parametrize the structures bent along σ by the bending angles θ_i. This provides a local deformation theory for such structures.

The fact that the set is non-empty also follows from doubling. Take the double of M minus a tubular neighborhood of σ to get a manifold with torus boundary. The incompressibility and acylindrical conditions on (M,σ) imply that this manifold in irreducible and atoroidal. By Thurston's geometrization theorem for Haken manifolds, the interior of this manifold has a complete hyperbolic structure which, by Mostow rigidity, is unique (see also [12]). By perturbing this structure one can get a cone manifold structure with sufficiently small cone angles on the double of M. By symmetry as above, this structure is geometrically a double as desired.

The next step is to control the global degeneration of the structures as the angles are changed. This depends on material in [17] and [18] (see also [4] and [14]) which is beyond the scope of this note. Roughly speaking, it involves the material used to prove the double limit theorem and to control the geometric and algebraic convergence of geometrically finite Kleinian groups in the proof of the geometrization of Haken manifolds.

Instead we point out that, although the techniques described before for controlling the degeneration of hyperbolic cone manifolds are not used in this proof, there are close connections between the limiting conditions on the bending angles and the 3 types of degenerations described for cone manifolds. Specifically, the fact that the convex structures have non-zero bending somewhere means that the doubled manifold will have volume bounded below, ruling out DEG 1. A non-trivial annulus a in M becomes a torus in the double which must intersect the singular locus since a must intersect σ. However, for a sequence of convex structures on M where $\chi_a(\Theta) \to 0$ the cone angles in the double along the components which intersect a are going to 2π. In the limit, the torus will be non-singular, an example of DEG 2. Similarly, a compressing disk m in M corresponds to a singular 2-sphere in the double. If

$\chi_m(\Theta) \to 2\pi$ this 2-sphere will become Euclidean (as a surface with cone points), another case of DEG 2. That the singular locus stays away from itself (DEG 3) is essentially a result of the convexity of the convex core and the boundedness of the lengths of the components σ_i.

Finally, one notes that global uniqueness follows from local uniqueness and Mostow rigidity for the structure where all of the angles θ_i are 0. This is because any two structures with the same angles could be deformed back to the this unique structure. Since the set $C(M, \sigma)$ has no branch points, the structures must have been the same all along.

$$\square$$

We note that, by the linearity of the angle conditions in Theorem 4, it is possible to decrease the angles θ_i so that the cone angles $2\theta_i$ decrease monotonically and linearly to zero, which is a complete structure. This proves the analog of Conjecture 1 in this context.

In [Bo2] Bonahon also proves the existence part of Theorem 4 for more general laminations. The conditions on the bending angles extend to similar conditions where the finite sums are replaced by integrals with respect to the transverse measure for the measured lamination. There is a further condition in the case when the manifold is an interval bundle over a surface i.e. when the corresponding Kleinian groups are quasi-Fuchsian or extended quasi-Fuchsian. However, the uniqueness part of the Theorem 4 is unknown for general laminations.

A local rigidity theorem (relative the bending measure) would suffice to prove the global 1-1 conjecture for general laminations. The doubling argument seems less useful here because the resulting structure on the double would be so complicated. Instead it would be preferable to understand boundary conditions on the convex core that would allow the proof of local rigidity to go through directly on the manifold with bent geodesic boundary. The boundary conditions would have to be self-adjoint to provide a Hodge theory, but also be set up so that the boundary term in the integration by parts argument (as sketched in the proof of Theorem 1) has the right sign. The lack of smoothness of the boundary and the complicated form of the equations makes this difficult and the problem is still open as of this writing.

Acknowledgements. The author would like to thank the Tanaguchi Foundation for its generous support, allowing him to attend the Tanaguchi Symposium in Finland. This discussion represents an expanded version of the talk given there.

References

1. Ahlfors,L., "Mobius transformations in several dimensions", Ordway Lecture Notes, Univ. of Minnesota, 1981.
2. Bers, L. "Spaces of Kleinian Groups", in *Maryland Conference in Several Complex Variables I.* Springer-Verlag Lecture Notes, No **155** (1970), 9-34.
3. Bleiler, S. and Hodgson, C., "Spherical space forms and Dehn filling", Topology **35** (1996), 809-833.

4. Bonahon, F., "Bouts des varietes hyperboliques de dimension 3", Annals of Math. **124** (1986), 3-92.

5. Bonahon, F., in preparation.

6. Calabi, E., "On compact riemannian manifolds with constant curvature, I , AMS Proceedings of Symposia in Pure Mathematics **3** (1961), 155-180.

7. Gromov, M. and Thurston, T., "Pinching constants for hyperbolic manifolds", Invent. Math. **89** (1987), 1-12.

8. Hodgson, C. and Kerckhoff, S., "Rigidity of hyperbolic cone manifolds and hyperbolic Dehn surgery", preprint.

9. Hodgson, C. and Kerckhoff, S., "The shape of Dehn surgery space, in preparation.

10. Kojima, S., "Deformations of hyperbolic 3-cone-manifolds", preprint.

11. Kra, I. "On spaces of Kleinian groups", Comm. Math. Helv. **47** (1972), 53-69.

12. Keen, L., Maskit,B., and Series, C., "Geometric finiteness and uniqueness for Kleinian groups with circle packing limit sets", preprint.

13. Maskit, B. "Self-Maps of Kleinian groups", Amer. J. Math. **93** (1971), 840-856.

14. Morgan, J. and Shalen, P., "Degenerations of hyperbolic structures, III: Actions of 3-manifold groups on trees and Thurston's compactness theorem", Ann. of Math. **127** (1988), 457-519.

15. Otal, J.-P. "Sur le coeur convexe d'une variete hyperbolic de dimension 3", preprint.

16. Thurston,W. *The Geometry and Topology of three-manifolds,* Princeton Univ. Math. Dept. Notes (1979).

17. Thurston, W. "Hyperbolic structures on 3-manifolds, II: Surface groups and 3-manifolds which fiber over the circle",preprint.

18. Thurston, W. "Hyperbolic structures on 3-manifolds, I: Deformations of acylindrical manifolds," Ann. of Math. **124** (1986), 203-246.

19. Weil, A., "Discrete subgroups of Lie groups, II", Annals of Math **75** (1962), 578-602.

DEPARTMENT OF MATHEMATICS, STANFORD UNIVERSITY, STANFORD, CA 94305, USA
E-mail address: spk@math.stanford.edu

Proceedings of
THE 37TH TANIGUCHI SYMPOSIUM ON
TOPOLOGY AND TEICHMÜLLER SPACES
held in Finland, July 1995
ed. by Sadayoshi KOJIMA *et al.*
©1996 World Scientific Publishing Co.
pp. 115–122

NONSINGULAR PARTS OF HYPERBOLIC 3-CONE-MANIFOLDS

SADAYOSHI KOJIMA

(Received January 18, 1996)

ABSTRACT. We prove that the nonsingular part of a compact hyperbolic 3-cone-manifold admits a complete hyperbolic structure provided that cone angles all are at most 2π, and that the automorphism group as a cone-manifold is finite.

§1. Introduction

A hyperbolic cone-manifold will be a riemannian manifold of constant negative sectional curvature with cone-type singularity along a codimension two geodesic submanifold. To each component of the singular set, associated is a cone angle. The cone angle takes value in positive real numbers, and might attain 2π. In this particular case, the singular set is not singular in fact, and simply a real geodesic submanifold. Hence the concept of the hyperbolic cone-manifold generalizes the hyperbolic manifold in the usual sense. Of course the hyperbolic cone-manifold is also a generalization of the hyperbolic orbifold with simple singularity. In dimension 3, the singular set forms a link in the ambient 3-manifold.

The purpose of this paper is to prove

Theorem (Nonsingular Parts are Hyperbolic). *Let C be a compact hyperbolic 3-cone-manifold and Σ the singular set. If each cone angle assigned to a component of Σ is $\leq 2\pi$, then the underlying space of $C - \Sigma$ admits a complete hyperbolic structure of finite volume.*

When cone angles all are equal to 2π, Σ is a disjoint union of simple closed geodesics by definition. The theorem in this special case, which claims that the complement of a geodesic link in a hyperbolic 3-manifold admits a complete hyperbolic structure, was proved ten years ago by Sakai [12, 7]. The proof we will give in §3 is independent

---

1991 *Mathematics Subject Classification.* Primary 57M50; Secondary 30F40.
Key words and phrases. cone-manifolds, hyperbolic 3-manifolds, Dehn filling.

of the one by Sakai, though both are appealing to Thurston's uniformization theorem for Haken manifolds [14, 9].

The theorem implies the following finiteness property of the automorphism group.

Corollary (Automorphism Groups are Finite). *Let C be a compact hyperbolic 3-cone-manifold and Σ the singular set. If each cone angle assigned to a component of Σ is $\leq 2\pi$, then the group of automorphisms of C as a cone-manifold up to isotopy is finite.*

The theorem and corollary were addressed in the first half of the author's talk focused on the global rigidity of hyperbolic 3-cone-manifolds under some conditions at the 37th Taniguchi Symposium. Actually they play an essential role in proving the rigidity, and this paper gives a detailed version of that part. The later half which involves a very different type of analysis will be discussed in [8].

The proof of the theorem is based heavily on the Hodgson-Kerckhoff local rigidity of hyperbolic 3-cone-manifolds [3], which has many applications such as Kerckhoff's [6] in this volume. We review their local rigidity and its topological consequences in the next section. Then we prove the theorem and corollary in §3.

The author would like to thank the Taniguchi Foundation for their generous support.

§2.   Local Rigidity and Dehn Filling

This section is to review the local rigidity of hyperbolic cone-manifolds established by Hodgson and Kerckhoff [3].

Let C be a compact *orientable* hyperbolic 3-cone-manifold. The singular set Σ forms a link of n components in C. C defines a nonsingular but incomplete hyperbolic structure on $C - \Sigma$ and we have an associated developing map $D : \widetilde{C - \Sigma} \to \mathbf{H}^3$ and a holonomy representation $\rho_0 : \Pi = \pi_1(C - \Sigma) \to \mathrm{PSL}_2(\mathbf{C})$. We call them a developing map and a holonomy representation of C.

To each component of Σ, associated is a cone angle α_j which takes value in $(0, \infty)$. Let m_j, $j = 1, 2, \cdots, n$, be an oriented meridional loop for each component. The image $\rho_0(m_j)$ of a meridian m_j by the holonomy is an elliptic element rotating $\mathbf{H}^3$ by α_j about the axis, though the angle of $\rho_0(m_j)$ makes sense only modulo 2π.

Lemma 1 (Holonomy Representations are Irreducible). *The action of $\rho_0(\Pi)$ has no global fixed points on the sphere at infinity.*

Proof. Choose a geodesic segment s joining two points on Σ but not lying on Σ, for instance the shortest common orthogonal to Σ. s lifts to the universal cover and is mapped by the developing map to $\mathbf{H}^3$. Notice that the developing map canonically extends over the completion of the universal cover $\widetilde{C - \Sigma}$. Also the covering projection $: \widetilde{C - \Sigma} \to C - \Sigma$ extends over the completion. Then for each end point of the developed image of the lift of s in $\mathbf{H}^3$, there is an associated unique geodesic passing through it that comes from a component of the preimage of Σ in the completion of the universal cover $\widetilde{C - \Sigma}$.

Thus we have two distinct such geodesics going through terminal points. Each of them is invariant by some loxodromic elements in $\rho_0(\Pi)$, and hence they admit no common fixed points on the sphere at infinity. In particular, there are no global fixed points of the action of $\rho_0(\Pi)$ on the sphere at infinity. $\square$

Removing a small open neighborhood of Σ from C, we get a compact hyperbolic manifold with boundary. Nearby deformations of such a compact hyperbolic manifold but a small neighborhood of the boundary are parameterized by holonomy representations [13].

Recall that the space of representations of the fundamental group Π in $\mathrm{PSL}_2(\mathbf{C})$ has a natural algebraic topology. There is a canonical projection to the set of conjugacy classes of representations $\mathcal{H}(\Pi) = \mathrm{Hom}(\Pi, \mathrm{PSL}_2(\mathbf{C}))/\mathrm{PSL}_2(\mathbf{C})$. This quotient set together with a quotient topology is globally non-Hausdorff in fact, but a small neighborhood of a conjugacy class represented by an irreducible representation in the above sense has a Hausdorff separation property. The local structure near such a representation is supposed to be identical to a corresponding part in the algebro geometric quotient of all $\mathrm{SL}_2(\mathbf{C})$-representations studied originally by Culler and Shalen [1] and more generally by Saito [11] in this volume. Hodgson-Kerckhoff's local rigidity [3] claims that the representations near the holonomy of a cone-manifold have even more a tractable structure.

To state the local rigidity more precisely, we recall some foundations of hyperbolic cone-manifolds. An orientation preservingly isometric transformation φ of $\mathbf{H}^3$ can be represented by a matrix Φ in $\mathrm{SL}_2(\mathbf{C})$. The complex length of φ is twice of an appropriate branch of the log of an eigenvalue of Φ. It measures how much φ translates $\mathbf{H}^3$ with twist along an invariant geodesic. Since there are two choices to be made, we adopt the following convention for choosing a complex length of an image of meridional element m_j by representations near ρ_0.

To each m_j, we choose a complex length of $\rho_0(m_j)$ by a multiple of its cone angle with the complex unit i, which we denote by $\mathcal{L}_{m_j}(\rho_0)$. It is of course equal to $i\alpha_j$. Then choose an orientation of the invariant geodesic ℓ_0 of $\rho_0(m_j)$ in $\mathbf{H}^3$ so that the rotational direction of m_j is counter clockwise. There are two ways to continuously extend this assignment of a complex length to that of transformation represented by $\rho(m_j)$ where ρ is close to ρ_0. Since ρ is close to ρ_0, the invariant geodesic ℓ of $\rho(m_j)$ is also close to ℓ_0 and inherits an orientation. The choice is made by selecting the sign of the real part. We choose it according to whether $\rho(m_j)$ translates the point on ℓ in positive or negative direction.

Thus we get a continuous mapping of the quotient space, which we denote by $\mathcal{L}_{m_j}$, defined on a neighborhood $\mathcal{D} \subset \mathcal{H}(\Pi)$ of a conjugacy class represented by ρ_0. Just arranging these maps, we get

$$\mathcal{L}_m : \mathcal{D} \to \mathbf{C}^n,$$

where $\mathcal{L}_m(\rho) = (\mathcal{L}_{m_1}(\rho), \cdots, \mathcal{L}_{m_n}(\rho))$.

Local Rigidity (Hodgson-Kerckhoff [3]). *If the cone angles assigned to the components of Σ all are $\leq 2\pi$, then $\mathcal{L}_m$ is a local diffeomorphism near the conjugacy class represented by ρ_0.*

In particular, the complex dimension of $\mathcal{H}(\Pi)$ near the conjugacy class represented by ρ_0 is equal to n, the number of components of Σ, and ρ_0 represents a smooth point.

We now introduce more topological parameterization due to Thurston [13], which is called the Dehn filling coefficients. Let us choose oriented longitudinal elements $\{l_j\}$ with respect to $\{m_j\}$. $\rho_0(l_j)$ leaves an axis ℓ_0 of $\rho_0(m_j)$ invariant. Then we choose a complex length of $\rho_0(l_j)$ so that its real part is positive or negative according to whether $\rho_0(l_j)$ moves a point on ℓ_0 in positive or negative direction, and its imaginary part lies in $[0, 2\pi)$. Extending this assignment for a neighborhood of ρ_0 as before again, we obtain a continuous map

$$\mathcal{L}_l : \mathcal{D} \to \mathbf{C}^n,$$

where $\mathcal{L}_l(\rho) = (\mathcal{L}_{l_1}(\rho), \cdots, \mathcal{L}_{l_n}(\rho))$.

Then consider the following system of equations,

$$p_j \, \mathcal{L}_{m_j} + q_j \, \mathcal{L}_{l_j} = 2\pi i,$$

in terms of $\mathbf{p} = (p_1, \cdots, p_n)$ and $\mathbf{q} = (q_1, \cdots, q_n)$. This is an equation defined over $\mathcal{D}$ and has an initial solution $\mathbf{p} = (2\pi/\alpha_1 \cdots, 2\pi/\alpha_n)$, $\mathbf{q} = 0$ at ρ_0. This solution extends to a continuous solution of the equation over $\mathcal{D}$. Collecting pairs (p_j, q_j)'s of the same components of $\mathbf{p}$ and $\mathbf{q}$, we call $((p_1, q_1), \cdots, (p_n, q_n))$ a generalized Dehn filing coefficient, and denote it for simplicity by $(\mathbf{p}, \mathbf{q})$.

Assigning Dehn filling coefficients to each representation, we get another mapping

$$\mathcal{DF} : \mathcal{D} \to (\mathbf{R}^2)^n.$$

To see this is again a diffeomorphism, let us see for each j how the perturbation of a complex length of $\mathcal{L}_{m_j} = i\alpha_j + \varepsilon$ affects on Dehn filling coefficients (p_j, q_j). Let us denote the corresponding change in the longitude by $\mathcal{L}_{l_j} = \delta$, where δ is a complex valued function of ε and is equal to $\mathcal{L}_{l_j}(\rho_0)$ when $\varepsilon = 0$. We split those into the real and imaginary parts: $\varepsilon = \varepsilon^r + i\varepsilon^i$, and $\delta = \delta^r + i\delta^i$. The Dehn filling coefficients are the appropriate solution of the equation

$$p_j(\alpha_j + \varepsilon^i) + q_j\delta^i = 2\pi,$$
$$p_j\varepsilon^r + q_j\delta^r = 0.$$

Fixing $p_j \approx 2\pi/\alpha_j$, and we get

$$\varepsilon^i = \frac{\delta^i}{\delta^r}\varepsilon^r + \frac{2\pi}{p_j} - \alpha_j.$$

Since δ^r is nonzero if $|\varepsilon| \ll 1$, small ε^r determines ε^i so as to satisfy the equation above. In particular, ε^r moves freely near 0 so that p_j is constant. On the other hand,

$$q_j = -\frac{\varepsilon^r}{\delta^r} p_j,$$

and δ^r is nonzero if $|\varepsilon| \ll 1$, q_j moves freely near 0 so that p_j is constant. This means that the variable change from the complex length to the Dehn filling coefficient is regular.

Hence we have another formulation of the local rigidity.

Local Rigidity (Hodgson-Kerckhoff [3]). *If the cone angles assigned to the components of Σ all are $\leq 2\pi$, then $\mathcal{DF}$ is a local diffeomorphism near the conjugacy class represented by ρ_0.*

It is now convenient to define the slope of the Dehn filling coefficient as the vector of ratios $\mathbf{q}/\mathbf{p} = (q_1/p_1, \cdots, q_n/p_n)$. We call the slope rational if each component is rational. To each representation near ρ_0, we assign a Dehn filling coefficient $(\mathbf{p}, \mathbf{q})$, and a hyperbolic manifold with boundary which is a deformation of C minus a small tubular neighborhood of Σ. Then the hyperbolic Dehn surgery theory in [13] says that there is a unique appropriate extension to fill the boundary, so that the filled piece to each end is either a solid torus with singular axis or a one point compactification of incomplete toral end according to whether the filling slope is rational or not. We call this general resultant a Dehn filled deformation of C. If each pair in the coefficient are relatively prime in particular, then the resultant is actually a nonsingular hyperbolic manifold. The topological type of the Dehn filled deformation depends only on its slope, but geometric type depends really on the Dehn filling coefficient.

Assigning to each small change of the representation a corresponding Dehn filled deformation, we get

Corollary 2. *There is a small neighborhood U of $((2\pi/\alpha_1, 0), \cdots, (2\pi/\alpha_n, 0))$ in $(\mathbf{R}^2)^n$ so that each representation in $\mathcal{DF}^{-1}(U)$ corresponds to a nearby Dehn filled deformation of C.*

The topological type of the Dehn filled deformation depends only on its slope, and we reformulate a topological consequence of the local rigidity in terms of slopes. Assigning the ratio $\mathbf{q}/\mathbf{p}$ to each coefficient, we get a projection π of $(\mathbf{R}^2 - (0, 0))^n$ to the space of slopes $\mathbf{S}$ which is homeomorphic $\mathbf{S}^1 \times \cdots \times \mathbf{S}^1$. The set of rational slopes $\mathbf{Q}$ are dense in $\mathbf{S}$.

By the local rigidity, the set of possible slopes for Dehn filled deformations of C contains at least an open set V in $\mathbf{S}$ containing $(0, \cdots, 0)$. On the other hand, given a rational slope $\mathbf{s} \in V$, we obtain a small one parameter family of Dehn filled deformations with a constant slope $\mathbf{s}$ corresponding to $U \cap \pi^{-1}(\mathbf{s})$. They are topologically the same. If we choose V small enough, then the only slopes in V whose corresponding Dehn filled deformations contain cone-manifolds with cone angle $\leq 2\pi$ consists of rational slopes in V containing at least one zero in some component. Contrarily, Dehn filled deformations of a rational slope in V each of whose component is nonzero consists of cone-manifolds with cone angles all $\geq 2\pi$.

Once we have a hyperbolic cone-manifold with cone angles all $\geq 2\pi$, then by the argument in Gromov and Thurston [2], we can modify a singular metric on it to be nonsingular with negative (not necessarily constant) sectional curvature. This metric property depends only on the topology of the Dehn filled deformations in fact, and also the topological type of the Dehn filled deformation depends only on the slope. Hence we use a topological terminology, Dehn surgery, to indicate the topological modification for a compact 3-manifold with toral boundary by sawing a solid torus along the slope. Then

Corollary 3. *There is an open neighborhood V of $(0, \cdots, 0)$ in S, so that the Dehn surgery along a rational slope in $V \cap (S^1 - \{0\})^n$ produces a manifold which admits a negatively curved metric.*

Hence many Dehn surgeries produce a manifold admitting a negatively curved metric. We say "MANY" surgeries in the next section if the set of slopes in question consists of rational ones in this open set in S.

§3.   Proof of Theorem and Corollary

Thurston's uniformization theorem for Haken manifolds [14, 9] provides a topological sufficient condition for a compact 3-manifold to admit a complete hyperbolic metric on its interior. We prove the theorem by verifying that the complement of the singular set satisfies this topological condition. Since we are concerned with only topological properties, the compactified piece will fit more to the terminologies in the 3-manifold theory. Hence we let E be the exterior of Σ in C, that is, the complement of a small tubular neighborhood of Σ in C. E is a compact 3-manifold with toral boundary. The topological sufficient condition in this special case is simply that E is irreducible, atoroidal and admits no Seifert fibrations. This is by virtue of the torus theorem, which claims that if there is a π_1-injective map from the torus, then either the target contains an incompressible torus or admits a Seifert fibration. The torus theorem for Haken manifolds which we need here was established by Jaco-Shalen [4] and Johannson [5].

Since we will use the local rigidity in several places, we assume that C is a compact orientable hyperbolic 3-cone-manifold where cone angles assigned all are at most 2π in the following four lemmas.

Lemma 4 (Exteriors are Irreducible). *E contains no essential 2-spheres.*

Proof. If not, then there is an essential 2-sphere S in E. By Corollary 3, "MANY" Dehn surgeries on E produce a manifold admitting a negatively curved metric. Since a negatively curved 3-manifold does not contain any essential 2-sphere, S bounds a 3-ball in the resultants. In particular, S separates C into two compact pieces P and Q. One of them yields a ball by each Dehn surgery. Choose a slope for each component of ∂E contained in Q small enough so that its Dehn surgered resultant is not homeomorphic to the ball. Then "MANY" Dehn surgeries on P must produce a ball. But we can see it impossible for example by computing the homology of the resultants. □

Lemma 5 (Exteriors have Incompressible Boundary). *∂E is incompressible.*

Proof. If not, there is a compressing disk for ∂E. Then by the irreducibilty, E is homeomorphic to the solid torus. However no Dehn surgery on the solid torus produces an aspherical manifold, and hence in particular a manifold which admits a negatively curved metric. This contradicts Corollary 3. □

Lemma 6 (Exteriors are atoroidal). *Every incompressible torus T in E is ∂-parallel.*

Proof. By Corollary 3, "MANY" Dehn surgeries on E produce a manifold admitting a negatively curved metric. Since a negatively curved 3-manifold is irreducible and contains no essential tori, T in E bounds a solid torus in the resultants. In particular, T separates E into two compact pieces P and Q.

Choose a slope for each component of ∂E contained in, say, Q small enough so that the resultant is not homeomorphic to the solid torus. Then "MANY" Dehn surgeries on P produce a solid torus.

We claim that there is an essential annulus between T and $P \cap \partial E$. To see this, recall a result of Wu [16] that an incompressible toral boundary of a 3-manifold without admitting compressible annuli to the other toral boundary will be compressible at most three Dehn surgered resultants on the other component. Hence if there is no essential annuli in P, then applying Wu's theorem for the first component of $P \cap \partial E$, we can choose a very small slope for that component so that T is still incompressible in the Dehn surgered resultant. Continue the same application in order of the components of $P \cap \partial E$, and we eventually find one of "MANY" slopes which produces the resultant where T is still incompressible. This is absurd. Hence $T \subset P$ must admit an essential annulus A which joins T with a component B of $P \cap \partial E$.

Choose a slope for each component of $P \cap \partial E$ other than B, if any, small enough so that the Dehn surgery resultant R still admit "MANY" Dehn surgeries on B producing a solid torus. Since there is an essential annulus A, R is a cable space. If the cabling is not trivial, infinitely many Dehn surgeries produce a solid torus in fact, but not "MANY" Dehn surgeries in our sense can produce a solid torus since the set of such slopes has only one accumulation point. Hence the cabling is trivial and R is homeomorphic to the torus times the interval.

If there are components other than B, their surgery cores lie in R as a nontrivial link L with prescribed longitude and meridian system. Then "MANY" Dehn surgeries on L and B must produce a solid torus. But we can see it impossible for example by computing the homology of the resultants. Hence there are no components of $P \cap \partial E$ other than B, and $P = R (\approx T^2 \times I)$. This means that T is ∂-parallel. $\square$

Lemma 7 (Exteriors are not Seifert Fibered). *E admits no Seifert fibrations.*

Proof. If E admits a Seifert fibration, all of Dehn surgeries produce either a Seifert fibered space or a graph manifold. In any case, the resultant admit no negatively curved metrics. This contradicts Corollary 3. $\square$

Proof of Theorem. When C is orientable, then the claim follows from the above four lemmas and Thurston's uniformization theorem for Haken manifolds. If C is not orientable, choose an orientable double cover $\tilde{C}$. Then its complement of the singular set admits a hyperbolic structure, and the covering translation restricted there acts as an isometry by Mostow rigidity [10]. Hence the quotient admits a hyperbolic structure. $\square$

Proof of Corollary. Any automorphism of C induces a homeomorphism of $C - \Sigma$. By Mostow rigidity [10], it is homotopic to an isometry with respect to the hyperbolic structure on $C - \Sigma$. Moreover by Waldhausen's theorem [15], this homotopy can be chosen by an isotopy. This isotopy extends to an isotopy of an automorphism of C. Hence the automorphism group of C is realized as a subgroup of the isometry group of $C - \Sigma$ which consists of maps extendible to automorphisms of C. Then since the isometry group of $C - \Sigma$ is finite, so is the automorphism group of C. $\square$

References

1. M. Culler and P. Shalen, *Varieties of group representations and splitting of 3-manifolds*, Ann. of Math., **117** (1983), 109 – 146.
2. M. Gromov and W. Thurston, *Pinching constants for hyperbolic manifolds*, Invent. Math., **89** (1987), 1 – 12.
3. C. Hodgson and S. Kerckhoff, *Rigidity of hyperbolic cone-manifolds and hyperbolic Dehn surgery*, preprint, (1993).
4. W. Jaco and P. Shalen, *Seifert fibered spaces in 3-manifolds*, Mem. Amer. Math. Soc., **220** (1979).
5. K. Johannson, *Homotopy equivalences of 3-manifolds with boundaries*, Lecture Notes in Math, vol 761, Springer-Verlag, 1979.
6. S. Kerckhoff, *Deformations of hyperbolic cone manifolds*, These Proceedings, 101–114.
7. S. Kojima, *Isometry transformations of hyperbolic 3-manifolds*, Top. and its Appl., **29** (1988), 297 – 307.
8. S. Kojima, *Deformations of hyperbolic 3-cone-manifolds*, preprint.
9. J. Morgan, *On Thurston's uniformization theorem for three-dimensional manifolds*, The Smith Conjecture (J. Morgan and H. Bass, eds), Academic Press, 1984, pp. 37 – 125.
10. G. D. Mostow, *Quasi-conformal mappings in n-space and the rigidity of hyperbolic space forms*, Publ. IHES, **34** (1973), 53 – 104.
11. K. Saito, *Character variety of representations of a finitely generated group in SL_2*, These Proceedings, 253–264.
12. T. Sakai, *Geodesic knots in a hyperbolic 3-manifold*, Kobe J. Math., **8** (1991), 81 – 87.
13. W. Thurston, *The geometry and topology of 3-manifolds*, Lecture Notes, Princeton University, 1977/78.
14. W. Thurston, *Three dimensional manifolds, Kleinian groups and hyperbolic geometry*, Bull. Amer. Math. Soc., **6** (1982), 357 – 381.
15. F. Waldhansen, *On irreducible 3-manifold which are sufficiently large*, Ann. of Math., **87** (1968), 56 – 88.
16. Y. Q. Wu, *Incompressibility of surfaces in surgered 3-manifolds*, Topology, **31** (1992), 271 – 279.

DEPARTMENT OF MATHEMATICAL AND COMPUTING SCIENCES, TOKYO INSTITUTE OF TECHNOLOGY, OHOKAYAMA, MEGURO, TOKYO 152 JAPAN

E-mail address: sadayosi@is.titech.ac.jp

Proceedings of
THE 37TH TANIGUCHI SYMPOSIUM ON
TOPOLOGY AND TEICHMÜLLER SPACES
held in Finland, July 1995
ed. by Sadayoshi KOJIMA *et al.*
©1996 World Scientific Publishing Co.
pp. 123–148

LEFSCHETZ FIBRATIONS OF GENUS TWO
- A TOPOLOGICAL APPROACH -

YUKIO MATSUMOTO

(Received April 11, 1996)

ABSTRACT. In this paper, we will study Lefschetz fibrations with the general fiber a Riemann surface Σ_2 of genus two, from C^∞-point of view. It will be proved that the isomorphism classes of Lefschetz fibrations of genus ≥ 2 are determined by their global monodromies. Certain 'fractional signature' concentrated on singular fibers will be discussed. Some examples of Lefschetz fibrations of genus 2 will be given with special emphasis on their global monodromies. In particular, examples of two non-isomorphic Lefschetz fibrations whose total spaces are mutually homeomorphic will be given.

1. Introduction

Lefschetz fibrations naturally appear as blow-ups (at the base points) of pencils of curves in algebraic surfaces. Although they are studied usually in the context of algebraic geometry, there are several approaches from the differential topological viewpoint. In fact, Lefschetz's original work on his pencils [15] was very topological. More recently Moishezon [24,p.162] defined Lefschetz fibrations in the C^∞-category, and studied their global monodromies in the case of fiber genus 1. A concordance classification of Lefschetz fibrations over the 2-sphere was made by Harer [11], in which he exploited the Hatcher-Thurston's presentation of the mappping class group $\mathcal{M}_g$.

Roughly speaking, a Lefschetz fibration in Moishezon's sense is a fibration of a smooth (real) 4-dimensional manifold over a surface with the general fiber another closed orientable surface, which may admit certain singular fibers.

We are interested in Lefschetz fibrations in his sense, because they provide us with many important smooth 4-manifolds as the total spaces, and the fibering structures seem to make it tractable to study the diffeomorphism types of such 4-manifolds.

1991 *Mathematics Subject Classification.* Primary 57M99; Secondary 30F99, 32S50.

When the genus of the general fiber is 1, the isomorphism classes of Lefschetz fibrations have been completely classified, [24], [17]. The next step to study would be the case of genus 2. However, this case is already difficult mainly because of the complicated structure of the mapping class group $\mathcal{M}_2$.

The purpose of this paper is to prove some basic results on Lefschetz fibrations of genus $g \geq 2$, and to give several results and examples for genus 2. New feature of our examples is explicit description of their global monodromies. Also we emphasize the correspondence between the defining relations of $\mathcal{M}_2$ and Lefschetz fibrations over the 2-sphere. Cf. [11].

In §2 of this paper, we prove that *the global monodromy of a Lefschetz fibration determines the fibering structure up to isomorphism*. This result is valid for arbitrary genus $g \geq 2$. In §3 we use Meyer's signature theorem to define certain 'fractional signature' concentrated on singular fibers of genus 2. This fractional signature can be defined for more general singular fibers than those of the Lefschetz types, and behaves additively at splitting of a singular fiber into simpler ones. In §4 we give some examples of Lefschetz fibrations of genus 2 and their global monodromies. Our final examples would be worth mentioning here: *there are two Lefschetz fibrations of genus 2 over the 2-sphere with 40 singular fibers such that their total spaces are homeomorpic to each other but the fibering structures are not isomorphic.* The phenomena of this kind never take place in Lefschetz fibrations of genus 1, [24],[17], where such phenomena appear only when 'multiple fibers' are inserted, [14], [9],[5]. In §5, we will prove Persson's result [27] on the number of singular fibers from monodromy theoretic viewpoint.

Algebro-geometric papers [12],[29],[26],[27],[28] were very useful to the author to learn about Lefschetz fibrations of genus 2. The author is grateful to T.Ashikaga, T.Fukui and Y.Kametani for useful information and comments. He was informed of the papers [28] and [27] by Ashikaga and Fukui, respectively.

Throughout the paper we will work in the C^∞-Category.

The following notations will constantly be used:

Σ_g: the closed oriented surface of genus g,

$\mathrm{Diff}^+(\Sigma_g)$: the group of all orientation preserving self-diffeomorphisms of Σ_g with the C^∞-topology,

$\mathrm{Diff}_0^+(\Sigma_g)$: the subgroup of $\mathrm{Diff}^+(\Sigma_g)$ consisting of all self-diffeomorphisms isotopic to the identity,

$\mathcal{M}_g = \mathrm{Diff}^+(\Sigma_g)/\mathrm{Diff}_0^+(\Sigma_g)$: the mapping class group of genus g.

2. Global Monodromy

We will begin with a precise definition of a Lefschetz fibration. (See [24, p.162], [17, p.594].)

Definition 2.1. Let M and B be compact oriented (not necessarily closed) C^∞-manifolds of dimensins 4 and 2, respectively. A C^∞-map $f : M \to B$ is called a *Lefschetz fibration of genus g* if the follwing conditions are satisfied:

(a) $\partial M = f^{-1}(\partial B)$;

(b) there is a finite set of points $b_1, b_2, \cdots, b_n$ (called the *critical values of f*) in $IntB$ $(= B - \partial B)$ such that $f|f^{-1}(B - \{b_1, \cdots, b_n\}) : f^{-1}(B - \{b_1, \cdots, b_n\}) \to B - \{b_1, \cdots, b_n\}$ is a C^∞-fiber bundle with the fiber diffeomorphic to Σ_g;

(c) for each i $(1 \le i \le n)$, there exits a single point $p_i \in f^{-1}(b_i)$ such that

    (1) $(df)_p : T_p(M) \to T_{f(p)}(B)$ is onto for any $p \in f^{-1}(b_i) - \{p_i\}$,

    (2) about p_i (resp. b_i), there exist local complex coordinates z_1, z_2 with $z_1(p_i) = z_2(p_i) = 0$ (resp. local complex coordinate ξ with $\xi(b_i) = 0$), so that f is locally written as $\xi = f(z_1, z_2) = z_1 z_2$;

(d) no fiber contains a (-1)-sphere, i.e., a smoothly embedded 2-sphere with self-intersection number -1.

We call a fiber $f^{-1}(b)$ a *singular fiber* if $b \in \{b_1, \cdots, b_n\}$ or else a *general fiber*. Also we call M the *total space*, B the *base space*, and f the *projection*.

In Condition (c) (2), we tacitly require the orientation of M (resp. B) to coincide with the canonical orientation determined by the local complex coordinates z_1, z_2 (resp. local complex coorninate ξ) about the critical point p_i (resp. b_i). Note that we do not assume any global complex structure on M. In Moishezon's definition [24], the irreducibility of a singular fiber is always assumed, while in the definition above, it is not necessarily so. Thus we need additional condition (d).

Remark. Harer's definition [11] of a Lefschetz fibration does not require the coincidence of the canonical orientation determined by (z_1, z_2) and the orientaion of M. We would like to call Harer's Lefschetz fibration an 'achiral' Lefschetz fibration. In what follows however we will confine ourselves to Moishezon's 'chiral' Lefschetz fibrations as difined above for simplicity.

Definition 2.2. ([24, p.164]) Lefschetz fibrations $f : M \to B$ and $f' : M' \to B'$ are said to be *isomorphic* if there exist orientation preserving diffeomorphisms $H : M \to M'$ and $h : B \to B'$ such that $f'H = hf$.

It is obvious that isomorphic Lefschetz fibrations are of the same genus and have the same numbers of critical values.

Let $f : M \to B$ be a Lefschetz fibration of genus g over a *connected* base space B. Take a base point $b_0 \in IntB - \{b_1, \cdots, b_n\}$, and identify the general fiber $F_0 = f^{-1}(b_0)$ with Σ_g by an orientaion preserving diffeomorphism $\Phi : \Sigma_g \to F_0$.

Let $l : [0, 1] \to B - \{b_1, \cdots, b_n\}$ be a loop with $l(0) = l(1) = b_0$. Since the pull-back l^*f of the fiber bundle $f|f^{-1}(B - \{b_1, \cdots, b_n\}) : f^{-1}(B - \{b_1, \cdots, b_n\}) \to B - \{b_1, \cdots, b_n\}$ is a trivial bundle over $[0, 1]$, there exists a map $\varphi : [0, 1] \times \Sigma_g \to M - f^{-1}(\{b_1, \cdots, b_n\})$ such that

(1) $f(\varphi(t, p)) = l(t)$ for all $(t, p) \in [0, 1] \times \Sigma_g$;

(2) the map $\varphi_t : \Sigma_g \to F_t (= f^{-1}(l(t)))$ defined by $\varphi_t(p) = \varphi(t, p)$ is an orientation preserving diffeomorphism;

(3) $\varphi_0 = \Phi : \Sigma_g \to F_0$.

Noticing that $F_1 = F_0$, we have a diffeomorphism $\Phi^{-1}\varphi_1 : \Sigma_g \to \Sigma_g$, whose isotopy class $[\Phi^{-1}\varphi_1]$ $(\in \mathcal{M}_g)$ is determined by the homotopy class of l, $\{l\}$, and Φ. This class

$[\Phi^{-1}\varphi_1]$ is called the *monodromy* associated with $\{l\}$ and Φ. We fix the identification $\Phi : \Sigma_g \to F_0$ for a while. Then by sending $\{l\}$ to $[\Phi^{-1}\varphi_1]$ we obtain a map

$$\rho : \pi_1(B - \{b_1, \cdots, b_n\}, b_0) \to \mathcal{M}_g.$$

The map ρ becomes a homomorphism (called the *monodromy representaion*) if $\mathcal{M}_g$ is assumed, by convention, to act on Σ_g from the *right*. (This is because the product ll' of loops traverses first l and then l'.) From now on we will adopt this convention.

Notice that if the identification $\Phi : \Sigma_g \to F_0$ is changed for another diffeomorphism, the monodromy representaion ρ is changed to $\alpha^{-1}\rho\alpha$, where α is an element of $\mathcal{M}_g$.

Let $f : M \to B$ and $f' : M' \to B'$ be Lefschetz fibrations of the same genus g with the sets of critical values $\{b_1, \cdots, b_n\}$, $\{b'_1, \cdots, b'_n\}$, and with base points b_0, b'_0 respectively. We assume that the base spaces B and B' are connected. Let $\rho : \pi_1(B - \{b_1, \cdots, b_n\}, b_0) \to \mathcal{M}_g$ (resp. $\rho' : \pi_1(B' - \{b'_1, \cdots, b'_n\}, b'_0) \to \mathcal{M}_g$) be the monodromy representation of $f : M \to B$ (resp. $f' : M' \to B'$) associated with some fixed identification $\Phi : \Sigma_g \to f^{-1}(b_0)$ (resp. $\Phi' : \Sigma_g \to (f')^{-1}(b'_0)$).

Definition 2.3. The monodromy representaions ρ and ρ' are said to be *equivalent* if there exist an element $\alpha \in \mathcal{M}_g$ and an orientaion preserving homeomorphism $h : (B, \{b_1, \cdots, b_n\}, b_0) \to (B', \{b'_1, \cdots, b'_n\}, b'_0)$ such that

$$\alpha^{-1}\rho\alpha = \rho' h_\#$$

where $h_\# : \pi_1(B - \{b_1, \cdots, b_n\}, b_0) \to \pi_1(B' - \{b'_1, \cdots, b'_n\}, b'_0)$ is the isomorphism induced by h.

The following theorem is the main result of this section.

Theorem 2.4. *Let $f : M \to B$ and $f' : M' \to B'$ be Lefschetz fibrations of genus $g(\geq 2)$ over connected base spaces. Then they are isomorphic if and only if their monodromy representaions are equivalent.*

The corresponding result for $g = 1$ holds if the base spaces have non-empty boundary or the sets of critical values are non-empty, [24,p.169], [17]. Although the details are quite different, the main idea of the following proof is due to Moishozon's argument for $g = 1$, [24,p.169-174]. A new ingredient for $g \geq 2$ is the contractibility of $\mathrm{Diff}_0^+(\Sigma_g)$.

Proof of Theorem 2.4. The 'only if' part is obvious. We will prove the 'if' part.

Let $h : (B, \{b_1, \cdots, b_n\}, b_0) \to (B', \{b'_1, \cdots, b'_n\}, b'_0)$ be the homeomorphism giving the equivalence of the monodromy representations of f and f'. By deforming h by an isotopy, we may assume that h is a diffeomorphism. Then replacing the Lefschetz fibration $f' : M' \to B'$ with $h^{-1}f' : M' \to B$, which is isomorphic to $f' : M' \to B'$, we may assume that $B = B'$, $\{b_1, \cdots, b_n\} = \{b'_1, \cdots, b'_n\}$, $b_0 = b'_0$ and $h = id_B$. Moreover, by changing the identification $\Phi' : \Sigma_g \to (f')^{-1}(b'_0)$ if necessary, we may take the element α (giving the conjugation between ρ and ρ') to be 1. Thus the proof of Theorem 2.4 is reduced to showing that *Lefschetz fibrations $f : M \to B$ and $f' : M' \to B$ with the same connected base space B, the same critical values $\{b_1, \cdots, b_n\}$,* and the same base point b_0, are isomorphic if their monodromy representations are identical:

$$\rho = \rho' : \pi_1(B - \{b_1, \cdots, b_n\}, b_0) \to \mathcal{M}_g.$$

Case (1) $\{b_1, \cdots, b_n\} = \emptyset$, i.e., the case of fiber bundles.

If in addition $\partial B \neq \emptyset$, it is easy to see that there exists a bundle isomorphism

$$H : M \to M'$$

such that $f'H = f$. In fact, B is decomposed into a handlebody

$$B = D \cup H_1 \cup \cdots \cup H_k, \qquad (b_0 \in D, k \geq 0),$$

and the isomorphism H is first constructed over the 2-disk D by extending the diffeomorphism $\Phi'\Phi^{-1} : f^{-1}(b_0) \to (f')^{-1}(b_0)$ to $f^{-1}(D) \to (f')^{-1}(D)$, and then it is extended over the 1-handles $H_1, \cdots, H_k$ using the hypothesis $\rho = \rho'$.

If $\partial B = \emptyset$, delete an open disk $Int\Delta$ from $B - \{b_0\}$. Then there is a bundle isomorphism $H' : f^{-1}(B - Int\Delta) \to (f')^{-1}(B - Int\Delta)$ by what we have observed above. Earle and Eells [7, p.21] proves that $\mathrm{Diff}_0^+(\Sigma_g)$ is contractible for $g \geq 2$. This implies that the isomorphism $H'|f^{-1}(\partial\Delta) : f^{-1}(\partial\Delta) \to (f')^{-1}(\partial\Delta)$ extends to an isomorphism $f^{-1}(\Delta) \to (f')^{-1}(\Delta)$, giving a required bundle isomorphism $H : M \to M'$.

Case (2) $\{b_1, \cdots, b_n\} \neq \emptyset$.

Let p_1 (resp. q_1) be the critical point of $f : M \to B$ (resp. $f' : M' \to B'$) over the first critical value b_1. By Condition (c) (2) in the definition of a Lefschetz fibration, there exist local complex coordinates z_1, z_2 about p_1 in M (resp. z_1', z_2' about q_1 in M') with $z_1(p_1) = z_2(p_1) = 0$ (resp. $z_1'(q_1) = z_2'(q_1) = 0$) and local complex coordinate ξ in B with $\xi(b_1) = 0$ (resp. ξ' in B with $\xi'(b_1) = 0$) so that with these coordinates f (resp. f') is locally written as

$$\xi = f(z_1, z_2) = z_1^2 + z_2^2$$
$$\xi' = f'(z_1', z_2') = (z_1')^2 + (z_2')^2.$$

Note that we have made coordinate changes on the original z_1, z_2 (resp. z_1', z_2') which appeared in Condition (c) (2) of Definition 2.1. We will throughout fix these coordinates (z_1, z_2), (z_1', z_2').

Let N be a small neighborhood of the first critical value b_1 in B. By the disk theorem, there exists an isotopy $h_t : B \to B$ $(0 \leq t \leq 1)$ such that

(1) h_t fixes b_1 and all points outside N, $0 \leq t \leq 1$,

(2) $h_0 = id_B$, and

(3) $\xi'(h_1(b)) = \xi(b)$ for all points b in a smaller neighborhood $N' \subset N$ of b_1.

Then considering $h_1 f : M \to B$ instead of $f : M \to B$ we may assume from the outset that $\xi = \xi'$ about b_1. The same remark can be applied to each b_i, $i = 1, 2, \cdots, n$.

Let D_1 be a 2-disk centered at b_1 and of radius ϵ (a small positive number):

$$D_1 = \{b \in N | |\xi(b)| \leq \epsilon\}.$$

Let $D_2, \cdots, D_n$ be similar disks centered at $b_2, \cdots, b_n$, respectively. We put

$$B_0 = B - Int(D_1 \cup \cdots \cup D_n),$$

and

$$M_0 = f^{-1}(B_0), \quad M_0' = (f')^{-1}(B_0).$$

Then $f|M_0$ and $f'|M_0'$ are fiber bundles with the identical monodromy: $\rho = \rho'$. Thus by Case (1), there exists a bundle isomorphism $H : M_0 \to M_0'$, which commutes with f and f'.

Now in the coordinate neighborhood $(U; z_1, z_2)$ of the critical point p_1 over b_1, consider a C^∞-function $\varphi : U \cap f^{-1}(D_1) \to \mathbf{R}$ defined by

$$\varphi(z_1, z_2) = -\Re(z_1^2 + z_2^2) = -x_1^2 - x_2^2 + y_1^2 + y_2^2$$

where $z_1 = x_1 + \sqrt{-1}y_1$, $z_2 = x_2 + \sqrt{-1}y_2$. The function φ is a Morse function of index 2 whose 'left-hand disk' D_L (cf. [22]) is

$$D_L = \{(z_1, z_2)|\ x_1^2 + x_2^2 \leq \epsilon, y_1 = y_2 = 0\}.$$

This disk D_L is attached to $f^{-1}(\epsilon)$ along the 'left-hand' circle

$$S_L = \partial D_L = \{(z_1, z_2)|\ x_1^2 + x_2^2 = \epsilon, y_1 = y_2 = 0\}.$$

We repeat the same construction in the coordinate neighborhood $(U'; z_1', z_2')$ of $q_1 \in M'$ to get a Morse function $\varphi' = -(x_1')^2 - (x_2')^2 + (y_1')^2 + (y_2')^2$, the disk $D_L' = \{(z_1', z_2')|(x_1')^2 + (x_2')^2 \leq \epsilon, y_1' = y_2' = 0\}$, and the circle $S_L' = \{(z_1', z_2')|(x_1')^2 + (x_2')^2 = \epsilon, y_1' = y_2' = 0\}$, where $z_1' = x_1' + \sqrt{-1}y_1'$, $z_2' = x_2' + \sqrt{-1}y_2'$. The disk D_L' is attached to $(f')^{-1}(\epsilon)$ along S_L'.

The Picard-Lefschetz theorem [15] states that the monodromy of the Σ_g-bundle $f|f^{-1}(\partial D_1) : f^{-1}(\partial D_1) \to \partial D_1$ is a negative Dehn twist of $f^{-1}(\epsilon)$ about the circle S_L, because S_L represents the 'vanishing cycle' associated with the critical point p_1 on the singular fiber $f^{-1}(b_1)$. The same thing can be said on the bundle $f'|(f')^{-1}(\partial D_1) : (f')^{-1}(\partial D_1) \to \partial D_1$. Here note that by Condition (d) in Definition 2.1, S_L and S_L' are not null-homotopic in the respective fibers, $f^{-1}(\epsilon)$ and $(f')^{-1}(\epsilon)$

Therefore, by Nielsen [25] (see also [20]), the bundle isomorphism $H : M_0 \to M_0'$ sends the homotopy class of S_L in $f^{-1}(\epsilon)$ to that of S_L' in $(f')^{-1}(\epsilon)$. Then by the Baer-Epstein theorem [8 §2], $H|f^{-1}(\epsilon) : f^{-1}(\epsilon) \to (f')^{-1}(\epsilon)$ is C^∞-isotopic to a diffeomorphism $f^{-1}(\epsilon) \to (f')^{-1}(\epsilon)$ which sends S_L onto S_L'. Also by rotating S_L' along itself by an isotopy within $(f')^{-1}(\epsilon)$, we can make $H|f^{-1}(\epsilon)$ satisfy

$$(2.1) \qquad z_i'(H(p)) = z_i(p), \ i = 1, 2, \ \text{for } \forall p \in S_L.$$

Since these isotopies of $H|f^{-1}(\epsilon)$ extends to a 'vertical' isotopy of $H : M_0 \to M_0'$, we may assume (2.1) from the beginning.

We need a lemma.

Lemma 2.5. *There exists a neighborhood V of S_L in $U \cap f^{-1}(\partial D_1)$ and an isotopy $K_t : (f')^{-1}(\partial D_1) \to (f')^{-1}(\partial D_1)$, $0 \leq t \leq 1$, satisfying the following conditions:*

(1) K_t is vertical, namely, $f'K_t = f'$ for $0 \leq t \leq 1$;
(2) $K_0 = $ the identity of $(f')^{-1}(\partial D_1)$;
(3) $z_i'(K_1 H(p)) = z_i(p)$, $i = 1, 2$, for all $p \in V$.

Proof. Take a small neighborhood V_0 of S_L in $f^{-1}(\partial D_1)$ so that $V_0 \cap f^{-1}(\epsilon)$ is a tubular neighborhood of S_L in the general fiber $f^{-1}(\epsilon)$. Note that we may assume the tubular neighborhood $V_0 \cap f^{-1}(\epsilon)$ is diffeomorphic to a product $S_L \times$ (interval). This tubular neighborhood is given in V_0 by the equation $z_1^2 + z_2^2 = \epsilon$, namely

$$\begin{cases} x_1^2 + x_2^2 - y_1^2 - y_2^2 = \epsilon \\ x_1 y_1 + x_2 y_2 = 0. \end{cases}$$

Thus it is parametrized by θ and u as

$$\begin{cases} x_1 = \sqrt{\epsilon + u^2}\cos\theta, & x_2 = \sqrt{\epsilon + u^2}\sin\theta \\ y_1 = -u\sin\theta, & y_2 = u\cos\theta. \end{cases}$$

Here θ and u belong to $[0, 2\pi]$ and $[-\delta, \delta]$ (for some $\delta > 0$), and parametrize S_L and the interval, respectively.

Similarly a tubular neighborhood of S_L' in $(f')^{-1}(\epsilon)$ is parametrized by θ and v as

$$\begin{cases} x_1' = \sqrt{\epsilon + v^2}\cos\theta, & x_2' = \sqrt{\epsilon + v^2}\sin\theta, \\ y_1' = -v\sin\theta, & y_2' = v\cos\theta, \end{cases}$$

where $(\theta, v) \in [0, 2\pi] \times [-\delta, \delta]$.

Recall that $H(S_L) = S_L'$ and that $H|f^{-1}(\epsilon) : f^{-1}(\epsilon) \to (f')^{-1}(\epsilon)$ preserves the orientations. Property (2.1) of H implies that H preserves the values of θ on S_L. Then by the uniqueness of a tubular neighborhood [10] applied to S_L' in $(f')^{-1}(\epsilon)$, there exists an isotopy $k_t : (f')^{-1}(\epsilon) \to (f')^{-1}(\epsilon)$, $0 \le t \le 1$ such that $k_0 = $ the identity, and

$$\theta(k_1 H(p)) = \theta(p), \qquad v(k_1 H(p)) = u(p),$$

for all $p \in V_0 \cap f^{-1}(\epsilon)$. This latter condition is equivalent to saying that $k_1(H|f^{-1}(\epsilon))$ preserves the values of z_i and z_i' on $V_0 \cap f^{-1}(\epsilon)$, $i = 1, 2$.

Let $\tilde{k}_t : (f')^{-1}(\partial D_1) \to (f')^{-1}(\partial D_1)$, $0 \le t \le 1$ be a vertical isotopy extending k_t with $\tilde{k}_0 = $ the identity. From the corresponding property of $k_1 H$, it follows that the composition $\tilde{k}_1 H$ preserves the values of z_i and z_i' on $V_0 \cap f^{-1}(\epsilon)$, $i = 1, 2$.

Now downstairs (on B), we take an open interval I_β on ∂D_1 defined by

$$I_\beta = \{b \in \partial D_1 | |\arg(\xi)| < \beta\},$$

where β is a small positive number. The inverse image $\mathcal{U} := (f')^{-1}(I_\beta)$ is an open neighborhood of $(f')^{-1}(\epsilon)$ in $(f')^{-1}(\partial D_1)$.

Let us construct a vector field X on $\mathcal{U}$ 'lifting' $\frac{\partial}{\partial \psi}$, namely, a vector field which satisfies

$$(df')_q(X_q) = (\frac{\partial}{\partial \psi})_b, \qquad q \in \mathcal{U},$$

where $b = f'(q)$ and $\psi = \arg(\xi)$. Let $\Gamma_X(q, \psi)$ be the integral curve of X starting from $q \in (f')^{-1}(\epsilon)$; $\Gamma_X(q, 0) = q$ for all $q \in (f')^{-1}(\epsilon)$. The argument $\psi = \arg(\xi)$ plays the role of the time parameter here, and we have

$$f'(\Gamma_X(q, \psi)) = \epsilon e^{\sqrt{-1}\psi} \in \partial D_1.$$

Γ_X gives a diffeomorphism

$$\Gamma_X : (f')^{-1}(\epsilon) \times (-\beta, \beta) \to \mathcal{U}.$$

Let V_0' denote the image $\tilde{k}_1 H(V_0)$ of V_0. Then V_0' is a neighborhood of S_L' in $(f')^{-1}(\partial D_1)$. Using the diffeomorphisms Γ_X and $\tilde{k}_1 H$, we can construct an embedding

$$E : (V_0' \cap (f')^{-1}(\epsilon)) \times (-\beta, \beta) \to \mathcal{U}$$

so that it satisfies

(2.2) $z_i((\tilde{k}_1 H)^{-1} E(q, \psi)) = z_i'(\Gamma_X(q, \psi)), \quad i = 1, 2,$

for all $(q, \psi) \in (V_0' \cap (f')^{-1}(\epsilon)) \times (-\beta, \beta)$.

Since we are working in regions, where f and f' are determined by z_i and z_i', $(i = 1, 2)$, and since $\tilde{k}_1 H$ commutes with f and f', we see that

$$f'(E(q, \psi)) = \epsilon e^{\sqrt{-1}\psi} \in \partial D_1.$$

Moreover, since $\tilde{k}_1 H$ preserves the values of z_i, z_i' $(i = 1, 2)$ on $V_0 \cap f^{-1}(\epsilon)$, we have $E(q, 0) = q$ for $q \in V_0' \cap (f')^{-1}(\epsilon)$.

Let Y be a vector field on $\mathcal{U}$ which lifts $\frac{\partial}{\partial \psi}$ and coincides with $\frac{d}{d\psi} E(q, \psi)$ in a neighborhood of $E(S_L' \times (-\beta, \beta))$.

Let $\Gamma_Y : (f')^{-1}(\epsilon) \times (-\beta, \beta) \to \mathcal{U}$ be the diffeomorphism defined by the integral curves of Y, so that for all $q \in (f')^{-1}(\epsilon)$,

$$\begin{cases} \Gamma_Y(q, 0) = q, \\ f'(\Gamma_Y(q, \psi)) = \epsilon e^{\sqrt{-1}\psi}. \end{cases}$$

Obviously Γ_Y coincides with E in a neighborhood of $E(S_L' \times (-\beta, \beta))$.

We define an isotopy $J_t : \mathcal{U} \to \mathcal{U}, 0 \le t \le 1$ so that it satisfies

$$J_t(\Gamma_Y(q, \psi)) = \Gamma_X(\tilde{q}, \psi),$$

where $\tilde{q} \in (f')^{-1}(\epsilon)$ is the solution of the following equation (see Figure 2.1):

$$\Gamma_Y(q, (1 - t)\psi) = \Gamma_X(\tilde{q}, (1 - t)\psi).$$

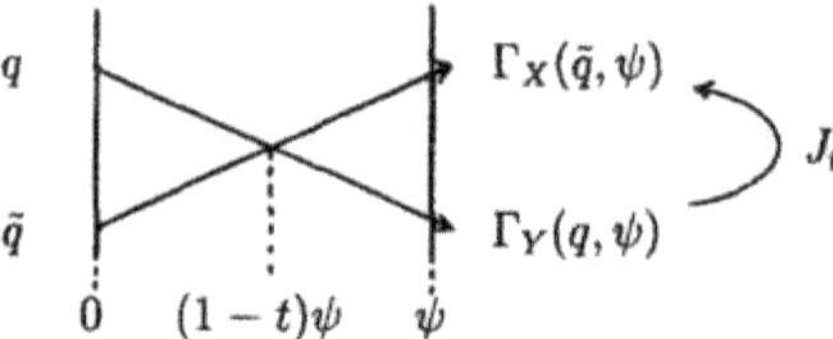

FIGURE 2.1

Obviously $\tilde{q}$ depends on (q, ψ, t) in a C^∞- manner. Therefore, $J_t, 0 \le t \le 1$, is a smooth isotopy of $\mathcal{U}$. It is easy to see that J_t is vertical and $J_0 = id_{\mathcal{U}}$.

Assertion. *J_1 satisfies*

$$z_i'(J_1 \tilde{k}_1 H(p)) = z_i(p), \quad i = 1, 2,$$

for all p in a neighborhood $V(\subset V_0)$ of S_L.

Taking this assertion for granted, we finish the proof of Lemma 2.5 as follows:
Let $\tilde{J}_t : (f')^{-1}(\partial D_1) \to (f')^{-1}(\partial D_1)$, $0 \le t \le 1$, be a vertical isotopy satisfying

(1) $\tilde{J}_0 =$ the identity of $(f')^{-1}(\partial D_1)$, and
(2) $\tilde{J}_t$ coincides with J_t on $(f')^{-1}(I_{\beta/2})$.

Then the required isotopy $K_t : (f')^{-1}(\partial D_1) \to (f')^{-1}(\partial D_1)$, $0 \le t \le 1$, is defined
to be the composition

$$K_t = \tilde{J}_t \tilde{k}_t, \qquad 0 \le t \le 1.$$

Conditions (1),(2) on K_t are immediately checked. Condition (3) is verified by applying Assertion above.

Proof of Lemma 2.5 is now completed. $\square$

Proof of Assertion. The proof uses property (2.2) of E. If $t = 1$, we have $\bar{q} = q$ (see Figure 2.1). Thus

$$J_1(\Gamma_Y(q, \psi)) = \Gamma_X(q, \psi).$$

If $q \in \tilde{k}_1 H(V_1 \cap f^{-1}(\epsilon))$ with $V_1(\subset V_0)$ a sufficiently small open neighborhood of S_L in $f^{-1}(\partial D_1)$, $\Gamma_Y(q, \psi)$ coincides with $E(q, \psi)$. Therefore, for $(q, \psi) \in \tilde{k}_1 H(V_1 \cap f^{-1}(\epsilon)) \times (-\beta, \beta)$, we have

$$
\begin{aligned}
z_i((\tilde{k}_1 H)^{-1} \Gamma_Y(q, \psi)) &= z_i((\tilde{k}_1 H)^{-1} E(q, \psi)) \\
&= z_i'(\Gamma_X(q, \psi)) && \text{by (2.2)} \\
&= z_i'(J_1 \Gamma_Y(q, \psi)), && i = 1, 2
\end{aligned}
$$

In other words, for all points p in the open set

$$(\tilde{k}_1 H)^{-1}(\Gamma_Y(\tilde{k}_1 H(V_1 \cap f^{-1}(\epsilon)) \times (-\beta, \beta))),$$

we have

$$z_i(p) = z_i'(J_1 \tilde{k}_1 H(p)), \quad i = 1, 2$$

as asserted. $\square$

We continue the proof of Theorem 2.4.

Extend the isotopy $K_t : (f')^{-1}(\partial D_1) \to (f')^{-1}(\partial D_1)$ of Lemma 2.5 to a vertical isotopy $\tilde{K}_t : M_0' \to M_0'$ and put $\tilde{H}_t = \tilde{K}_t H : M_0 \to M_0'$, $0 \le t \le 1$. Then replacing $H : M_0 \to M_0'$ with $\tilde{H}_1 : M_0 \to M_0'$, we may assume from the outset that

$$H(S_L) = S_L',$$

and that

$$z_i'(H(p)) = z_i(p), \quad i = 1, 2, \quad \text{for } p \in V,$$

where V is a neighborhood of S_L in $f^{-1}(\partial D_1)$.

We think of the coordinate neighborhoods $(U; z_1, z_2)$ and $(U'; z_1', z_2')$ as neighborhoods of the left-hand disks D_L and D_L', respectively. Assuming U to be small enough, define an embedding

$$\tau : f^{-1}(D_1) \cap U \to (f')^{-1}(D_1) \cap U'$$

by $z_i'(\tau(p)) = z_i(p)$, $i = 1, 2$, for $p \in f^{-1}(D_1) \cap U$. If U' is chosen appropriately, the map τ is a diffeomorphism of $f^{-1}(D_1) \cap U$ onto $(f')^{-1}(D_1) \cap U'$. Also if U is so small that $f^{-1}(\partial D_1) \cap U \subset V$, then τ coincides with H on $f^{-1}(\partial D_1) \cap U$, because both τ and H preserve the values of $(z_1, z_2), (z_1', z_2')$ there. Thus $H : M_0 \to M_0'$ extends to a diffeomorphism

$$H \cup \tau : M_0 \cup U \to M_0' \cup U',$$

which commutes with f and f'.

Now descending again, in a neighborhood of D_1, we construct a vector field $\mathcal{X}$ which satisfies the following conditions:

(1) Let $I_\beta(\subset \partial D_1)$ be the interval considered in the proof of Lemma 2.5. If $b \in I_\beta$, the vector $\mathcal{X}_b$ is perpendicular to ∂D_1 with respect to the metric $|d\xi|^2$ and points inward;

(2) if $b \in \partial \bar{I}_\beta$, then $\mathcal{X}_b = 0$;

(3) if b is on the real axis $\Im(\xi) = 0$, then $\mathcal{X}_b$ is tangent to the real axis pointing in the negative direction;

(4) the 'time 1 map' associated with $\mathcal{X}$ sends I_β to the curve C as shown in Figure 2.2, where C is C^∞-tangent to ∂D_1 at $\partial \bar{I}_\beta$ and cuts the real axis orthogonaly at a negative α, $|\alpha|$ being small enough.

FIGURE 2.2

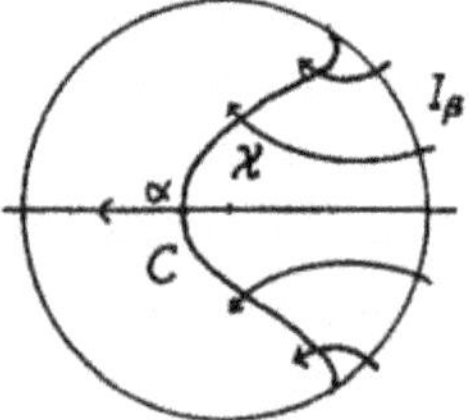

Let W be the closed region in D_1 bouded by $C \cup I_\beta$. We want to lift the vector field $\mathcal{X}|W$ to a vector field $\mathcal{Y}$ on $f^{-1}(W) - \{p_1\}$.

Upstairs, in U, we considered the left-hand disk D_L associated with the Morse function $\varphi = -\Re(z_1^2 + z_2^2)$. Now we need a right-hand disk $D_R = \{(z_1, z_2)|x_1 = x_2 = 0, y_1^2 + y_2^2 \leq |\alpha|\}$, too. This D_R is attached to $f^{-1}(\alpha)$. Note that $f(D_L) = [0, \epsilon], f(D_R) = [\alpha, 0]$, and that $U \cap f^{-1}(W)$ is a neighborhood of $D_L \cup D_R$ in $f^{-1}(W)$.

Since $(df)_p : T_p(M) \to T_{f(p)}(B)$ is onto for any $p \in f^{-1}(W) - \{p_1\}$, we can lift $\mathcal{X}|W$ to a vector field $\mathcal{Y}$ on $f^{-1}(W) - \{p_1\}$ so that $\mathcal{Y}_p$ is tangent to $D_L \cup D_R$ if $p \in D_L \cup D_R - \{p_1\}$, and so that $\mathcal{Y}$ is perpendicular to $f^{-1}(C) \cup f^{-1}(I_\beta)$ with respect to a suitable metric on M

Let $U_1(\subset U)$ be a 'saturated' neighborhood of $D_L \cup D_R$ with respect to $\mathcal{Y}$, namely, U_1 is a union of $D_L \cup D_R$ and the full trajectories of $\mathcal{Y}$ which leave $f^{-1}(I_\beta)$ and reach $f^{-1}(C)$.

Similarly to the above, since $(df')_q : T_q(M') \to T_{f'(q)}(B)$ is onto for any $q \in (f')^{-1}(W) - \{q_1\}$, we can lift $\mathcal{X}|W$ to a vector field $\mathcal{Z}$ on $(f')^{-1}(W) - \{q_1\}$ so that $\mathcal{Z}$ is perpendicular to $(f')^{-1}(C) \cup (f')^{-1}(I_\beta)$ with respect to a suitable metric. We require in addition that $\mathcal{Z}$ should coincide with $d\tau(\mathcal{Y})$ on $\tau(U_2)$, where $U_2 \subset U_1$ is a smaller saturated neighborhood of $D_L \cup D_R$.

Let $\Gamma_{\mathcal{Y}}(\tilde{p},t)$ (resp. $\Gamma_{\mathcal{Z}}(\tilde{q},t)$) denotes the integral curve of $\mathcal{Y}$ (resp. $\mathcal{Z}$) which starts from $\tilde{p} \in f^{-1}(\overline{I}_\beta)$ (resp. $\tilde{q} \in (f')^{-1}(\overline{I}_\beta)$), t being the time parameter.

We define a diffeomorphism

$$H' : M_0 \cup f^{-1}(W) \to M_0' \cup (f')^{-1}(W)$$

by

$$H'(p) = (H \cup \tau)(p)$$

if $p \in M_0 \cup U_2$, or by

$$H'(p) = \Gamma_{\mathcal{Z}}(H(\tilde{p}),t)$$

if $p = \Gamma_{\mathcal{Y}}(\tilde{p},t) \in f^{-1}(W) - (D_L \cup D_R)$.

The well-definedness and the commutativity with f and f' are easily verified, and we obtain an isomorphism $H' : f^{-1}(B_0 \cup W) \to (f')^{-1}(B_0 \cup W)$. Note that the difference between $f^{-1}(B_0 \cup W)$ and $f^{-1}(B_0 \cup D_1)$ is a trivial Σ_g-bundle over a 2-disk $\overline{D_1 - W}$. Similarly the difference between $(f')^{-1}(B_0 \cup W)$ and $(f')^{-1}(B_0 \cup D_1)$ is a trivial Σ_g-bundle over the same 2-disk $\overline{D_1 - W}$. Then by the contractibility of $\mathrm{Diff}_0^+(\Sigma_g)$ for $g \geq 2$, H' can be extended to an isomorphism between $f^{-1}(B_0 \cup D_1)$ and $(f')^{-1}(B_0 \cup D_1)$.

Repeating the above argument for each of $D_2, \cdots , D_n$, we finally obtain an isomorphism of $f : M \to B$ to $f' : M' \to B$ as required.

This completes the proof of Theorem 2.4. $\square$

As for the existence of a Lefschetz fibration of genus $g \geq 1$ we obtain

Theorem 2.6. *Let B be a compact connected oriented surface, $\{b_0, b_1, \cdots , b_n\}$ a finite set of points in $IntB$. Let $\lambda : \pi_1(B - \{b_1, \cdots , b_n\}, b_0) \to \mathcal{M}_g$ be a homomorphism. Then there exists a Lefschetz fibration $f : M \to B$ whose monodromy representation is equivalent to λ if and only if $\lambda(\gamma_i)$ is the negative Dehn twist about an essential simple closed curve on Σ_g, $i = 1, \cdots , n$, where γ_i is a loop on B based at b_0 which surrounds exclusively b_i.*

Outline of Proof. If $\{b_1, \cdots , b_n\} = \emptyset$, the proof belongs to the standard bundle theory, so we assume $\{b_1, \cdots , b_n\} \neq \emptyset$.

Note that there are $[\frac{g}{2}] + 1$ equivalence classes, $\mathcal{C}_0, \mathcal{C}_1, \cdots , \mathcal{C}_{[g/2]}$, of essential simple closed curves on Σ_g under the action of self-homeomorphisms, where $\mathcal{C}_0$ is the class of non-separating curves, and $\mathcal{C}_j$ is the class of curves separating Σ_g into surfaces of genera j and $g - j$, $(1 \leq j \leq [g/2])$. Accordingly, by Theorem 2.4, there are exactly $[\frac{g}{2}] + 1$ isomorphism classes of Lefschetz fibrations of genus g over the 2-disk with a single singular fiber over the center. Cf.[20]. We choose a representative $f_j : M_j \to D$ from each isomorphism class and call it a 'standard model'. Let D_i be a small 2-disk centered at b_i. We can construct a Σ_g-bundle over $B_0 := B - \bigcup_{i=1}^n IntD_i$ whose monodromy representation is $\lambda|B_0$. Then we fill in the hole over each D_i with a copy of an appropriate standard model over a 2-disk, using a suitable gluing bundle isomorphism over the circle so that the vanishing cycle of the inserted standard model coincides with the circle about which the negative Dehn twist $\lambda(\gamma_i)$ occurs.

This completes the outline of the proof of Theorem 2.6. $\square$

The author was asked by G.Tian the following problem (in Beijing, 1994), which would be very interesting to investigate:

Problem 2.7. Given a $(C^\infty$-$)$Lefschetz fibration over a closed surface, is it isomorphic to a holomorphic Lefschetz fibration?

3. Fractional Signature

In this section we use Meyer's theory [21] on the signature of surface bundles over surfaces to define the 'fractional signature' of singular fibers of genus 2. First we will extend the notion of Lefschetz fibrations to a more general class of fiber spaces, in which our argument of this section will apply more naturally.

Definition 3.1. Let M and B be compact oriented (not necessarily closed) C^∞-manifolds of dimensions 4 and 2 , respectively. A C^∞-map $f : M \to B$ is called a *locally analytic fibration of genus g* if it satisfies Conditions (a) and (b) of Definition 2.1 and the following Condition (c') instead of (c):
(c') for each i $(1 \le i \le n)$ and at each point $p \in f^{-1}(b_i)$, the germ (f,p) is conjugate via (not necessarily orientation preserving) diffeomorpisms to the germ at 0 of a holomorphic function $\mathbb{C}^2 \to \mathbb{C}$. Moreover, there exists at least one critical point of f on the fiber $f^{-1}(b_i)$

Remarks 1) In Condition (c') of the above definition, if the local diffeomorphisms giving the conjugations of the germs are always orientation preserving, we would like to call such a fibration a *locally holomorphic fibration*.
2) An 'achiral' Lefschetz fibration of Harer[11] and a 'good torus fibration' considered in [30] are special cases of locally analytic fibrations.

Let F and F' be (singular or non-singular) fibers in locally analytic fibrations $f : M \to B$ and $f' : M' \to B'$ respectively.

Definition 3.2. F and F' are said to be *topologically equivalent* $(F \sim F')$ if there exist neighborhoods $N(\subset B)$ and $N'(\subset B')$ of points $b = f(F)$ and $b' = f'(F')$, respectively, and orientation preserving homeomorphisms $H : (f)^{-1}(N) \to (f')^{-1}(N')$ and $h : N \to N'$ so that $h(b) = b'$ and $f'H = hf$.

Let $S_g^{loc.an}$ denote the set of equivalence classes, classified under $\sim$, of all (singular or non-singular) fibers that appear in locally analyic fibrations of genus g. From now on we will mainly be concerned with the case of genus 2. For $g = 2$ we have the following theorem, in which $\frac{1}{5}\mathbb{Z}$ denotes the additive group $\{\frac{m}{5} \in \mathbb{Q} \mid m \in \mathbb{Z}\}$.

Theorem 3.3. *There exists a function*

$$\sigma : S_2^{loc.an} \to \frac{1}{5}\mathbb{Z}$$

with the following properties:

(1) *if F is a non-singular fiber, then $\sigma(F) = 0$;*

(2) *if $f : M \to B$ is a locally analytic fibration of genus 2 over a closed surface B (M is necessarily closed), and if $F_1, \cdots, F_n$ are all the singular fibers in this fibration, then $\mathrm{Sign}(M) = \sum_{i=1}^n \sigma(F_i)$.*

The *signature* $\mathrm{Sign}(M)$ ($\in \mathbb{Z}$) of a compact connected oriented 4-manifold M is defined, of course, to be the number of positive eigenvalues of the intersecton form $I : H_2(M) \times H_2(M) \to \mathbb{Z}$ minus the number of negative eigenvalues. Note that $\mathrm{Sign}(M)$ makes sense even if $\partial M \neq \emptyset$.

The above theorem implies that if the base space is closed, the signature of the total space of a locally analytic fibration of genus 2 is 'concentrated' on singular fibers, and that each singular fiber behaves as if it carried its own 'fractional signature'. The corresponding fact for genus 1 has been observed in [16], [18]. For genus ≥ 3, we cannot expect this kind of concentration of the signature, because for each $g \geq 3$ there are Σ_g-bundles over a closed surface whose total space have non-zero signature, [1], [13], [21].

Proof of Theorem 3.3. Meyer [21,p.257] proved that if $g = 2$, five times his 'signature cocycle' $z \in C^2(\mathcal{M}_2; \mathbb{Z})$ is a coboundary. As a consequence, there exists a function

$$\phi : \mathcal{M}_2 \to \frac{1}{5}\mathbb{Z}$$

satisfying the following conditions:

(1) $\phi(1) = 0$;
(2) $\phi(\alpha^{-1}) = -\phi(\alpha)$;
(3) $\phi(\gamma^{-1}\alpha\gamma) = \phi(\alpha)$, i.e., ϕ is a class function;
(4) $z(\alpha, \beta) = \phi(\alpha) + \phi(\beta) - \phi(\alpha\beta)$.

Let $f : M \to B$ be a locally analytic fibration over a closed surface B. Let $\{b_1, \cdots, b_n\}$ be the set of critical values, D_i a small 2-disk on B centered at b_i and $F_i = f^{-1}(b_i)$. We put B_0 for $B - \bigcup_{i=1}^n IntD_i$, and M_0 for $M - \bigcup_{i=1}^n f^{-1}(IntD_i)$. We give an orientation to ∂D_i from the inside of D_i, and let $\alpha_i \in \mathcal{M}_2$ be the monodromy of the bundle $f^{-1}(\partial D_i) \to \partial D_i$ in this direction of ∂D_i. Then Meyer's signature theorem [21] states that

$$\mathrm{Sign}(M_0) = -\sum_{i=1}^n \phi(\alpha_i)$$

By the Novikov additivity [2], we have

$$\mathrm{Sign}(M) = \mathrm{Sign}(M_0) + \sum_{i=1}^n \mathrm{Sign}(f^{-1}(D_i)).$$

Thus by defining $\sigma(F_i)$ by

$$\sigma(F_i) := -\phi(\alpha_i) + \mathrm{Sign}(f^{-1}(D_i)),$$

we have the desired formula $\mathrm{Sign}(M) = \sum_{i=1}^n \sigma(F_i)$. $\square$

Definition 3.4. Let $f : M \to D$ be a locally analytic fibration of genus g over a 2-disk D with a single singular fiber F. Suppose there is another locally analytic fibration $f' : M' \to D$ of genus g over the same 2-disk, with the set of singular fibers $\{F_1', F_2', \cdots, F_r'\}$. Moreover, suppose there exists an orientation preserving diffeomorphism $H : M \to M'$ which commutes with f and f' on the boundary: $(f'|\partial M') \circ (H|\partial M) = f|\partial M$. Then we say that the singular fiber F *splits* into the set of singular fibers $F_1', F_2', \cdots, F_r'$. Sometimes we will write this splitting as

$$F \to F_1' + F_2' + \cdots + F_r'.$$

It is easy to see that if a singular fiber F_i of a localy analytic fibration $f : M \to B$ splits into $F_{i1}, F_{i2}, \cdots, F_{ir}$, then we can change the projection f within an arbitrarily small neighborhood of F_i (without changing M and B) so that the new projection f' has singular fibers $F_{i1}, F_{i2}, \cdots, F_{ir}$ in place of F_i.

Remark. The splitting defined above might better be called *splitting in a weak sense*. Compare the splitting in a stronger sense which will be discussed in [19], where singular fibers split through certain perturbation of the projecton map.

Theorem 3.5. *Suppose a singular fiber F of genus 2 splits into $F_1', F_2', \cdots, F_r'$, then*

$$\sigma(F) = \sum_{i=1}^{r} \sigma(F_i').$$

Proof. We use the same notation as in Defintion 3.4. Let $D_1, D_2, \cdots, D_r$ be disjoint small 2-disks in $IntD$ centered at $b_1, b_2, \cdots, b_r$, respectively, where $b_i = f'(F_i')$, $i = 1, 2, \cdots, r$. We put M_0' for $M' - \bigcup_{i=1}^{r} Int((f')^{-1}(D_i))$. We give orientations to the boundaries $\partial D, \partial D_1, \cdots, \partial D_r$ from the inside of $D, D_1, \cdots, D_r$, respectively, with which we calculate the monodromies $\alpha, \alpha_1, \cdots, \alpha_r (\in \mathcal{M}_2)$ of the restricted Σ_2-bundles over $\partial D, \partial D_1, \cdots, \partial D_r$. Then by Meyer's theorem,

$$\mathrm{Sign}(M_0') = \phi(\alpha) - \sum_{i=1}^{r} \phi(\alpha_i).$$

By the Novikov additivity,

$$\mathrm{Sign}(M) = \mathrm{Sign}(M')$$

$$= \mathrm{Sign}(M_0') + \sum_{i=1}^{r} \mathrm{Sign}((f')^{-1}(D_i))$$

$$= \phi(\alpha) - \sum_{i=1}^{r} \phi(\alpha_i) + \sum_{i=1}^{r} \mathrm{Sign}((f')^{-1}(D_i)),$$

from which we have

$$\sigma(F) = \mathrm{Sign}(M) - \phi(\alpha)$$

$$= -\sum_{i=1}^{r} \phi(\alpha_i) + \sum_{i=1}^{r} \mathrm{Sign}((f')^{-1}(D_i))$$

$$= \sum_{i=1}^{r} \sigma(F_i').$$

This completes the proof. $\square$

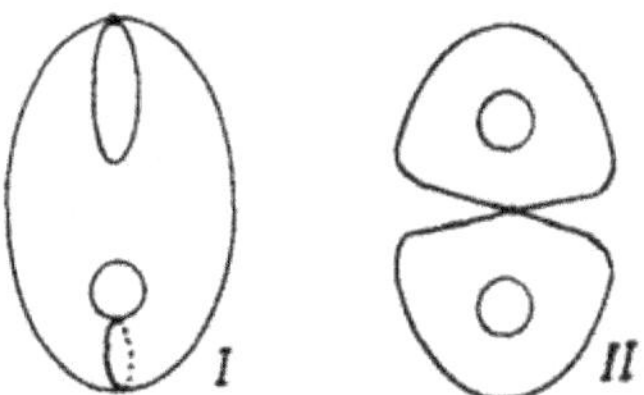

FIGURE 3.1. Lefschetz singular fibers I and II.

There are exactly two topological types of Lefschetz singular fibers of genus 2, which we would like to call *singular fibers of type I and type II*, respectively. A singular fiber of type I is obtained by pinching a *non-separating* simple closed curve on Σ_2 to a point, while a singular fiber of type II is obtained by pinching a *separating* simple closed curve . Their monodromies are negative Dehn twists about the respective curves.

We will say that a Lefschetz fibration $f : M \to B$ is of type $aI + bII$ if it contains singular fibers of type I to the number a, and those of type II to the number b.

The fractional signature of Lefschetz singular fibers of type I and II are given by the following

Proposition 3.6.

$$\sigma(I) = -\frac{3}{5}, \quad \sigma(II) = -\frac{1}{5}.$$

Proof. Example A in §4 below shows the existence of a Lefschetz fibration of $\mathbb{C}P_2\#13\overline{\mathbb{C}P_2}$ (the complex projective plane blown up 13 times) over the 2-sphere S^2 whose type is $20I$. By Theorem 3.3 , we have

$$\mathrm{Sign}(\mathbb{C}P_2\#13\overline{\mathbb{C}P_2}) = 20\sigma(I).$$

Since $\mathrm{Sign}(\mathbb{C}P_2\#13\overline{\mathbb{C}P_2}) = -12$, we obtain $\sigma(I) = -\frac{3}{5}$.

Example B in §4 shows the existence of a Lefschetz fibration of $S^2 \times T^2\#4\overline{\mathbb{C}P_2}$ ($S^2 \times T^2$ blown up 4 times) over S^2 whose type is $6I + 2II$. Thus by Theorem 3.3,

$$\mathrm{Sign}(S^2 \times T^2\#4\overline{\mathbb{C}P_2}) = 6\sigma(I) + 2\sigma(II).$$

Substituting the known values $\mathrm{Sign}(S^2 \times T^2\#4\overline{\mathbb{C}P_2}) = -4$ and $\sigma(I) = -\frac{3}{5}$, we obtain $\sigma(II) = -\frac{1}{5}$. $\square$

Corollary 3.7. *The value of Meyer's function* $\phi : \mathcal{M}_2 \to \frac{1}{5}\mathbb{Z}$ *on the negative Dehn twist about a non-separating simple closed curve is equal to* $\frac{3}{5}$, *and that on the negative twist about a separating simple closed curve is equal to* $-\frac{4}{5}$.

This corollary is proved by the definition of σ together with the easily computed values of Sign(fibered neighborhood of I) $= 0$ and Sign (fibered neighborhood of II) $= -1$.

It would be interesting to think of a singular fiber of genus 2 as an analogue of an elementary particle in physics. The fractional signature could be interpreted as the electric charge; it is conserved at splitting. Furthermore there is an analogue of mass.

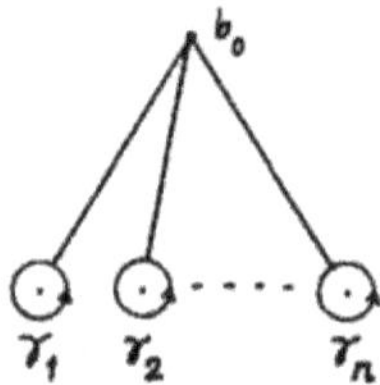

FIGURE 4.1

Following [27], we will call it the *Euler contribution*. This can be defined for singular fibers of arbitrary genus.

Definition 3.8. (Cf. [27]) Let F be a (singular or non-singular) fiber in a locally analytic fibration of genus g. The *Euler contribution* of F, denoted by $\varepsilon(F)$, is defined as follows:

$$\varepsilon(F) := e(F) - e(\Sigma_g),$$

where $e(\cdot)$ denotes the Euler characteristic.

$\varepsilon(F)$ is invariant under topological equivalence. If F is non-singular, then $\varepsilon(F) = 0$. It is easy to see that if $\{F_1, F_2, \cdots, F_n\}$ is the totality of the singular fibers in a locally analytic fibration $f : M \to B$, then

$$e(M) = e(B)e(\Sigma_g) + \sum_{i=1}^{n} \varepsilon(F_i).$$

Here B may have nonempty boundary.

Remark. $\varepsilon(F)$ is always non-negative. Also $\varepsilon(F) = 0$ if and only if F is nonsingular, provided that $g \neq 1$.

Suppose a singular fiber F splits as $F \to F_1 + F_2 + \cdots + F_r$. Then , as is easily seen,

$$\varepsilon(F) = \sum_{i=1}^{r} \varepsilon(F_i).$$

A Lefschetz singular fiber has Euler contribution 1.

4. Examples of Lefschetz Fibrations of Genus 2

Let $f : M \to B$ be a Lefschetz fibration of genus g. Suppose that the base space B is the 2-disk D or the 2-sphere S^2. Then the monodromy representation $\rho : \pi_1(b - \{b_1, \cdots, b_n\}, b_0) \to \mathcal{M}_g$ can be given by an n-tuple

$$(4.1) \qquad\qquad\qquad (g_1, g_2, \cdots, g_n)$$

where $g_i = \rho(\gamma_i)$, $i = 1, 2, \cdots, n$, and $\gamma_1, \gamma_2, \cdots, \gamma_n$ are the loops drawn on B as in Figure 4.1.

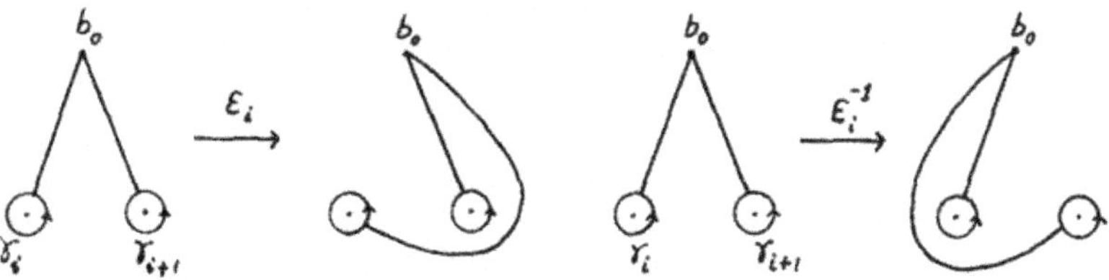

FIGURE 4.2

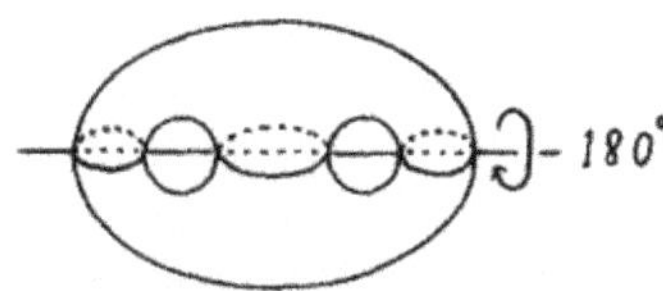

FIGURE 4.3. Hyperelliptic involution ι.

If we change the loops by the *i-th elementary homeomorphism* ε_i or its inverse ε_i^{-1} as shown in Figure 4.2, the above presentation (4.1) is changed into

$$(g_1, \cdots, g_{i-1}, g_{i+1}, g_{i+1}^{-1} g_i g_{i+1}, g_{i+2}, \cdots, g_n)$$

or

$$(g_1, \cdots, g_{i-1}, g_i g_{i+1} g_i^{-1}, g_i, g_{i+2}, \cdots, g_n),$$

respectively. We will call these transformations the *i-th elementary transformation* or its inverse.

The elementary transformatins do not change the *total monodromy*

$$g_1 g_2 \cdots g_n.$$

In the case $B = S^2$, the total monodromy is trivial.

We will return to the case of genus 2. The main purpose of this section is to give some examples of Lefschetz fibrations of genus 2. One can find many examples in [26], [29], but the new feature of ours is the explicit descriptions of their global monodromies.

Example A. (Lefschetz fibration of $CP_2 \# 13\overline{CP_2}$)

Let $\iota : \Sigma_2 \to \Sigma_2$ be the hyperelliptic involution (see Figure 4.3), and $\tau : S^2 \to S^2$ the 180° rotation of a 2-sphere about the axis through the poles.

The quotient space $S^2 \times \Sigma_g / \tau \times \iota$ has 12 singular points corresponding to the fixed points $Fix(\tau) \times Fix(\iota)$ of $\tau \times \iota$. Blowing up these singularities, we obtain a compact complex surface M^ι. There is a natural projection $f^\iota : M^\iota \to S^2/\tau \cong S^2$ induced by the projection to the first factor $S^2 \times \Sigma_2 \to S^2$, and M^ι has the structure of a locally analytic fibration over the 2-sphere. In fact, f^ι is a holomorphic fibration.

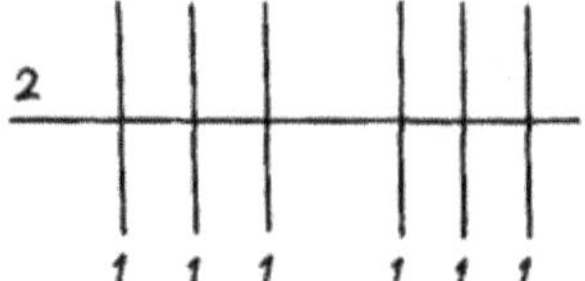

FIGURE 4.4. Singular fiber F^ι

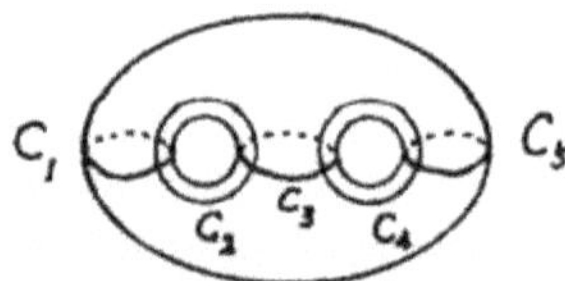

FIGURE 4.5. $\zeta_i = $ negative Dehn twist about C_i.

The fibration $f^\iota : M^\iota \to S^2$ contains two singular fibers over the north and the south poles of S^2. They are topologically equivalent, and the monodromy about each singular fiber is the hyperelliptic involution ι. We denote either singular fiber by F^ι *Structure of F^ι.*   The quotient space Σ_2/ι is homeomorphic to S^2 with 6 cusps corresponding to $Fix(\iota)$. When blowing up $S^2 \times \Sigma_2/\tau \times \iota$ to M, 6 spheres with self-intersecton number -2 are inserted at these cusps on Σ_2/ι. Thus the final shape of F^ι is as indicated in Figure 4.4, where arcs stand for smooth 2-spheres and the numbers attached to them are the multiplicities of f^ι about the irreducible components. (For a general method of constructing a singular fiber starting from its monodromy, see [20].)

Here we quote a result obtained by a computer calculation, from [19].

Fact 1. *F^ι splits into 10 Lefschetz singular fibers of type I. We may assume that this splitting takes place within a small neighborhood of F^ι, and that the new critical values $b_1, b_2, \cdots, b_{10}$ are in a small disk D centered at the critical value of F^ι. Then taking loops as in Figure 4.1, and changing them by elementary transformations and/or their inverses, we can arrange so that the monodromy representaion $\rho : \pi_1(D - \{b_1, \cdots, b_{10}\}, b_0) \to \mathcal{M}_2$ is given by*

$$(4.2) \qquad\qquad (\zeta_1, \zeta_2, \zeta_3, \zeta_4, \zeta_5, \zeta_5, \zeta_4, \zeta_3, \zeta_2, \zeta_1),$$

where, $\zeta_i, i = 1, \cdots, 5$ are the canonical generators of $\mathcal{M}_2$ as shown in Figure 4.5. (Cf.[3])

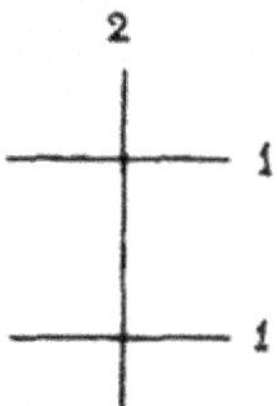

FIGURE 4.6

By Fact 1, we can change the projection $f^\iota : M^\iota \to S^2$ so that the resulting projection has 20 Lefschetz singular fibers $F_1, \cdots, F_{20}$ of type I. (We use the same notation f^ι to denote the new projection.) The associated monodromy representation is given by repeating the sequence (4.2) twice. The total monodromy is trivial, and this corresponds to the well known defining relation of the mapping class group $\mathcal{M}_2$:

$$(\zeta_1\zeta_2\zeta_3\zeta_4\zeta_5^2\zeta_4\zeta_3\zeta_2\zeta_1)^2 = 1.$$

In this sense the above relation 'produces' the Lefschetz fibration $f^\iota : M^\iota \to S^2$ of Exmple A, and uniquely up to isomorphism, by Theorem 2.4.

Proposition 4.1. *The total space M^ι is diffeomorphic to $\mathbb{C}P_2\#13\overline{\mathbb{C}P_2}$*

Proof. Let $\varphi : M^\iota \to \Sigma_2/\iota$ be the projection to the 2-shere $\Sigma_2/\iota(\cong S^2)$ induced by the projection to the second factor $S^2 \times \Sigma_2 \to \Sigma_2$. A general fiber of φ is a 2-sphere, and there are 6 singular fibers over $Fix(\iota)$. By a similar argument to the one done to obtain Figure 4.4, we see that each singular fiber consists of three 2-spheres as shown in Figure 4.6.

Claim. *Supose a singular fiber in a locally holomorphic fibration of genus $g(\geq 0)$ consists of s irreducuble components $\Theta_1, \Theta_2, \cdots, \Theta_s$ with the multiplicities $m_1, m_2, \cdots, m_s$, respectively. Then the self-intersection number $\Theta_i \cdot \Theta_i$ is given by the following formula:*

$$\Theta_i \cdot \Theta_i = -\frac{1}{m_i}\sum_{j\neq i} m_j\Theta_j \cdot \Theta_i.$$

This is well known. (Cf.[30].)

By Claim, the central sphere in Figure 4.6 has the self-intersection number -1. Thus it can be blown down. After that, we have two spheres, each having multiplicity 1, which cut each other transeversely in a point. By Claim again, either sphere has self-intersection number -1 and can be blown down. Proceeing in this way, we obtain a sphere bundle over the sphere,$S^2\tilde{\times}S^2$, after blowing down M^ι 12 times. Thus

$$M^\iota \cong S^2\tilde{\times}S^2\#12\overline{\mathbb{C}P_2} \cong \mathbb{C}P_2\#13\overline{\mathbb{C}P_2}.$$

This completes the proof of Proposition 4.1.   □

Example B. (Lefschetz fibration of $S^2 \times T^2\#4\overline{\mathbb{C}P_2}$)

Let $\omega : \Sigma_2 \to \Sigma_2$ be the involution shown in Figure 4.7, $\tau : S^2 \to S^2$ the 180° rotation as before.

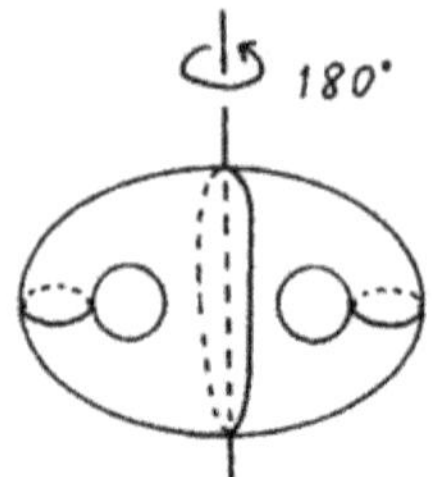

FIGURE 4.7. Involution ω.

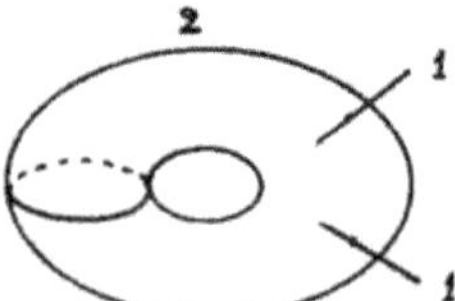

FIGURE 4.8. Singular fiber F^ω.

The quotient space $S^2 \times \Sigma_2/\tau \times \omega$ has 4 singular points. By blowing up these singularities, we obtain a compact complex surface M^ω and a holomorphic map $f^\omega : M^\omega \to S^2$ induced by the first projection $S^2 \times \Sigma_2 \to S^2$.

The fibration $f^\omega : M^\omega \to S^2$ has two singular fibers over the north and the south poles of S^2. They are topologically equivalent, and the monodromy around each of them is the involution ω. We denote either singular fiber by F^ω.

Structure of F^ω. The quotient space Σ_2/ω is homeomorphic to T^2 with two cusps. Blowing them up, we obtain the configuration of of F^ω as shown in Figure 4.8.

By a computer calculation we can prove the following fact (see [19]):

Fact 2. *F^ω splits into 4 Lefschetz singular fibers (3 of them have type I, and the rest has type II). We may assume that the new critical values are in a small disk D centered at the critical value of F^ω. Taking loops as in Figure 4.1 and changing them by elementary transformations and/or their inverses, we can arrange so that the monodromy representation $\rho : \pi_1(D - \{b_1, \cdots, b_4\}, b_0) \to \mathcal{M}_2$ is given by the quadruple of negative Dehn twits about the simple closed curves indicated below:*

(4.3)

By Fact 2, the projection $f^\omega : M^\omega \to S^2$ is changed so that the resulting projection has type $6I + 2II$.

The product of the negative Dehn twists about the simple closed curves (4.3) is ω, and juxtaposing two copies of the sequence (4.3) gives the global monodromy of the

Lefschetz fibration $f^\omega : M^\omega \to S^2$.

Proposition 4.2. *The total space M^ω is diffeomorphic to $S^2 \times T^2 \# 4\overline{CP_2}$.*

Proof. Let $\psi : M^\omega \to \Sigma_2/\omega (\cong T^2)$ be the projection induced by the projection to the second factor $S^2 \times \Sigma_2 \to \Sigma_2$. A general fiber of ψ is a 2-sphere, and there are two singular fibers over $Fix(\omega)$. Their shapes are the same as Figure 4.6. Blowing down the (-1)-spheres 4 times, we obtain a sphere bundle over T^2, $S^2 \tilde\times T^2$. Therefore,

$$M^\omega \cong S^2 \tilde\times T^2 \# 4\overline{CP_2} \cong S^2 \times T^2 \# 4\overline{CP_2}.$$

This completes the proof. $\square$

Example C. (Lefschetz fibration of $K3\#2\overline{CP_2}$)

It is known that a $K3$ surface is obtained by taking a double cover of CP_2 branched along a smooth complex curve C_6 of degree 6, (Cf. [6, p.15]). Let p_0 be a point in $CP_2 - C_6$. The Lefschetz pencil of complex lines through p_0 gives a fiber bundle $CP_2\#\overline{CP_2} \to CP_1$ when CP_2 is blown up at p_0. A fiber of this bundle is the 2-sphere, and generically it intersects C_6 transversely in 6 points. Let M be the 2-fold branched cover of $CP_2\#\overline{CP_2}$ branched along C_6. Obviously, M is diffeomorphic to $K3\#2\overline{CP_2}$. M inherits a Lefschetz fibration from $CP_2\#\overline{CP_2}$ whose general fiber is a double cover of a 2-sphere branched at 6 points that is a Riemann surface Σ_2 of genus 2.

To see the monodromy representation of $M \to CP_1$, we make some numerical calculations. (We used *Mathematica*.) First consider an affine complex curve C_6' in $\mathbb{C}^2$ defined by the following equation

$$y^6 - 6y - 10 + x = 0.$$

A complex line L_a defined by $x = a$(const.) is tangent to C_6' if and only if $a = a_i$, where

$$a_i = 10 + 5\exp(2\pi\sqrt{-1}(i+2)/5), \quad i = 1, \cdots, 5.$$

Take 0 as a base point and consider the five loops $l_1, l_2, \cdots, l_5$ on the complex x-axis (See Figure 4.9). The line L_0 over the base point 0 intersects C_6' in 6 points $y_1, y_2, \cdots, y_6$, where

$$y_1 = -1.1909\cdots$$
$$y_2 = -0.8938\cdots - \sqrt{-1} \times 1.0981\cdots$$
$$y_3 = 0.6663\cdots - \sqrt{-1} \times 1.4493\cdots$$
$$y_4 = 1.6458\cdots$$
$$y_5 = 0.6663\cdots + \sqrt{-1} \times 1.4493\cdots$$
$$y_6 = -0.8938\cdots + \sqrt{-1} \times 1.0981\cdots.$$

If the point a starts from 0, moves along the loop l_k, and comes back, then the points $\{y_1, \cdots, y_6\}$ are isotoped as indicated by Figure 4.10 according to the movement of the intersection points $L_a \cap C_6'$.

YUKIO MATSUMOTO

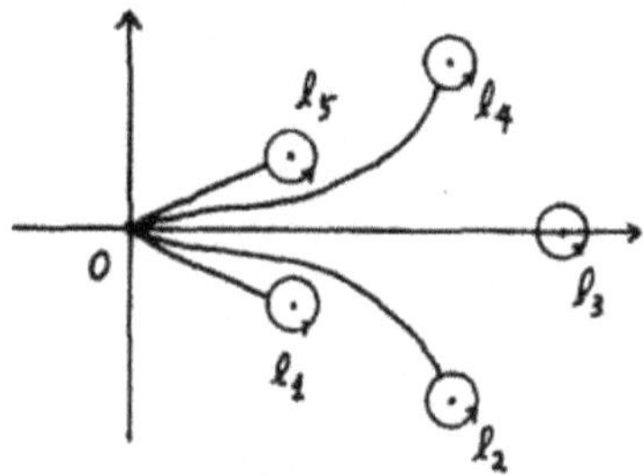

FIGURE 4.9. Loops $l_1, \cdots, l_5$

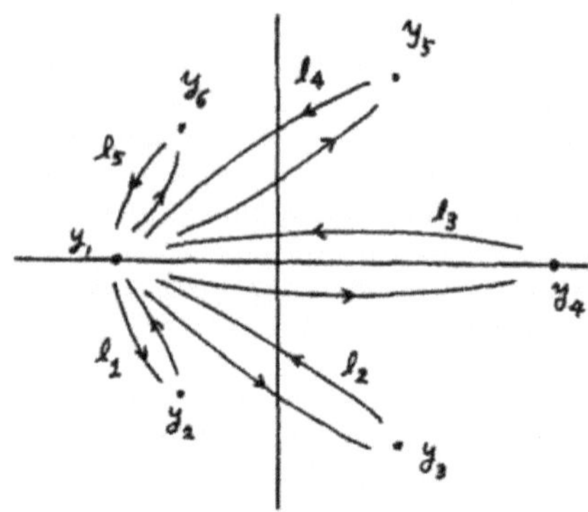

FIGURE 4.10

Let σ_k be the isotopy in L_0 corresponding to the elementary braid which permutes y_k and y_{k+1}, $k = 1, \cdots, 5$. Then the above movements of $\{y_1, \cdots, y_6\}$ (monodromy) are expressed as follows:

$$l_1 \mapsto \sigma_1$$
$$l_2 \mapsto \sigma_1^{-1}\sigma_2\sigma_1$$
$$l_3 \mapsto \sigma_1^{-1}\sigma_2^{-1}\sigma_3\sigma_2\sigma_1$$
$$l_4 \mapsto \sigma_1^{-1}\sigma_2^{-1}\sigma_3^{-1}\sigma_4\sigma_3\sigma_2\sigma_1$$
$$l_5 \mapsto \sigma_1^{-1}\sigma_2^{-1}\sigma_3^{-1}\sigma_4^{-1}\sigma_5\sigma_4\sigma_3\sigma_2\sigma_1.$$

Therfore, the loops $l_1' = l_1 l_2 l_3 l_4 l_5 l_4^{-1} l_3^{-1} l_2^{-1} l_1^{-1}$, $l_2' = l_1 l_2 l_3 l_4 l_3^{-1} l_2^{-1} l_1^{-1}$, $l_3' = l_1 l_2 l_3 l_2^{-1} l_1^{-1}$, $l_4' = l_1 l_2 l_1^{-1}$, and $l_5' = l_1$ correspond to σ_5, σ_4, σ_3, σ_2, and σ_1, respectively.

Let L_∞ be a line in CP_2 through p_0. We identify $CP_2 - L_\infty$ with (z, y)-plane. Let $\pi : \mathbb{C}^2 \to \mathbb{C}^2$ be a 6-fold branched cover from (z, y)-plane to (x, y)-plane defined by $(z, y) \mapsto (z^6, y)$. The inverse image $\pi^{-1}(C_6')$ can be regarded as the affine part of C_6. In the z-axis, we have 6 sets of 5 loops $\{l_{1j}', \cdots, l_{5j}'\}_{j=1,\cdots,6}$, which are the inverse image $\pi^{-1}\{l_1', \cdots, l_5'\}$.

The projection $(z, y) \mapsto z$ gives rise to the projection $CP_2 \# \overline{CP_2} \to CP_1$. The latter lifts to the projection $M \to CP_1$, whose monodromy along l_{kj}' is the Dehn twist ζ_{6-k} of Σ_2, because ζ_{6-k} is the lift of σ_{6-k} to the 2-fold branched cover($= \Sigma_2$) of CP_1 branched at $\{y_1, \cdots, y_6\}$.

Thus we have proved that the monodromy representation of the Lefschetz fibration $M \to CP_1$ is given by repeating the sequence

$$(\zeta_5, \zeta_4, \zeta_3, \zeta_2, \zeta_1)$$

6 times, or if we prefer, the sequence

$$(\zeta_1, \zeta_2, \zeta_3, \zeta_4, \zeta_5)$$

6 times. (Note that ζ_{6-k} is conjugate to ζ_k by the involution ω of Figure 4.7.)

The total monodromy $(\zeta_1\zeta_2\zeta_3\zeta_4\zeta_5)^6$ is trivial. This is of course one of the defining relations of $\mathcal{M}_2$, (cf.[3, p.184]).

This completes the construction of Example C.

Proposition 4.3. *In $\mathcal{M}_2$, the following relation holds:*

$$(\zeta_1\zeta_2\zeta_3\zeta_4)^5 = \iota,$$

where ι is the hyperelliptic involution (Figure 4.3).

Proof. We set $a = \zeta_1\zeta_2\zeta_3\zeta_4\zeta_5$, and $b = \zeta_1^2\zeta_2\zeta_3\zeta_4\zeta_5$. Then by the defining relation, $a^6 = 1$. Obviously, $\zeta_1 = ba^{-1} = ba^5$. Inspecting the right action of a on Σ_2, we can verify $\zeta_i = a^{i-1}\zeta_1 a^{1-i}$, $i = 2, 3, 4, 5$. Substitute ba^5 for ζ_1 to obtain

$$\zeta_2 = aba^4, \quad \zeta_3 = a^2ba^3, \quad \zeta_4 = a^3ba^2, \quad \zeta_5 = a^4ba.$$

Rewrite the known relaton $\zeta_1\zeta_2\zeta_3\zeta_4\zeta_5^2\zeta_4\zeta_3\zeta_2\zeta_1 = \iota$ (cf. [3, p.188]) using the above expresson of ζ_i in terms of a and b, then we obtain $a^{-1}(ba^4)^5a = \iota$. Since ι belongs to the center of $\mathcal{M}_2$, we have $(ba^4)^5 = \iota$. We will further rewrite ba^4 as follows:

$$ba^4 = b \cdot a^4 ba \cdot a^{-1}b^{-1} = b\zeta_5 a^{-1}b^{-1}$$
$$= b(\zeta_5 \cdot \zeta_5^{-1}\zeta_4^{-1}\zeta_3^{-1}\zeta_2^{-1}\zeta_1^{-1})b^{-1} = b(\zeta_4^{-1}\zeta_3^{-1}\zeta_2^{-1}\zeta_1^{-1})b^{-1}$$

Then we have $b(\zeta_1\zeta_2\zeta_3\zeta_4)^{-5}b^{-1} = \iota$. Using the fact that ι belongs to the center and that $\iota^2 = 1$, we finally obtain $(\zeta_1\zeta_2\zeta_3\zeta_4)^5 = \iota$ as asserted. $\square$

Example D. (Two non-isomorphic Lefschetz fibrations whose total spaces are homeomorphic.)

In $\mathcal{M}_2$ we have the following relations

$$(\zeta_1\zeta_2\zeta_3\zeta_4\zeta_5^2\zeta_4\zeta_3\zeta_2\zeta_1)^4 = 1$$
$$(\zeta_1\zeta_2\zeta_3\zeta_4)^{10} = 1$$

The first relation is an immediate consequence of $(\zeta_1\zeta_2\zeta_3\zeta_4\zeta_5^2\zeta_4\zeta_3\zeta_2\zeta_1)^2 = 1$, and the second one follows from Proposition 4.3. These two relations produce Lefschetz fibrations of genus 2 $M_1 \to S^2$, $M_2 \to S^2$, respectively. Both fibrations have type $40I$. If a Lefschetz singular fiber has monodromy ζ_i around it, its vanishing cycle is C_i (Figure 4.5). Thus in M_1 and M_2, the circles $C_1, \cdots, C_4$ (at least) on the general fiber over the base point p_0 are null homotopic. This implies that M_1 and M_2 are simply connected. They have the same signature -24 (by Theorem 3.3 and Proposition 3.6) and the Euler number 36 (since a Lefschetz singular fiber has Euler contribution 1). The intersection forms of M_1 and M_2 are odd by Rochlin's theorem.

Therefore, M_1 and M_2 are homotopy equivalent smooth 4-manifolds [23], and by Freedman's theorem [9], homeomorphic to $5\mathbb{C}P_2\#29\overline{\mathbb{C}P_2}$. However, they are not isomorphic as two Lefschetz fibrations, because the monodromy representation ρ_1 of $M_1 \to S^2$ is onto (its image is generated by $\{\zeta_1, \zeta_2, \zeta_3, \zeta_4, \zeta_5\}$) while ρ_2 of $M_2 \to S^2$ is not (its image is a subgroup generated by $\{\zeta_1, \zeta_2, \zeta_3, \zeta_4\}$, a proper subgroup of $\mathcal{M}_2$).

Question. Is M_1 diffeomorphic to M_2 ?

Remarks 1) Y.Kametani pointed out that M_1 is diffeomorphic to an elliptic surface. Thus M_1 is not diffeomorphic to $5\mathbb{C}P_2\#29\overline{\mathbb{C}P_2}$ by Donaldson's theorem [6].
2) Considering 'fiber-connected sums' we obtain Lefschetz fibrations of genus 2, $nM_1 \to S^2$ and $nM_2 \to S^2$, both of wich have type $40nI$ ($n = 1, 2, \cdots$). If n is odd, the total spaces nM_1 and nM_2 are homeomorpic, but they are not isomorpic as Lefschetz fibrations.

5. Persson's Theorem

In this final section we will give a monodromy theoretic proof to a result of Persson [27, p.135].

Theorem 5.1. *(Persson) Suppose a Lefschetz fibration $M \to B$ of genus 2 has type $aI + bII$, and suppose the base space B is closed. Then $a + 2b \equiv 0 \ (mod.10)$.*

Proof. The abelianization $\mathcal{M}_2^{ab}$ of $\mathcal{M}_2$ is isomorphic to $\mathbb{Z}/10$, (see [3, p.184]). Under the identification $\mathcal{M}_2^{ab} = \mathbb{Z}/10$, ζ_i corresponds to 1. Since $(\zeta_1\zeta_2\zeta_1)^4$ is a negative Dehn twist about an essential separating simple closed curve on Σ_2, the monodromy about a singular fiber of type II corresponds to 2 in $\mathcal{M}_2^{ab}$. Thus the total monodromy of $M \to B$ corresponds to $a + 2b$.

Since the product $\gamma_1\gamma_2\cdots\gamma_n$ of the loops in Figure 4.1 is null-homologous on $B -$ {critical values}, the abelianized total monodromy is 0. Thus we have $a + 2b \equiv 0 \pmod{10}$. $\square$

(Added on July 18,1996.) Recently, T.Fuller [31] answered the question at the end of section 4, showing that M_1, M_2 and $5\mathbb{C}P_2\#29\overline{\mathbb{C}P_2}$ are mutually non-diffeomorphic.

References

1. M.F.Atiyah, *The signature of fiber-bundles*, Global Analysis ed. by D.C.Spencer and S.Iyanaga, Univ. Tokyo Press, and Princeton Univ. Press, (1969) 73 –84.
2. M.F.Atiyah and I.M.Singer, *The index of elliptic operaters: III*, Ann. Math., **87** (1968), 546–604.
3. J.Birman, *Braids, links, and mapping class groups*, Princeton Univ. Press, Princeton, N.J. USA. (1974).
4. S.K.Donaldson, *An application of gauge theory to four dimensional topology*, J. Diff. Geom **18** (1983), 279–315.
5. S.K.Donaldson, *La topologie différentielle des surfaces complexes*, C.R.Acad.Sci **301** (1985), 317–320.
6. S.K.Donaldson and P.B.Kronheimer, *The geometry of four-manifolds*, Clarendon Press, Oxford (1990).
7. C.J.Earle and J.Eells *A fiber bundle description of Teichmüller theory*, J. Diff. Geom. **3** (1969), 19–43.
8. D.B.A.Epstein, *Curves on 2-manifolds and isotopies*, Acta Math. **115** (1966), 83-107.
9. M.H.Freedman, *The topology of four-dimensional manifolds*, J. Diff. Geom. **17** (1982), 357–454.
10. V.Guilman and A.Pollack, *Differential Topology*, Prentice-Hall, (1974).
11. J.Harer, *Pencils of curves in 4-manifolds* Thesis, UCB. 1979.
12. E.Horikawa, *On algebraic surfaces with pencils of curves of genus 2*, Complex analysis and algebraic geometry, Cambridge Univ. Press (1977), 79–90.
13. K.Kodaira, *A certain type of irregular algebraic surfaces*, J.Analyse Math. **19** (1967), 207–215.
14. K.Kodaira, *On homotopy K3 surfaces*, Essays on topology and related topics (Mémoires dédiés à George de Rham), ed. by A.Haefliger and R.Narasimhan, Springer Verlag (1970), 58–69.
15. S.Lefschetz, *L'analysus situs et la géométrie algébrique*, Gauthier-Villar (1930).
16. Y.Matsumoto, *On 4-manidolds fibered by tori, II*, Proc.Japan Acad. **59** (1983), 100–103.
17. Y.Matsumoto, *Diffeomorphism types of elliptic surfaces*, Topology, **25** (1986), 549–563.
18. Y.Matsumoto, *Topology of torus fibrations*, Sugaku Expositins **2** (1989), 55–73.
19. Y.Matsumoto, *Splitting of certain singular fibers of genus 2*, in preparation.
20. Y.Matsumoto and J.M.Montesinos-Amilibia, *Pseudo-periodic homeomorphisms and degeneration of Riemann surfaces*, Bull.A.M.S. **30** (1994), 70-75.
21. W.Meyer, *Die Signatur von Flächenbündeln*, Math.Ann., **201** (1973), 239-264.
22. J.Milnor, *Lectures on the h-cobordism theorem*, Princeton Univ. Press, (1965).
23. J.Milnor and D.Husemoller, *Symmetric bilinear forms*, Springer Verlag, (1973).
24. B.Moishezon, *Complex surfaces and connected sums of complex projective planes*, Lecture Lotes in Math. 603 Springer Verlag (1977).
25. J.Nielsen. *Surface transformation classes of algebraically finite type*, Mat.-Fys. Medd. Danske Vid. Selsk. **21** (1944), 3–89: and Collected Papers vol **2**, Birkhäuser (1986).
26. U. Persson, *Chern invariants of surfaces of general type*, Compositio Mathematica **43** (1981), 3–58.

27. U.Persson, *Genus two fibrations revisited*, Lecture Notes Math. 1507, Springer Verlag (1992), 133–144.
28. M.Reid, *Problems on pencils of small genus*, preprint.
29. G.Xiao, *Surfaces fibrées en courbes de genre deux*, Lecture Notes in Math. 1137 Springer Verlag (1985).
30. Y.Matsumoto, *Good torus fibrations*, Contemporary Math. **35** (1984), 375–397.
31. T.Fuller, *Diffeomorphism types of genus 2 Lefschetz fibrations*, preprint, University of Texas, Austin, 1996.

GRADUATE SCHOOL OF MATHEMATICAL SCIENCES, UNIVERSITY OF TOKYO, 3-8-1, KOMABA, MEGURO-KU, TOKYO 153, JAPAN

E-mail address: yukiomat@ms.u-tokyo.ac.jp

Proceedings of
THE 37TH TANIGUCHI SYMPOSIUM ON
TOPOLOGY AND TEICHMÜLLER SPACES
held in Finland, July 1995
ed. by Sadayoshi KOJIMA *et al.*
©1996 World Scientific Publishing Co.
pp. 149–158

A GEOMETRIC APPROACH TO THE COMPLEX OF CURVES ON A SURFACE

YAIR N. MINSKY

(Received March 11, 1996)

1. The complex of curves

Geometry has played an important role in the study of Teichmüller space and its group of symmetries, the Mapping Class Group of a surface. In [13], Harvey introduced a simplicial complex associated to a surface called the *complex of curves*, which was intended to encode the asymptotic geometry of Teichmüller space in analogy with similar constructions for symmetric spaces. This complex was then studied by Harer [11, 12] from a cohomological point of view and by Ivanov [16, 15] with applications to the structure of the mapping class group.

Harvey defined his complex as follows. Let S be a finite genus surface with finitely many punctures, and let "essential simple curve" on S denote a non-trivial homotopy class of simple curves which are also non-peripheral, i.e. cannot be deformed to a puncture. The essential simple curves are the vertices of the complex, and k-simplices are sets of $k + 1$ curves that can be realized disjointly. It is immediate from the definition that this complex is finite dimensional, but also that it is usually locally infinite – there are usually infinitely many homotopically distinct curves in the complement of a simple curve.

In joint work with Howard Masur, we have been studying Harvey's complex from the point of view of hyperbolic metric spaces in the sense of Gromov and Cannon (see below). Part of our motivation is an attempt to explain the presence and absence of negative-curvature properties of the Teichmüller space, whereas further long-term motivation includes an improved understanding of the mapping class group, and possible applications to Kleinian groups.

For a few simple cases (such as the torus) we will see in Section 3 that this complex is 0-dimensional, and in those cases we will make an adjustment to the definition that yields a 1-complex. In any case let $K(S)$ denote the 1-skeleton of the complex.

To state our main theorem, we need Gromov's notion of δ-*hyperbolicity*, or "negative curvature" (in the large) in Cannon's terminology, which may be given in terms of the following "thin triangle" property: Let X be a geodesic metric space. For any three points $x, y, z \in X$ let $[xy], [xz], [yz]$ denote geodesics between pairs (not necessarily unique). If $[xz]$ is always contained in a δ-neighborhood of $[xy] \cup [yz]$ for some fixed $\delta \geq 0$, we say that X is δ-hyperbolic. This property encodes many of the most characteristic large-scale properties of spaces such as hyperbolic space $\mathbf{H}^n$, or infinite simplicial trees (see [6, 10, 9]).

Theorem (Masur-Minsky [19]). *The 1-complex $K(S)$ endowed with the metric giving length 1 to edges is δ-hyperbolic for some $\delta(S) \geq 0$*

Remark. 1. Note that, from the point of view of this theorem, it does not matter if we include the entire complex or just its 1-skeleton, since if we endow each simplex with a standard (say Euclidean) metric then the full complex is quasi-isometric to its 1-skeleton. The property of δ-hyperbolicity is invariant under quasi-isometry, although the specific value of δ is not. 2. The definition of δ-hyperbolicity often includes the additional assumption of *propriety*, i.e. that balls of finite radius are compact. This is clearly false for $K(S)$, and one might emphasize the distinction by saying that $K(S)$ is an improper hyperbolic space.

This article is an exposition of the ideas and motivations behind this theorem. We will discuss a sketch of the proof, and note some remaining questions and future plans.

2. Geometry of Teichmüller space

A persistent, but incomplete, analogy has long been recognized between the geometry of the Teichmüller space $\mathcal{T}(S)$ and that of δ-hyperbolic spaces. For example, $\mathcal{T}(S)$ admits a boundary at infinity similar to that of hyperbolic space (See e.g. [8, 23]), and the Teichmüller geodesic flow on its quotient, the moduli space, is ergodic (see [22]). The analogy was exploited, for example, in Bers [1], Kerckhoff [18], and Wolpert [29]).

On the other hand, Masur [20] first showed that the Teichmüller metric cannot be negatively curved in a local sense, and more recently Masur-Wolf [24] showed that it is not δ-hyperbolic. Although the Weil-Petersson metric on $\mathcal{T}(S)$ has negative sectional curvatures, they are not bounded away from zero [21, 30].

The connection between the complex of curves and the geometry of $\mathcal{T}(S)$ is based on the collar lemma (see e.g. [17, 5]) which implies in particular that there is a universal $\epsilon_0 > 0$ such that any collection of simple geodesics of length less than ϵ_0 on a hyperbolic surface are pairwise disjoint. In other words, vertices in $K(S)$ are on a common simplex precisely when the curves they represent can be made simultaneously short for some point in $\mathcal{T}(S)$. If we define the "thin region" of an essential curve α as $T(\alpha) = \{\sigma \in \mathcal{T}(S) : \ell_\sigma(\alpha) < \epsilon_0\}$ (here ℓ_σ denotes hyperbolic length in the metric σ) then $K(\alpha)$ is therefore the *nerve* of the family of regions $\{T(\alpha)\}$. (For a reinterpretation of this in the case of tori and spheres, see section 3).

It has often been noted that these regions where some curve is short are precisely the places where the geometry of $\mathcal{T}(S)$ is non hyperbolic. This is related to the fact that the presence of a short curve α decomposes the surface into pieces that are geometrically nearly independent, thus giving the region $T(\alpha)$ the approximate geometry of a product (see [21, 30, 26]). A product of infinite-diameter spaces cannot be δ-hyperbolic. This is related to the fact that the mapping class group of a higher-genus surface is known not to be δ-hyperbolic because it contains high rank abelian subgroups, generated by elements whose supports lie on disjoint subsurfaces.

On the other hand, outside the thin regions there is evidence of hyperbolic behavior. For example, we show in [27] that a Teichmüller geodesic that stays away from $\cup_\alpha T(\alpha)$ admits a closest-point retraction that satisfies a contraction property (see Section 4.2) which characterizes such projections in a hyperbolic space.

In this context, one interpretation of our main theorem is that $T(S)$ is *relatively hyperbolic* with respect to the family of regions $T(\alpha)$. This terminology is due to Farb [7], who showed for example that a Kleinian group with finite-volume but non-compact quotient is relatively hyperbolic with respect to the abelian subgroups corresponding to cusps. In other words, all the non-hyperbolicity in $T(S)$ is "restricted to" the regions $T(\alpha)$, but the way in which these regions are arranged in $T(S)$ is itself hyperbolic.

3. The low dimensional cases

In a small number of cases, the complex of curves is either trivial or "classically" known, and our Theorem holds vacuously or is fairly easy to prove. If S is a sphere with between 0 and 3 holes, then $K(S)$ is empty, and hyperbolicity for the empty set is a matter of terminology.

If S is a torus, once-punctured torus or 4-times punctured sphere, then any two essential simple curves must intersect, so that $K(S)$ has no edges under the general definition. However, with a small adjustment in the definition K becomes a sensible and familiar 1-complex: Let (α, β) be an edge whenever α and β have the lowest possible intersection number. For the tori this is 1, and for the 4-punctured sphere this is 2.

For the rest of this section assume that S is a torus (with 0 or 1 holes). The sphere case is similar. Choosing an identification of $H_1(S)$ with $\mathbf{Z}^2$, the essential simple curves on S are well known to be in one-to-one correspondence with the slopes of elements in $\mathbf{Z}^2$, namely the rational numbers p/q including $1/0 = \infty$. Thus the 0-skeleton K_0 is identified with $\mathbf{Q} \cup \infty$ in the circle $S^1 = \mathbf{R} \cup \infty$, and the identification is unique up to the action of $\mathrm{PSL}_2(\mathbf{Z}) = Isom(\mathbf{Z}^2)$ on the circle.

The full 1-complex K may then be embedded as an ideal triangulation of the unit disk bounded by S^1 ("ideal" means that the vertices are on the boundary) by realizing each edge as a hyperbolic geodesic in the Poincaré model of $\mathbf{H}^2$. To see that this is an embedding, we note first that the unsigned intersection number of curves represented by p/q and r/s (written in lowest terms) is simply the absolute value of the determinant $|ps - qr|$. One may easily check this for $p/q = \infty$ and apply $\mathrm{PSL}_2(\mathbf{Z})$ to get the general case. (For the 4-punctured sphere, twice the determinant is the intersection number.) Thus we note that there is an edge from ∞ to 0, and that no other edge crosses this one – for if $|ps - qr| = 1$ the signs of p/q and r/s must agree. Applying $\mathrm{PSL}_2(\mathbf{Z})$ we see that no two edges intersect, so the complex embeds. We further check that ∞ is connected by edges to exactly the integers $n/1$, and that

152   YAIR N. MINSKY

$n/1, m/1$ are connected exactly when $n = m \pm 1$. We therefore have a triangulation
around ∞, and hence around every vertex – see figure 1. This famous picture, familiar
in the study of modular forms, is also known as the Hatcher-Thurston complex for
the torus; see [14, 4]. Note also that up to combinatorial equivalence there is only
one ideal triangulation of the disk.

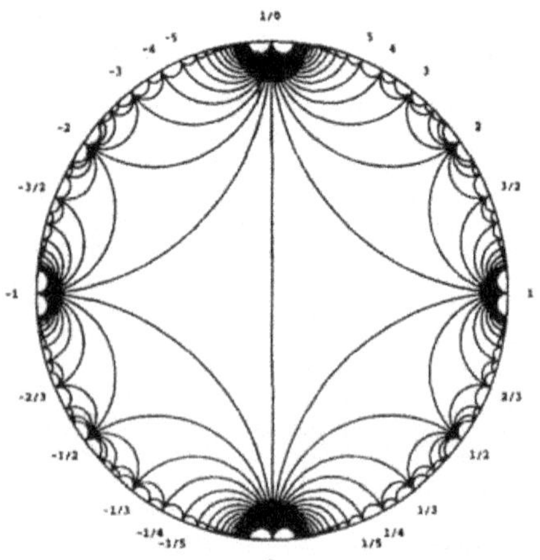

FIGURE 1. The complex of curves on the torus, embedded in the unit disk

Let us now prove K is hyperbolic. We begin by giving a description of the geodesics
in K.

Because of its embedding in the disk, K has the property that any edge e separates
K into two components. Given two vertices x, y let $E(x, y)$ denote the set of edges
that separate x from y. Then not just geodesics but every continuous path in K from
x to y must pass through some vertex (or both) of every edge $e \in E(x, y)$.

If $E(x, y)$ is empty then x and y are vertices of the same triangle and the geodesic
between them is an edge. From now on assume $E(x, y) \neq \emptyset$. We will check that any
geodesic from x to y is in the $\frac{1}{2}$-neighborhood of $E(x, y)$; in fact we will show a bit
more.

It is easy to see that $E(x, y)$ admits an order where $e < f$ if e separates x from the
interior of f. Successive elements in this order share a vertex, called a *pivot*, and the
set of pivots is thus ordered as well. If p is a pivot let $w(p)$ (its "width") be the number
of edges in $E(x, y)$ incident to p. (Remark: the sequence of numbers $\{w(p)\}$ is exactly
the "left-right" sequence that appears in the classification of homeomorphisms of the
torus, and is closely related to continued-fraction expansions for the slopes of curves.)

For $e \in E(x, y)$ consider the ways in which a geodesic $[xy]$ may cross e. If e is
incident to no pivots then $E(x, y) = \{e\}$ and x and y are opposite vertices of a
quadrilateral consisting of two triangles in K. Thus $d(x, y) = 2$ and there are two
geodesics, going around the quadrilateral in the two possible ways.

If e is incident to a pivot p, let $e_1 < \cdots < e_{w(p)}$ be the sequence of edges incident
to p (including e), and let v_j be the vertex of e_j different from p.

A shortest path from x to y must enter the configuration $\cup e_i$ at one vertex or the other of e_1 and exit through $e_{w(p)}$. (Note that x may be equidistant from p and v_1 or closer to v_1, but not closer to p, and similarly for y and $p, v_{w(p)}$.) If $w(p) \geq 4$ then the path must pass through p and not around through $v_2, \ldots, v_{w(p)-1}$, but if $w(p) = 2$ or 3 there may be (finitely many) choices. In any case a shortest path cannot pass through any vertices on the other side of an edge (v_j, v_{j+1}) from p, since any such digression would be longer than simply traversing the edge.

Thus we see that although geodesics may not be unique, they must pass through the block of edges around each pivot in one of a finite number of well-defined ways, and in particular they pass entirely through vertices of edges in $E(x, y)$ and hence must remain in a $\frac{1}{2}$-neighborhood of $E(x, y)$.

Now consider a triple of points x, y, z and the triangle $[xyz]$ formed by geodesics $[xy], [yz]$ and $[xz]$. We must show that this triangle satisfies the thin triangle property.

Let e be an edge of $E(x, z)$. If e does not separate x from y then either e separates y from z, or y is a vertex of e. Thus $E(x, z)$ is contained in the union of $E(x, y) \cup E(y, z)$ together with the edges incident to y. It follows that the vertices of any geodesic $[xz]$ are distance 1 from $[xy] \cup [yz]$ and thus that $[xz]$ itself is at most distance 3/2 from $[xy] \cup [yz]$. Thus K is 3/2-hyperbolic.

Let us also remark that the interpretation of $K(S)$ as the nerve of the family of regions $\{T(\alpha)\}$ in Section 2 is still valid, with appropriate definitions. The Poincaré disk in which we embedded $K(S)$ is in fact isometric to the Teichmüller space of the torus (or punctured torus, or 4-punctured sphere). It is not hard to show that there is a constant $\epsilon_1 > 0$ such that for any hyperbolic punctured torus, there are at most two curves of length ϵ_1 or less. If there are two, they must have intersection number 1. Thus $K(S)$ is the nerve of the family of regions $T(\alpha)$ defined using ϵ_1 in place of ϵ_0, and in fact two such regions $T(\alpha), T(\beta)$ meet in a point of tangency in the interior of the edge (α, β) in our embedding of $K(S)$. A similar statement can be made for the 4-punctured sphere and, using extremal length, for the unpunctured torus.

4. Sketch of the proof in the general case

For surfaces S not covered in the previous section, we return to Harvey's original definition for $K(S)$, and the proof becomes more complicated. There are two main components: (a) control of distance in K via nesting of train-tracks, and (b) a projection lemma using Teichmüller geometry.

Short distances in K are relatively easy to understand. That is, $d(\alpha, \beta) = 1$ exactly when α and β are disjoint. $d(\alpha, \beta) = 2$ when α and β intersect, but are both disjoint from an essential curve γ. Finally, $d(\alpha, \beta) > 2$ whenever α and β *bind* S; that is, the components of $S \setminus (\alpha^* \cup \beta^*)$ are disks and once-punctured disks (where α^* and β^* are geodesic representatives in some hyperbolic metric). However, distinguishing between distances of 3 or more requires a further idea.

4.1. Train-tracks and nested polyhedra

We briefly recall Thurston's notion of a *train-track* (see Penner-Harer [28] for a complete treatment). A train-track is a smoothly embedded 1-complex τ in S (considered up to isotopy) with edges, called *branches*, meeting tangently at the vertices, which are called *switches*. A simple closed curve can be *carried* on τ if it is homotopic to a non-degenerate smoothly immersed curve in τ, and in that case the degree of covering of each branch gives a positive measure on the branches which satisfies the *switch condition*: at each switch, total incoming measure equals total outgoing measure.

Further topological constraints are also imposed on τ to guarantee that curves carried on it are homotopically essential and determined uniquely by their measures, and that each branch is traversed by some closed curve (recurrence). One can further demand (birecurrence) that τ can be realized in a hyperbolic metric on S in such a way that all edges are long and nearly geodesic. A finite number of such train-tracks suffice to carry all the essential simple closed curves in S.

The set of real positive weights on τ satisfying the switch conditions is, modulo scaling, a convex polyhedron $P(\tau)$ in the projective space $P\mathbf{R}^B$ where B is the set of branches of τ. The rational points in $P(\tau)$ correspond to simple curves or unions of simple curves.

We can make the following observation, where a *maximal* train track is one all of whose complementary components are triangles or punctured monogons.

Lemma. *Let τ be maximal and let a vertex α of $K(S)$ determine a measure in the interior of $P(\tau)$. If $d(\alpha, \beta) = 1$ then β is carried in τ.*

In other words, a 1-neighborhood of $int(P(\tau))$ is in $P(\tau)$. Here "interior" means the set of measures positive on every branch. A sketch of the proof is this: Assume that τ is realized nearly geodesically and α and β are geodesics, in some hyperbolic metric on S. Since α traverses every branch of τ, β cannot intersect τ essentially since it is disjoint from α. Since τ is maximal any segment of β in the complement of τ can be pushed onto τ and we may conclude that β is carried on τ.

Now we recall that a train-track τ' may be *carried* on τ if it can be homotoped so that all its edges are immersed in τ. Then every curve carried on τ' is carried on τ, and $P(\tau') \subset P(\tau)$. In view of the Lemma, if we now have a sequence $\tau_1, \dots, \tau_n$ such that $P(\tau_{j+1}) \subset int(P(\tau_j))$, then if α is carried on τ_n and β is not carried on τ_1, we may conclude $d(\alpha, \beta) \geq n$.

In particular one may use this idea to show that $\mathrm{diam}(K) = \infty$, a fact which may not have been obvious before, in view of the local infiniteness of the complex.

A generalization of this observation to non-maximal train tracks allows us to go the other way – given two curves far apart we can find a sequence of train-tracks whose polyhedra are nested (in a slightly weaker way) and separate the two curves.

4.2. A projection lemma

In a δ-hyperbolic space X, geodesics have the following *contraction property*. Any geodesic L in X admits a projection $\pi_L : X \to L$ taking each point in X to a closest point on L, and π is contracting in this sense: $\mathrm{diam}(\pi(B_x(d(x, L))))$ is uniformly bounded for all $x \in X$, where $B_x(R)$ is an R-neighborhood of x. This is easy to see for example when X is a tree, or $\mathbf{H}^n$.

Conversely, one may prove that if X contains a family of paths that connect any two points in X and each admitting projections with this contraction property (with uniform bounds), then the paths are quasi-geodesics and X is δ-hyperbolic.

Our proof of hyperbolicity for K uses this approach, showing in particular that a family of paths in K induced by the family of Teichmüller geodesics in Teichmüller space has the contraction property.

For any point σ in $\mathcal{T}(S)$, let $\Phi(\sigma)$ denote the set of curves of minimal extremal lengths in σ. It is not hard to see that this is a finite set, of uniformly bounded diameter in $K(S)$. With this, we can convert any Teichmüller geodesic L to a path in K, simply by choosing some element of $\Phi(t)$ for each $t \in L$ (we do not require paths to be continuous, since hyperbolicity only requires precision within some bounded constant). By slight abuse of notation, we call this path $\Phi \circ L$.

To describe the projection from $K(S)$ to $\Phi \circ L$, we use the notion of *balance*. The Teichmüller geodesic L is determined by a holomorphic quadratic differential q on S, whose horizontal and vertical foliations are respectively stretched and shrunk as we move along L in one direction. Thus for every curve α and point on L there is a horizontal and vertical length with respect to these foliations. As long as α is not completely vertical or horizontal, there is a unique point $t_\alpha \in L$ where its horizontal and vertical lengths are equal, and we say α is balanced at t_α. Our "projection" map is then defined by taking α to a point in $\Phi(t_\alpha)$.

To prove that this projection has the contraction property, we need to prove something like this. Let α be a point in $\Phi \circ L$ and let β, γ be two other points in $K(S)$ which are far from α but relatively close to each other (that is, $d(\beta, \gamma)/d(\beta, \alpha)$ is bounded by a suitable constant), and suppose that β projects to α. We would like to know that the projection of γ is a bounded distance from α.

The basic idea is that, as we move along L, the transition of any curve μ from mostly vertical to mostly horizontal occurs in a bounded distance as measured in $K(S)$ (although not as measured in Teichmüller distance), once we reach the point where μ does not consist of nearly vertical segments filling up almost the entire surface. As soon as the nearly vertical part of μ misses some definite horizontal arc in S (as measured in the quadratic differential metric), we show that a subsurface filled by the vertical parts begins to steadily shrink, making a bounded number of topological reductions until it is a disk, and the boundary components of the resulting sequence of nested surfaces yield a bounded path in $K(S)$.

Using this argument, we show that if β is balanced and γ is far from balanced (say more vertical than horizontal), we can move a bounded distance to a point where β is almost completely horizontal and γ is still almost completely vertical, and both fill up the whole surface.

At this point we use the train-track machinery to reach a contradiction. The distance of β from α implie the existence of a long sequence of nested train-track polyhedra which contain β, and the bound on $d(\beta, \gamma)$ then traps γ deep inside a long subsequence of these polyhedra (the actual proof is complicated by the fact that the train-tracks may not be maximal). It follows that there is a train-track which simultaneously carries both β and γ, but which intersects any short curve on the surface many times. In particular the intersection number of β and γ with each other is considerably smaller than their intersection numbers with any short curve. We then show that this contradicts the fact that one of them is nearly horizontal and the other nearly vertical. We conclude that in fact β and γ are balanced a bounded distance apart in $\Phi \circ L$.

With the contraction property thus established, we conclude simultaneously that $K(S)$ is hyperbolic, and that the class of paths determined by Teichmüller geodesics are quasi-geodesics in $K(S)$. We note that this fact is also easy to verify in the simpler cases discussed in Section 3.

5. Further comments and questions

δ-hyperbolicity is useful often because of the control one obtains on local geometry using *asymptotic* data. For example, two geodesic segments whose endpoints are some fixed distance apart remain a bounded distance apart along their whole lengths (we call them "fellow travelers"). Using this idea, there is a canonical *boundary at infinity* ∂X for a hyperbolic space X, consisting of equivalence classes of infinite geodesic rays, so that two geodesics whose endpoints at infinity are the same are fellow travelers.

One is thus led to ask, for example, what is the boundary at infinity for $K(S)$. In the case of the torus it is not hard to see that the boundary may be identified with the irrational numbers. In general, one can show that a sequence of curves α_n whose distance from a fixed base point is going to infinity must (up to subsequence) converge to a maximal geodesic lamination, and that if two sequences converge to distinct laminations then they determine different points at infinity. However when two sequences converge to the same geodesic lamination one has to consider the possibility that they converge to it in two very different ways -- in particular, it may be that the lamination supports two distinct projective classes of transverse measures, and that as measured laminations the sequences have different limits. Thus the question remains whether the boundary can be identified with the space of maximal geodesic laminations, or maximal geodesic laminations with projective measure classes, or something else.

For a proper δ-hyperbolic space X, the union $X \cup \partial X$ in a natural topology is compact. The same is not true in our case, because of the lack of local finiteness. This in fact indicates a basic difficulty in applying hyperbolicity to study the complex of curves: namely, what good is it to know that two geodesics are a bounded distance apart, if a ball of bounded distance still contains an infinite number of vertices? Note that this question is nicely resolved in the case of the complex for the torus, where we had an explicit description for geodesics in spite of the fact that the link of every vertex was infinite. In particular we note that for every pivot on the path between two vertices, if the width w of the pivot was sufficiently large (at least 4) then the geodesic was forced to travel through, and not around, the pivot. This idea generalizes to higher genus. That is, given one geodesic passing through some vertex x, if its entry and exit points are sufficiently far apart in the link of x (which we note is itself a complex of curves for a subsurface) then we can show that any fellow travelling geodesic must pass through x as well. This opens the way for an inductive study of $K(S)$, since as soon as two geodesics are known to pass through the same vertex, they both induce paths in the complex corresponding to the link of that vertex. In an upcoming paper we hope to develop this idea, and apply it to a study of the mapping class group.

Harvey's complex arises naturally in the study of Kleinian groups, where one of the intriguing open problems is Thurston's Ending Lamination Conjecture, which proposes that a hyperbolic 3-manifold is uniquely determined by its topological type and a list of *end invariants* encoding the asymptotic geometry of its ends. For example, if N is a geometrically infinite hyperbolic 3-manifold homotopy-equivalent to a surface S then the ending laminations of N describe the limits of sequences of closed curves on S whose geodesic representatives in N leave every compact set in N (see Bonahon [2]).

A solution to Thurston's conjecture would require some way to convert the asymptotic information in the ending laminations to more precise geometric information. For example, one may ask to understand the structure of the set K_L of vertices of $K(S)$ whose geodesic lengths in N are at most L, for some fixed $L > 0$. In view of the hyperbolicity theorem one might seek to show that K_L is quasi-geodesic. This result was shown in [25] for the case that S is a punctured torus, and played a role in our proof of the Ending Lamination Conjecture for that case. (Bowditch [3] has also considered the relation between $K(S)$ and representations in $\mathrm{PSL}_2(\mathbf{C})$ when S is a punctured torus, showing for example that a set defined similarly to K_L is connected in $K(S)$, even if the representation is not discrete).

In the general case there is some hope that a similar analysis of K_L can go through, and become a first step towards a solution of the conjecture.

Finally we remark that we do not know a purely combinatorial proof of the hyperbolicity theorem which does not use Teichmüller theory. In addition, except in the low-dimensional cases the constant δ is completely non-constructive, as it depends on bounds obtained from compactness arguments on spaces of quadratic differentials. It would be very interesting to find a proof which places some bounds on δ, or at least describes its asymptotic dependence on the topological type of S.

References

1. L. Bers, *An extremal problem for quasiconformal mappings and a theorem by Thurston*, Acta Math. **141** (1978), 73–98.
2. F. Bonahon, *Bouts des variétés hyperboliques de dimension 3*, Ann. of Math. **124** (1986), 71–158.
3. B. Bowditch, *Markoff triples and quasifuchsian groups*, University of Southampton Preprint #249, 1995.
4. B. Bowditch and D. B. A. Epstein, *Natural triangulations associated to a surface*, Topology **27** (1988), 91–117.
5. P. Buser, *Geometry and Spectra of Compact Riemann Surfaces*, Birkhäuser, 1992.
6. J. Cannon, *The theory of negatively curved spaces and groups*, Ergodic theory, symbolic dynamics, and hyperbolic spaces (Trieste, 1989), Oxford Univ. Press, 1991, pp. 315–369.

7. B. Farb, *Relatively Hyperbolic and Automatic Groups with Applications to Negatively Curved Manifolds*, Ph.D. thesis, Princeton University, 1994.

8. A. Fathi, F. Laudenbach, and V. Poenaru, *Travaux de Thurston sur les surfaces*, vol. 66-67, Asterisque, 1979.

9. E. Ghys and P. de la Harpe, *Sur les groupes hyperboliques d'aprés Mikhael Gromov*, Birkhäuser, 1990.

10. M. Gromov, *Hyperbolic groups*, Essays in Group Theory (S. M. Gersten, editor), MSRI Publications no. 8, Springer-Verlag, 1987.

11. J. Harer, *Stability of the homology of the mapping class group of an orientable surface*, Ann. of Math. **121** (1985), 215–249.

12. ———, *The virtual cohomological dimension of the mapping class group of an orientable surface*, Invent. Math. **84** (1986), 157–176.

13. W. J. Harvey, *Boundary structure of the modular group*, Riemann Surfaces and Related Topics: Proceedings of the 1978 Stony Brook Conference (I. Kra and B. Maskit, eds.), Ann. of Math. Stud. 97, Princeton, 1981.

14. A. E. Hatcher and W. P. Thurston, *A presentation for the mapping class group*, Topology **19** (1980), 221–237.

15. N. V. Ivanov, *Automorphisms of complexes of curves and of Teichmüller spaces*, Preprint.

16. ———, *Complexes of curves and the Teichmüller modular group*, Uspekhi Mat. Nauk **42** (1987), 55–107.

17. L. Keen, *Collars on Riemann surfaces*, Discontinuous groups and Riemann surfaces (Proc. Conf., Univ. Maryland 1973), Ann. of Math. Studies 79, Princeton, 1974, pp. 263–268.

18. S. Kerckhoff, *The Nielsen realization problem*, Ann. of Math. **117** (1983), 235–265.

19. H. Masur and Y. Minsky, *Geometry of the complex of curves I: Hyperbolicity*, In preparation.

20. H. A. Masur, *On a class of geodesics in Teichmüller space*, Ann. of Math. **102** (1975), 205–221.

21. ———, *The extension of the Weil-Petersson metric to the boundary of Teichmüller space*, Duke Math. J. **43** (1976), 623–635.

22. ———, *Transitivity properties of the horocyclic and geodesic flows on moduli space*, J. D'analyse Math. **39** (1981), 1–10.

23. ———, *Two boundaries of Teichmüller space*, Duke Math. J. **49** (1982), no. 1, 183–190.

24. H. A. Masur and M. Wolf, *Teichmüller space is not Gromov hyperbolic*, MSRI preprint No. 011-94, 1994.

25. Y. Minsky, *The classification of punctured-torus groups*, Preprint.

26. ———, *Extremal length estimates and product regions in Teichmüller space*, Stony Brook IMS Preprint #1994/11, to appear in Duke Math. J.

27. ———, *Quasi-projections in Teichmüller space*, Stony Brook IMS Preprint #1994/16, to appear in J. für die Reine und Angew. Math.

28. R. Penner with J. Harer, *Combinatorics of train tracks*, Annals of Math. Studies no. 125, Princeton University Press, 1992.

29. S. A. Wolpert, *Geodesic length functions and the Nielsen problem*, J. Differential Geom. **25** (1987), 275–296.

30. ———, *The hyperbolic metric and the geometry of the universal curve*, J. Differential Geom. **31** (1990), 417–472.

STATE UNIVERSITY OF NEW YORK, STONY BROOK, NY 11794
E-mail address: yair@math.sunysb.edu

Proceedings of
THE 37TH TANIGUCHI SYMPOSIUM ON
TOPOLOGY AND TEICHMÜLLER SPACES
held in Finland, July 1995
ed. by Sadayoshi KOJIMA *et al.*
©1996 World Scientific Publishing Co.
pp. 159–186

A LINEAR REPRESENTATION
OF THE MAPPING CLASS GROUP OF ORIENTABLE SURFACES
AND CHARACTERISTIC CLASSES OF SURFACE BUNDLES

SHIGEYUKI MORITA

(Received March 26, 1996)

1. Introduction

Let Σ_g be a closed orientable surface of genus g and let $\mathcal{M}_g$ be its mapping class group. Namely it is the group of isotopy classes of orientation preserving diffeomorphisms of Σ_g. In this paper, we always assume that $g \geq 2$. $\mathcal{M}_g$ acts on the Teichmüller space $\mathcal{T}_g$ of Σ_g properly discontinuously and the quotient space $\mathbf{M}_g = \mathcal{T}_g/\mathcal{M}_g$ is the moduli space of Riemann surfaces of genus g. We have a natural isomorphism

$$H^*(\mathcal{M}_g; \mathbf{Q}) \cong H^*(\mathbf{M}_g; \mathbf{Q}).$$

In the papers [22, 23, 24, 33], certain cohomology classes $e_i \in H^{2i}(\mathcal{M}_g; \mathbf{Z})$ have been defined and it was proved in [22, 23] that the homomorphism

$$\Phi : \mathbf{Q}[e_1, e_2, \cdots] \longrightarrow \varinjlim_{g \to \infty} H^*(\mathcal{M}_g; \mathbf{Q})$$

is injective, where the right hand side denotes the stable cohomology group (see [10]) of the mapping class group.

Let H denote the first integral homology group $H_1(\Sigma_g; \mathbf{Z})$ of Σ_g. Then there is defined a skew symmetric bilinear form

$$H \otimes H \to \mathbf{Z}$$

which is induced from the intersection number. The natural action of $\mathcal{M}_g$ on H preserves this form so that if we fix a symplectic basis of H, then we have a representation

$$\rho_0 : \mathcal{M}_g \longrightarrow \mathrm{Sp}(2g; \mathbf{Z})$$

where $\mathrm{Sp}(2g; \mathbf{Z})$ denotes the Siegel modular group of genus g. $\mathrm{Sp}(2g; \mathbf{Z})$ is a discrete subgroup of $\mathrm{Sp}(2g; \mathbf{R})$ and the maximal compact subgroup of $\mathrm{Sp}(2g; \mathbf{R})$ is isomorphic to the unitary group $\mathrm{U}(g)$. Hence we have the Chern classes

$$c_i \in H^{2i}(\mathrm{Sp}(2g; \mathbf{Z}); \mathbf{Z}) \quad (i = 1, \cdots)$$

which are induced from those of the universal g-dimensional complex vector bundle over the classifying space of $\mathrm{Sp}(2g; \mathbf{R})$. The pull back to the classifying space of $\mathrm{Sp}(2g; \mathbf{Z})$ of this universal bundle is *flat* as a real $2g$-dimensional vector bundle so that all of their Pontrjagin classes vanish. Thus any Chern class c_{2i} of even index can be expressed as a polynomial in the Chern classes $c_1, c_3, \cdots$ of odd indices (see [24] for details). Moreover it is a classical result of Borel [4, 5] that the stable cohomology of $\mathrm{Sp}(2g; \mathbf{Z})$ is a polynomial algebra generated by these Chern classes:

$$\lim_{g \to \infty} H^*(\mathrm{Sp}(2g; \mathbf{Z}); \mathbf{Q}) \cong \mathbf{Q}[c_1, c_3, c_5, \cdots].$$

Now it was proved in [22, 23, 24, 33] that, for any i, the pull back of the Chern class c_{2i-1} by the homomorphism ρ_0 above is expressible as a polynomial in $e_1, e_3, \cdots, e_{2i-1}$ whose leading term is a non-zero multiple of e_{2i-1}. Hence the image of the homomorphism

$$\rho_0^* : \lim_{g \to \infty} H^*(\mathrm{Sp}(2g; \mathbf{Z}); \mathbf{Q}) \longrightarrow \lim_{g \to \infty} H^*(\mathcal{M}_g; \mathbf{Q})$$

is exactly equal to $\mathbf{Q}[e_1, e_3, \cdots] \subset \lim_{g \to \infty} H^*(\mathcal{M}_g; \mathbf{Q})$.

The above fact can be interpreted geometrically as follows. The cohomology classes e_i serve as the characteristic classes of oriented differentiable fibre bundles with fibre Σ_g because the classifying space $\mathrm{BDiff}^+\Sigma_g$ of such bundles is an Eilenberg-MacLane space $K(\mathcal{M}_g, 1)$ by the result of Earle and Eells [6]. In particular, if we have a differentiable family of Riemann surfaces of genus g, then we have their e_i classes and they express how the given family is twisted topologically. These classes also detect how the complex structure is varied along the base space of the family. It turns out that these two roles of e_i are almost the same each other as is manifested in the fact that there is a natural continuous mapping

$$\mathrm{BDiff}^+\Sigma_g \longrightarrow \mathbf{M}_g$$

which is a $\mathbf{Q}$-equivalence.

Now if $\pi : E \to X$ is an oriented Σ_g-bundle over a C^∞ manifold X, then we have the associated holonomy homomorphism

$$h : \pi_1 X \longrightarrow \mathcal{M}_g$$

which describes the structure of the given bundle completely. The composite homomorphism $\rho_0 \circ h : \pi_1 X \to \mathrm{Sp}(2g; \mathbf{Z})$ expresses how the fundamental group $\pi_1 X$ of the base space acts on the first cohomology group of the fibres. Hence we can conclude that the characteristic classes e_{odd} of *odd* indices measure how the bundle is twisted *cohomologically*. This is the geometrical meaning of e_{odd} classes.

In the case where $\pi : E \to X$ is a family of Riemann surfaces of genus g, there is another explanation of the e_{odd} classes. Namely, in such a case, we can construct a g-dimensional complex vector bundle η over X by defining the fibre η_x over $x \in X$ to be

$$\eta_x = H^0(E_x; \Omega^1)$$

where $E_x = \pi^{-1}(x)$ and $H^0(E_x; \Omega^1)$ is the space of holomorphic differentials on the Riemann surface E_x. Hence we have the Chern classes of η. At the universal space level, these are nothing but the universal Chern classes c_i mentioned above (up to signs). It follows that the characteristic classes e_{odd} are equivalent to the Chern classes of η.

The above consideration raises the following problems concerning the characteristic classes e_{even} of *even* indices:

(i) Can the classes e_{even} be detected by a certain finite dimentional linear representation of $\mathcal{M}_g$?

(ii) What is the geometrical meaning of e_{even} classes ?

The purpose of the present paper is to give answers to both of the above two questions.

Roughly speaking, they are given as follows. The representation ρ_0 is induced from the natural action of $\mathcal{M}_g$ on H which is nothing but the abelianizaton of $\pi_1\Sigma_g$. Next to the abelianization, it is natural to consider the *first nilpotent* quotient

$$N_1 = \pi_1\Sigma_g/\Gamma_2(\pi_1\Sigma_g)$$

which is the factor group of $\pi_1\Sigma_g$ modulo its two fold commutator subgroup $\Gamma_2(\pi_1\Sigma_g)$ (see §2 for details). $\mathcal{M}_g$ acts on N_1 by outer automorphisms and it induces a representation

$$\rho_1 : \mathcal{M}_g \longrightarrow U \rtimes \mathrm{Sp}(2g; \mathbf{Z})$$

of $\mathcal{M}_g$ into certain semi-direct product. Here U is an $\mathrm{Sp}(2g; \mathbf{Z})$-module defined to be $U = \frac{1}{2}\Lambda^3 H/H$ where $\Lambda^3 H$ denotes the third exterior power of H and it contains H as a natural submodule (see [30] and §2 below for details). It turns out that ρ_1 is in fact a linear representation of $\mathcal{M}_g$ (Lemma 2.2). If we restrict ρ_1 to the Torelli group $\mathcal{I}_g(= \mathrm{Ker}\,\rho_0)$, then we obtain Johnson's homomorphism $\tau_1 : \mathcal{I}_g \to \Lambda^3 H/H$ given in [11]. The topological meaning of ρ_1 can be described as follows. Namely if $\pi : E \to X$ is an oriented Σ_g-bundle with holonomy $h : \pi_1 X \to \mathcal{M}_g$, then the composition $\rho_1 \circ h : \pi_1 X \to U \rtimes \mathrm{Sp}(2g; \mathbf{Z})$ expresses how $\pi_1 X$ acts on the first nilpotent quotient of the fundamental group of the fibres. The representation ρ_1 is also important in the work of Hain [8] in the context of algebraic geometry.

Now let $H^\bullet(U; \mathbf{Q})^{\mathrm{Sp}}$ be the subalgebra of $H^\bullet(U; \mathbf{Q})$ consisting of $\mathrm{Sp}(2g; \mathbf{Z})$-invariant rational cohomology classes of U. Then it is easy to see that there is a natural homomorphism $H^\bullet(U; \mathbf{Q})^{\mathrm{Sp}} \to H^\bullet(U \rtimes \mathrm{Sp}(2g; \mathbf{Z}); \mathbf{Q})$. If we combine this homomorphism with $\rho_1^\bullet : H^\bullet(U \rtimes \mathrm{Sp}(2g; \mathbf{Z}); \mathbf{Q}) \to H^\bullet(\mathcal{M}_g; \mathbf{Q})$, then we obtain

$$\rho_1^\bullet : H^\bullet(U; \mathbf{Q})^{\mathrm{Sp}} \longrightarrow H^\bullet(\mathcal{M}_g; \mathbf{Q})$$

which is actually defined at the *cocycle level*. By making use of the classical representation theory, we can describe the group $H^\bullet(U; \mathbf{Q})^{\mathrm{Sp}}$ explicitly (see §3).

The following is the main result of the present paper.

Theorem. *Let $\mathcal{M}_g$ be the mapping class group of Σ_g and let $\rho_1 : \mathcal{M}_g \to U \rtimes \mathrm{Sp}(2g; \mathbf{Z})$ be the representation of $\mathcal{M}_g$ induced from its action on the first nilpotent quotient of $\pi_1\Sigma_g$. Then, for any i, there exists an element $\beta_i \in H^{2i}(U; \mathbf{Q})^{\mathrm{Sp}}$ such that $\rho_1^\bullet(\beta_i) = e_i$, where $e_i \in H^{2i}(\mathcal{M}_g; \mathbf{Q})$ is the i-th characteristic class of $\mathcal{M}_g$ with rational coefficients. In particular, the classes e_{2i} of even indices can be detected by the finite dimensional linear representation ρ_1 of $\mathcal{M}_g$.*

The kernel of the representation ρ_1 coinsides with the subgroup $\mathcal{K}_g$ of $\mathcal{M}_g$ which is the one generated by Dehn twists along *separating* simple closed curves on Σ_g (see [13, 30]). Hence as an immediate corollary to the above theorem, we obtain

Corollary. *The characteristic classes $e_i \in H^{2i}(\mathcal{M}_g; \mathbf{Q})$ $(i = 1, 2, \cdots)$ vanish on the subgroup $\mathcal{K}_g$. Hence, if the holonomy group of an oriented Σ_g-bundle is contained in $\mathcal{K}_g$, then its characteristic class e_i is zero for all i.*

Now we summarize the contents of the present paper briefly. In §2, we review the construction of the homomorphism $\rho_1 : \mathcal{M}_g \to U \rtimes \mathrm{Sp}(2g; \mathbf{Z})$ as well as a series of representations ρ_i $(i = 1, 2, \cdots)$ from our paper [30]. The main results will be given in §3 in detail, where we also give an explicit description of $H^*(U; \mathbf{Q})^{\mathrm{Sp}}$ and $H^*(\Lambda^3 H; \mathbf{Q})^{\mathrm{Sp}}$. §4 is devoted to a description of the representation ρ_2. Details of the ingredient of this section will be given in a separate paper [31]. In §5, we give an extension of a result of Hain [9] which determines the kernel of the homomorphism $\tau_1^* : H^2(U; \mathbf{Q}) \to H^2(\mathcal{I}_g; \mathbf{Q})$. In §6 we construct a cocycle for the universal Euler class $e \in H^2(\mathcal{M}_{g,*}; \mathbf{Q})$ (see §2 and [24] for the definition). Finally in §7, we give a proof of the main theorem (Theorem 3.4).

The above follows the author's talk at the Taniguchi symposium "Topology and Teichmüller spaces" which was held in July 1995, at Katinkulta, Finland, rather closely. After that, in a joint work with N. Kawazumi, we obtained a complete description of the homomorphism $\rho_1^* : H^*(U; \mathbf{Q})^{\mathrm{Sp}} \to H^*(\mathcal{M}_g; \mathbf{Q})$. More precisely, by combining our results with those of Kawazumi given in [14, 15, 16], we proved that the image of the above homomorphism is *exactly* equal to the subalgebra generated by the characteristic classes e_i. Details will be given in our forthcoming paper [17].

However we think that our original approach would still have its own meaning because it is directly related to the cohomological structure of the Torelli group while our main technical argument in [17] is based on a procedure to evaluate the cocycles on the moduli space directly.

The author would like to thank M. Furuta for correspondences at a very early stage (around the latter half of 80's) of the present work, R. Hain for many enlightening and useful discussions and K. Masuda for a helpful suggestion related to the description of $H^*(U; \mathbf{Q})^{\mathrm{Sp}}$. It is also a pleasure to thank C.-F. Bödigheimer, N. Kawazumi, M. Kontsevich and E. Looijenga for stimulating discussions.

2. Linear representations of $\mathcal{M}_g$

In this section, we review our previous results in [30] (see also [28, 29]). In particular we recall the definition of the representation

$$\rho_1 : \mathcal{M}_g \longrightarrow U \rtimes \mathrm{Sp}(2g; \mathbf{Z})$$

which was mentioned in the introduction and describe an enhancement of it. We also prove that ρ_1 is in fact a finite dimensional linear representation (Lemma 2.2).

We first introduce two types of the mapping class groups other than the usual one. Thus let $\mathcal{M}_{g,1}$ be the mapping class group of Σ_g relative to an embedded disc $D \subset \Sigma_g$ and let $\mathcal{M}_{g,*}$ be the mapping class group of Σ_g relative to the base point $* \in D$ of Σ_g. As is well known, these three types of the mapping class groups are related by the following two natural extensions

$$1 \longrightarrow \pi_1 \Sigma_g \longrightarrow \mathcal{M}_{g,*} \longrightarrow \mathcal{M}_g \longrightarrow 1$$

$$0 \longrightarrow \mathbf{Z} \longrightarrow \mathcal{M}_{g,1} \longrightarrow \mathcal{M}_{g,*} \longrightarrow 1$$

where the second central extension corresponds to the Euler class $e \in H^2(\mathcal{M}_{g,*}; \mathbf{Z})$ and the infinite cyclic group $\mathbf{Z} \subset \mathcal{M}_{g,1}$ is generated by the Dehn twist along a simple closed curve on $\Sigma_g \setminus \operatorname{Int} D$ which is parallel to ∂D.

These mapping class groups can be naturally considered as certain (outer) automorphism groups of the fundamental groups. More precisely, by a classical result of Dehn and Nielsen and its generalizations, we can write

$$\mathcal{M}_g \cong \operatorname{Out}_+ \pi_1 \Sigma_g, \quad \mathcal{M}_{g,*} \cong \operatorname{Aut}_+ \pi_1 \Sigma_g$$

$$\mathcal{M}_{g,1} = \{\varphi \in \operatorname{Aut} \pi_1 \Sigma_g{}^0; \varphi(\zeta) = \zeta\}$$

where $\Sigma_g{}^0 = \Sigma_g \setminus \operatorname{Int} D$ and ζ is the element of the free group $\pi_1 \Sigma_g{}^0$ corresponding to the boundary curve so that $\pi_1 \Sigma_g = \pi_1 \Sigma_g{}^0 / \langle \zeta \rangle$.

For any group π, let $\{\Gamma_k(\pi)\}_{k \geq 0}$ denote the lower central series of π. Namely $\Gamma_0(\pi) = \pi$ and $\Gamma_k(\pi)$ is inductively defined to be $[\pi, \Gamma_{k-1}(\pi)]$ for all $k \geq 1$. We set $N_k(\pi) = \pi / \Gamma_{k+1}(\pi)$ and call it the k-th nilpotent quotient of π. By the definition, $N_0(\pi)$ is nothing but the abelianization of π. Obviously the lower central series is preserved by any automorphism of the group π. Hence we have a series of representations

$$\rho_k : \operatorname{Aut} \pi \longrightarrow \operatorname{Aut} N_k(\pi) \quad (k = 0, 1, \cdots)$$

and similarly for the outer automorphisms.

Now we apply this procedure to the mapping class groups $\mathcal{M}_{g,1}, \mathcal{M}_{g,*}$ and $\mathcal{M}_g$. Let us write the nilpotent quotients of $\pi_1 \Sigma_g{}^0$ and $\pi_1 \Sigma_g$ simply by

$$N_k^0 = N_k(\pi_1 \Sigma_g{}^0), \quad N_k = N_k(\pi_1 \Sigma_g).$$

In particular, we have $N_0^0 = N_0 = H$ and the first nilpotent quotients N_1^0, N_1 are described by the central extensions

$$0 \longrightarrow \Lambda^2 H \longrightarrow N_1^0 \longrightarrow H \longrightarrow 0$$

$$0 \longrightarrow \Lambda^2 H / \mathbf{Z} \longrightarrow N_1 \longrightarrow H \longrightarrow 0$$

where the subgroup $\mathbf{Z} \subset \Lambda^2 H$ is generated by the symplectic class $\omega_0 \in \Lambda^2 H$. Namely $\omega_0 = \sum_i x_i \wedge y_i$ where $x_1, \cdots, x_g, y_1, \cdots, y_g$ is any symplectic basis of H.

We have representations

$$\rho_k^0 : \mathcal{M}_{g,1} \longrightarrow \operatorname{Aut} N_k^0, \quad \rho_k : \mathcal{M}_{g,*} \longrightarrow \operatorname{Aut} N_k$$

$$\bar{\rho}_k : \mathcal{M}_g \longrightarrow \operatorname{Out} N_k$$

for all k. Here we attach the symbols "0" and "$^-$" to distinguish the three cases. However we will occasionally omit them if there will be no fear of confusion and this remark also applies to another symbol τ which will appear below. The first one ρ_0 is nothing but the classical representation of the mappig class group onto $\mathrm{Sp}(2g; \mathbf{Z})$ in all three cases. As was shown in [30], $\mathrm{Aut}\, N_k^0$ can be embedded into a certain linear algebraic group defined over $\mathbf{Q}$ as a discrete subgroup and the same proof yields the same result for the group $\mathrm{Aut}\, N_k$. It follows that ρ_k^0 and ρ_k are finite dimensional linear representations. We will see below that $\bar{\rho}_1$ is also linear.

These representations induce a decreasing filtration of the mapping class groups by setting

$$\mathcal{M}_{g,1}(k) = \mathrm{Ker}\, \rho_{k-1}^0, \quad \mathcal{M}_{g,*}(k) = \mathrm{Ker}\, \rho_{k-1}$$
$$\mathcal{M}_g(k) = \mathrm{Ker}\, \bar{\rho}_{k-1}.$$

The cases for $k = 1$ correspond to the Torelli group which is the subgroup of the mapping class group consisting of all elements acting trivially on the homology of the surface. We use the notations $\mathcal{M}_{g,1}(1) = \mathcal{I}_{g,1}, \mathcal{M}_{g,*}(1) = \mathcal{I}_{g,*}$ and $\mathcal{M}_g(1) = \mathcal{I}_g$ for them.

Let

$$\mathcal{L}^0 = \bigoplus_{k \geq 1} \mathcal{L}_k^0, \quad \mathcal{L} = \bigoplus_{k \geq 1} \mathcal{L}_k$$

be the graded Lie algebra of $\pi_1 \Sigma_g{}^0$ and $\pi_1 \Sigma_g$, respectively, which is associated with the lower central series; $\mathcal{L}_k^0 = \Gamma_{k-1}(\pi_1 \Sigma_g{}^0)/\Gamma_k(\pi_1 \Sigma_g{}^0)$ and $\mathcal{L}_k = \Gamma_{k-1}(\pi_1 \Sigma_g)/\Gamma_k(\pi_1 \Sigma_g)$. Clearly $\mathcal{L}_1^0 = H$ and it is well known that $\mathcal{L}^0$ is a free graded Lie algebra generated by elements of H (see [20]). Let I be the ideal of $\mathcal{L}^0$ generated by $\omega_0 \in \Lambda^2 H = \mathcal{L}_2^0$. Then a result of Labute [20] says that there is a natural isomorphism $\mathcal{L} \cong \mathcal{L}^0/I$.

Now define two $\mathrm{Sp}(2g; \mathbf{Z})$-modules $\mathcal{H}_k^0$ and $\mathcal{H}_k$ by

$$\mathcal{H}_k^0 = \mathrm{Ker}\Big(\mathrm{Hom}(H, \mathcal{L}_{k+1}^0) \to \mathcal{L}_{k+2}^0 \Big)$$

$$\mathcal{H}_k = \mathrm{Ker}\Big(\mathrm{Hom}(H, \mathcal{L}_{k+1}) \to \mathcal{L}_{k+2} \Big).$$

Here the homomorphism $\mathrm{Hom}(H, \mathcal{L}_{k+1}^0) \to \mathcal{L}_{k+2}^0$ is defined as follows. We have a canonical isomorphism $\mathrm{Hom}(H, \mathcal{L}_{k+1}^0) \cong H \otimes \mathcal{L}_{k+1}^0$ which is induced from the Poincaré duality $H^* \cong H$. Then the required one is the composition of this isomorphism followed by the natural mapping $H \otimes \mathcal{L}_{k+1}^0 \to \mathcal{L}_{k+2}^0$ which is given by the bracket operation of the Lie algebra $\mathcal{L}^0$. The homomorphism $\mathrm{Hom}(H, \mathcal{L}_{k+1}) \to \mathcal{L}_{k+2}$ is defined similarly. As was shown in [31], if we restrict the representations ρ_k^0 and ρ_k to the subgroups $\mathcal{M}_{g,1}(k)$ and $\mathcal{M}_{g,*}(k)$, respectively, then we obtain homomorphisms

$$\tau_k^0 : \mathcal{M}_{g,1}(k) \longrightarrow \mathcal{H}_k^0, \quad \tau_k : \mathcal{M}_{g,*}(k) \longrightarrow \mathcal{H}_k$$

for all $k \geq 1$. We call them Johnson's homomorphisms of order k because the definition of τ_k^0 was originally given in [12]. In our paper [31] we treated only the case of τ_k^0. However almost the same argument yields the case of τ_k. In particular, the fact that $\mathrm{Im}\, \tau_k$ is contained in $\mathcal{H}_k$ can be proved exactly the same way as the case of τ_k^0. We remark here that we have *shifted* the indices of the lower central series as well as those of ρ and τ appropriately in the above description.

It is clear that the above constructions for the two types of the mapping class groups $\mathcal{M}_{g,1}$ and $\mathcal{M}_{g,*}$ are almost the same each other. However the relation between them is rather delicate. We have an obvious mapping $\mathcal{M}_{g,1}(k) \to \mathcal{M}_{g,*}(k)$ and it is easy to see that we have a central extension

$$0 \longrightarrow \mathbf{Z} \longrightarrow \mathcal{M}_{g,1}(k) \longrightarrow \mathcal{M}_{g,*}(k) \longrightarrow 1$$

for $k = 0, 1, 2$. Also it is easy to see that $\mathcal{M}_{g,1}(k) \to \mathcal{M}_{g,*}(k)$ is injective for all $k \geq 3$. The determinations of $\mathrm{Im}\tau_1$ given in [11], $\mathrm{Im}\tau_2$ in [26] and $\mathrm{Im}\tau_3$ in [3, 9] imply that $\mathcal{M}_{g,1}(k) \cong \mathcal{M}_{g,*}(k)$ for $k = 3, 4$. However it has been an open question whether the mapping $\mathcal{M}_{g,1}(k) \to \mathcal{M}_{g,*}(k)$ is surjective (and hence an isomorphism) for general k (cf. Remark (4.5) of Asada-Nakamura [3]). Here we remark that the filtration of $\mathcal{M}_{g,*}$ given in [3] is defined to be the image of that of $\mathcal{M}_{g,1}$ so that it may be different from ours.

Let us define J_k by

$$J_k = \mathrm{Ker}\Big(\mathrm{Hom}(H, I_{k+1}) \to I_{k+2}\Big)$$

where recall that $I = \oplus_k I_k$ is the ideal of $\mathcal{L}^0$ generated by ω_0 so that $I = \mathrm{Ker}(\mathcal{L}^0 \to \mathcal{L})$. Then we have the following commutative diagram

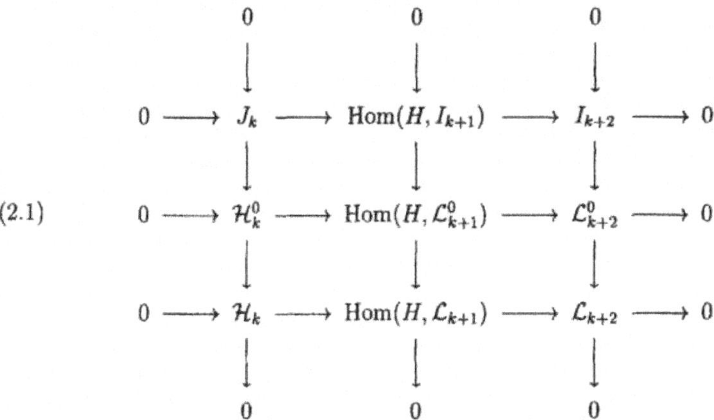

$$(2.1)$$

where all of the rows as well as the columns are exact.

Lemma 2.1. *Assume that* $\mathcal{M}_{g,1}(\ell) \cong \mathcal{M}_{g,*}(\ell)$ *for all* $3 \leq \ell \leq k$. *Then the homomorphism* $\mathcal{M}_{g,1}(k+1) \to \mathcal{M}_{g,*}(k+1)$ *is surjective (and hence is an isomorphism) if and only if*

$$\mathrm{Im}\tau_k^0 \cap J_k = \{0\}.$$

Proof. Immediate from the definitions. $\square$

However, very recently, Hain settles the above question affirmatively in [9] where he develops a beautiful theory about the structure of the Torelli group by using the mixed Hodge structures.

Now we consider the filtration $\{\mathcal{M}_g(k)\}_{k \geq 0}$ of the usual mapping class group $\mathcal{M}_g$. Observe that it can be also described as $\mathcal{M}_g(k) = \pi(\mathcal{M}_{g,*}(k))$ where $\pi : \mathcal{M}_{g,*} \to \mathcal{M}_g$ is the natural projection. Hence there is an exact sequence

$$1 \longrightarrow \pi_1 \Sigma_g \cap \mathcal{M}_{g,*}(k) \longrightarrow \mathcal{M}_{g,*}(k) \longrightarrow \mathcal{M}_g(k) \longrightarrow 1.$$

By a result of Asada-Kaneko [2] (see also [1]), we have $\pi_1\Sigma_g \cap \mathcal{M}_{g,*}(k) = \Gamma_{k-1}(\pi_1\Sigma_g)$ so that there is a natural injection $\mathcal{L}_k \subset \mathcal{H}_k$. Hence we can define Johnson's homomorphism

$$\overline{\tau}_k : \mathcal{M}_g(k) \longrightarrow \overline{\mathcal{H}}_k$$

of order k by setting $\overline{\mathcal{H}}_k = \mathcal{H}_k/\mathcal{L}_k$ and $\overline{\tau}_k(\overline{\varphi}) = \tau_k(\varphi)\mathrm{mod}\mathcal{L}_k$ where $\overline{\varphi} \in \mathcal{M}_g(k)$ is any element and $\varphi \in \mathcal{M}_{g,*}(k)$ is any of its lift.

Each of the graded modules

$$\bigoplus_{k\geq 1}\mathcal{H}_k^0, \quad \bigoplus_{k\geq 1}\mathcal{H}_k, \quad \bigoplus_{k\geq 1}\overline{\mathcal{H}}_k$$

has a natural structure of a graded Lie algebra and it is a very important problem to identify

$$\bigoplus_{k\geq 1}\mathrm{Im}\tau_k^0, \quad \bigoplus_{k\geq 1}\mathrm{Im}\tau_k, \quad \bigoplus_{k\geq 1}\mathrm{Im}\overline{\tau}_k$$

as their graded Lie subalgebras (see [31]).

In the papers [3, 8], Johnson's homomorphisms are described from different points of view.

Now we describe the representation ρ_1 after our paper [30]. Let H denote the first integral homology group $H_1(\Sigma_g; \mathbf{Z})$ of Σ_g as before and let $\Lambda^3 H$ be the third exterior power of H. Let $\frac{1}{2}\Lambda^3 H$ be $\Lambda^3 H \otimes \frac{1}{2}\mathbf{Z}$. It is naturally an $\mathrm{Sp}(2g; \mathbf{Z})$-module so that we can make the semi-direct product $\frac{1}{2}\Lambda^3 H \rtimes \mathrm{Sp}(2g; \mathbf{Z})$. The homomorphism ρ_1^0 is now described as

$$\rho_1^0 : \mathcal{M}_{g,1} \longrightarrow \frac{1}{2}\Lambda^3 H \rtimes \mathrm{Sp}(2g; \mathbf{Z}).$$

Here the restriction of ρ_1^0 to the Torelli group $\mathcal{I}_{g,1}$ is equal to Johnson's homomorphism $\tau_1^0 : \mathcal{I}_{g,1} \rightarrow \Lambda^3 H$ given in [11]. More precisely, we extended Johnson's homomorphism τ_1^0 from the Torelli group to the whole of the mapping class group $\mathcal{M}_{g,1}$ to obtain a *crossed* homomorphism

$$\tilde{k} : \mathcal{M}_{g,1} \longrightarrow \frac{1}{2}\Lambda^3 H.$$

Then the homomorphism ρ_1^0 above is given explicitly by $\rho_1^0(\varphi) = (\tilde{k}(\varphi), \rho_0(\varphi))$ ($\varphi \in \mathcal{M}_{g,1}$). It is easy to see that the representation ρ_1^0 is trivial on $\mathrm{Ker}(\mathcal{M}_{g,1} \rightarrow \mathcal{M}_{g,*}) \cong \mathbf{Z}$ so that we can write

$$\rho_1 : \mathcal{M}_{g,*} \longrightarrow \frac{1}{2}\Lambda^3 H \rtimes \mathrm{Sp}(2g; \mathbf{Z}).$$

Next we describe the representation $\overline{\rho}_1$. There is an $\mathrm{Sp}(2g; \mathbf{Z})$-equivariant mapping

$$i : H \longrightarrow \Lambda^3 H$$

defined by $i(u) = u \wedge \omega_0$ ($u \in H$). It is easy to see that i is injective and henceforth we identify the image $i(H) \subset \Lambda^3 H$ of i with H. $\frac{1}{2}\Lambda^3 H$ also contains H as a submodule so that we have the quotient $\frac{1}{2}\Lambda^3 H/H$. We simply write U for $\frac{1}{2}\Lambda^3 H/H$. $\mathrm{Sp}(2g; \mathbf{Z})$ acts on U naturally so that we can make the semi-direct product $U \rtimes \mathrm{Sp}(2g; \mathbf{Z})$. Now it is easy to see that the restriction of ρ_1 to $\mathrm{Ker}(\mathcal{M}_{g,*} \rightarrow \mathcal{M}_g) = \pi_1\Sigma_g$ is equivalent to the abelianization $\pi_1\Sigma_g \to H$ followed by the inclusion $i : H \to \Lambda^3 H$. Hence the representation $\bar{\rho}_1$ is given by

$$\bar{\rho}_1 : \mathcal{M}_g \longrightarrow U \rtimes \mathrm{Sp}(2g; \mathbf{Z})$$

whose first term is a crossed homomorphism $\tilde{k} : \mathcal{M}_g \to U$. It was actually proved in [30] that the rank of $H^1(\mathcal{M}_g; U)$ is equal to 1 and the crossed homomorphism $\tilde{k}$ above is a generator of this group (modulo possibly torsions).

Lemma 2.2. *The representation $\rho_1 : \mathcal{M}_g \longrightarrow U \rtimes \mathrm{Sp}(2g; \mathbf{Z})$ is a linear representation.*

Proof. Let us write $U_\mathbf{Q}$ for $U \otimes \mathbf{Q}$. Then $U_\mathbf{Q}$ is a finite dimensional vector space over $\mathbf{Q}$. Since U is a free abelian group, U can be considered as a subgroup (lattice) of $U_\mathbf{Q}$. Now U is naturally an $\mathrm{Sp}(2g; \mathbf{Z})$-module so that $\mathrm{Sp}(2g; \mathbf{Z})$ acts on $U_\mathbf{Q}$ linearly. In fact it is easy to see that $U_\mathbf{Q}$ is an irreducible rational representation of the algebraic group $\mathrm{Sp}(2g; \mathbf{Q})$. Hence we have the induced homomorphism

$$r : \mathrm{Sp}(2g; \mathbf{Z}) \longrightarrow \mathrm{GL}(U_\mathbf{Q})$$

which is injective. In general, it is well known that the semi-direct product $U_\mathbf{Q} \rtimes \mathrm{GL}(U_\mathbf{Q})$ can be embedded into $\mathrm{GL}(U_\mathbf{Q} \oplus \mathbf{Q})$ as the group of *affine transformations*. Then the composition

$$U \rtimes \mathrm{Sp}(2g; \mathbf{Z}) \longrightarrow U_\mathbf{Q} \rtimes \mathrm{GL}(U_\mathbf{Q}) \subset GL(U_\mathbf{Q} \oplus \mathbf{Q})$$

embedds the representation ρ_1 into a general linear group. This completes the proof. $\square$

3. Statement of the main result

The linear representation

$$\rho_1 : \mathcal{M}_g \longrightarrow U \rtimes \mathrm{Sp}(2g; \mathbf{Z})$$

whose definition was recalled in the previous section, induces a homomorphism in cohomology

$$\rho_1^* : H^*(U \rtimes \mathrm{Sp}(2g; \mathbf{Z}); \mathbf{Q}) \longrightarrow H^*(\mathcal{M}_g; \mathbf{Q}).$$

Let us prepare a general fact on the cohomology of a semi-direct product. Thus let G, N be any two groups and suppose that there is given a homomorphism $G \to \mathrm{Aut} N$ so that we can make the associated semi-direct product $N \rtimes G$. The group law of $N \rtimes G$ is given by $(\nu_1, \gamma_1)(\nu_2, \gamma_2) = (\nu_1 \gamma_1(\nu_2), \gamma_1 \gamma_2)$ $(\nu_i \in N, \gamma_i \in G)$.

Lemma 3.1. *In the situation as above, suppose a k-cocycle $c \in Z^k(N)$ of N is invariant under the action of G. Namely the equality*

$$c\big(\gamma(\nu_1), \cdots, \gamma(\nu_k)\big) = c(\nu_1, \cdots, \nu_k)$$

holds for all $\gamma \in G, \nu_i \in N$. Then the cochain $\tilde{c} \in C^k(N \rtimes G)$ defined by

$$\tilde{c}\big((\nu_1, \gamma_1), \cdots, (\nu_k, \gamma_k)\big) = c\big(\nu_1, \gamma_1(\nu_2), \gamma_1 \gamma_2(\nu_3), \cdots, \gamma_1 \cdots \gamma_{k-1}(\nu_k)\big)$$

is a cocycle of the semi-direct product $N \rtimes G$.

Proof. A straightforward computation shows that $\delta \tilde{c} = 0$. $\square$

The following proposition can be proved by applying the above Lemma to the semi-direct product $U \rtimes \mathrm{Sp}(2g; \mathbf{Z})$. Recall here that we have a canonical isomorphism $H^*(U_\mathbf{Q}) \cong \Lambda^* U_\mathbf{Q}^*$ where $U_\mathbf{Q} = U \otimes \mathbf{Q}$.

Proposition 3.2. *The representation $\rho_1 : \mathcal{M}_g \to U \rtimes \mathrm{Sp}(2g; \mathbf{Z})$ induces a homomorphism*

$$\rho_1^* : (\Lambda^k U_\mathbf{Q}^*)^{\mathrm{Sp}} \longrightarrow H^k(\mathcal{M}_g; \mathbf{Q})$$

which is defined at the cocycle level. More precicely, suppose that $c \in Z^k(U_\mathbf{Q}; \mathbf{Q})$ is a k-cocycle of $U_\mathbf{Q}$ which is given as a multi-linear alternating k-form

$$c : \overbrace{U_\mathbf{Q} \times \cdots \times U_\mathbf{Q}}^{k\ times} \longrightarrow \mathbf{Q}$$

and suppose that it is invariant under the action of $\mathrm{Sp}(2g; \mathbf{Z})$. Then the assignment

$$(\varphi_1, \cdots, \varphi_k) \mapsto \tilde{c}\big(\rho_1(\varphi_1), \cdots, \rho_1(\varphi_k)\big)$$
$$= c\big(\tilde{k}(\varphi_1), (\varphi_1)_* \tilde{k}(\varphi_2), \cdots, (\varphi_1 \cdots \varphi_{k-1})_* \tilde{k}(\varphi_k)\big) \ (\varphi_i \in \mathcal{M}_g)$$

defines a k-cocycle of $\mathcal{M}_g$ and it represents $\rho_1^*([c])$.

If we apply the same argument as above to the representation

$$\rho_1 : \mathcal{M}_{g,*} \longrightarrow \frac{1}{2}\Lambda^3 H \rtimes \mathrm{Sp}(2g; \mathbf{Z}),$$

then we obtain

Proposition 3.3. *The representation ρ_1 of $\mathcal{M}_{g,*}$ induces a homomorphism*

$$\rho_1^* : \big(\Lambda^*(\Lambda^3 H_\mathbf{Q})^*\big)^{\mathrm{Sp}} \longrightarrow H^*(\mathcal{M}_{g,*}; \mathbf{Q})$$

which is defined at the cocycle level.

Now, as mentioned before, it is easy to see that $U_\mathbf{Q}$ is an irreducible rational representation of $\mathrm{Sp}(2g; \mathbf{Q})$ which can be realized in $H_\mathbf{Q}^{\otimes 3}$ where $H_\mathbf{Q} = H \otimes \mathbf{Q}$. In fact, if we define a contraction $C : \Lambda^3 H \to H$ by $C(u \wedge v \wedge w) = 2\big((u \cdot v)w + (v \cdot w)u + (w \cdot u)v\big)$ $(u, v, w \in H)$, then it is well known (see [7]) that $\mathrm{Ker}\, C \otimes \mathbf{Q}$ is an irreducible rational representation of $\mathrm{Sp}(2g; \mathbf{Q})$ and the composition $\mathrm{Ker}\, C \subset \Lambda^3 H \to U$ is easily seen to be an isomorphism after tensoring with $\mathbf{Q}$. Hence a fundamental result of Weyl in classical representation theory implies that any $\mathrm{Sp}(2g; \mathbf{Q})$-invariant mapping

$$\Lambda^k U_\mathbf{Q} \longrightarrow \mathbf{Q}$$

is induced from some contraction $H_\mathbf{Q}^{\otimes 3k} \to \mathbf{Q}$ restricted to the subspace $\Lambda^k U_\mathbf{Q} \subset H_\mathbf{Q}^{\otimes 3k}$. Hence if k is odd, then $(\Lambda^k U_\mathbf{Q}^*)^{\mathrm{Sp}} = 0$ and for even k, it can be described as follows.

For each positive integer k, let $\mathcal{G}_{2k}$ denote the set of isomorphism classes of connected trivalent graphs with $2k$ vertices. As is well known $\mathcal{G}_{2k}$ can also be identified with the set of equivalence classes of pants decompositions of a closed orientable surface of genus $k + 1$. We set $\mathcal{G}$ to be the disjoint union of $\mathcal{G}_{2k}$ for all $k \geq 1$

$$\mathcal{G} = \coprod_{k \geq 1} \mathcal{G}_{2k}.$$

Let $\mathcal{G}_{2k}^0 \subset \mathcal{G}_{2k}$ be the subset consisting of those graphs which have no *loops*. Here a loop means an edge whose two endpoints are attached to the same vertex. We set $\mathcal{G}^0 = \coprod_{k \geq 1} \mathcal{G}_{2k}^0$. Now let

$$\mathbf{Q}[\alpha_\Gamma; \Gamma \in \mathcal{G}]$$

be the polynomial algebra over $\mathbf{Q}$ with a set of generators α_Γ corresponding to each element $\Gamma \in \mathcal{G}$. We define the degree of $\alpha_\Gamma (\Gamma \in \mathcal{G}_{2k})$ to be equal to $2k$. We also consider the polynomial algebra

$$\mathbf{Q}[\beta_\Gamma; \Gamma \in \mathcal{G}^0]$$

the degrees of whose generators are defined similarly. Here we use different symbol β_Γ instead of α_Γ because we consider $\mathbf{Q}[\beta_\Gamma; \Gamma \in \mathcal{G}^0]$ as a *twisted* subspace of $\mathbf{Q}[\alpha_\Gamma; \Gamma \in \mathcal{G}]$. Now a classical theorem of Weyl in the invariant theory implies that there exist natural *surjective* homomorphisms

$$\mathbf{Q}[\alpha_\Gamma; \Gamma \in \mathcal{G}] \longrightarrow \left(\Lambda^*(\Lambda^3 H_\mathbf{Q})^* \right)^{\mathrm{Sp}}$$

$$\mathbf{Q}[\beta_\Gamma; \Gamma \in \mathcal{G}^0] \longrightarrow \left(\Lambda^* U_\mathbf{Q}^* \right)^{\mathrm{Sp}}$$

which are isomorphisms in the range $6k \leq 2g$. More precisely, each homogenious element of degree $2k$ of the relevant polynomial algebra define a way of contraction

$$C_\Gamma : H_\mathbf{Q}^{\otimes 6k} \longrightarrow \mathbf{Q}$$

and its restrictions to the subspaces $\Lambda^{2k}(\Lambda^3 H_\mathbf{Q})$ or $\Lambda^{2k}(U_\mathbf{Q})$ of $H_\mathbf{Q}^{\otimes 6k}$ yield the above homomorphisms. This kind of use of graphs and representation theory appeared already in the works of Kontsevich [18, 19]. A detailed proof of these facts will be given in our forthcoming paper [17].

If we combine the above argument with Proposition 3.2 and Proposition 3.3, then we obtain homomorphisms

$$\Phi : \mathbf{Q}[\alpha_\Gamma; \Gamma \in \mathcal{G}] \longrightarrow H^*(\mathcal{M}_{g,*}; \mathbf{Q})$$

$$\Phi : \mathbf{Q}[\beta_\Gamma; \Gamma \in \mathcal{G}^0] \longrightarrow H^*(\mathcal{M}_g; \mathbf{Q})$$

which are defined at the cocycle level. More precisely, any (not necessarily connected) trivalent graph Γ with $2k$ vertices defines a $2k$-cocycle of the mapping class group $\mathcal{M}_{g,*}$ and if Γ has no loops, then it defines a $2k$-cocycle of $\mathcal{M}_g$.

Now we can state our main result.

Theorem 3.4. *The following diagram is commutative*

$$
\begin{array}{ccc}
\mathbf{Q}[\alpha_\Gamma; \Gamma \in \mathcal{G}] & \xrightarrow{\ \Phi\ } & H^*(\mathcal{M}_{g,*}; \mathbf{Q}) \\
\Big\uparrow & & \Big\uparrow{\scriptstyle \pi^*} \\
\mathbf{Q}[\beta_\Gamma; \Gamma \in \mathcal{G}^0] & \xrightarrow[\ \Phi\ ]{} & H^*(\mathcal{M}_g; \mathbf{Q})
\end{array}
$$

where the right vertical homomorphism is induced from the projection $\pi : \mathcal{M}_{g,*} \to \mathcal{M}_g$
*and the left vertical map is a "twisted" inclusion mentioned above. Moreover any
characteristic class* $e_i \in H^{2i}(\mathcal{M}_g; \mathbf{Q})(i = 1, 2, \cdots)$ *as well as the Euler class* $e \in$
$H^2(\mathcal{M}_{g,*} : \mathbf{Q})$ *belong to the images of the homomorphisms* Φ *so that* $\mathrm{Im}\Phi$ *contains
the subalgebra of the cohomology algebra of the mapping class groups generated by
these characteristic classes.*

Remark 3.5. As was mentioned in the introduction, in a joint work with Kawazumi,
we have determined the homomorphism Φ completely. More precisely, we proved that
$\mathrm{Im}\Phi$ is exactly equal to the subalgebra generated by the characteristic classes e_i and
e. Moreover we obtained explicit formulae to express the classes $\Phi(\alpha_\Gamma)$ as well as
$\Phi(\beta_\Gamma)$ as polynomials in e_i, e. Details will be given in [17].

4. Description of the representation ρ_2 of the mapping class group

In this section we describe the representation

$$
\rho_2 : \mathcal{M} \longrightarrow \mathrm{Aut} N_2
$$

(see §2 for the definition) for all of the three types of the mapping class groups
$\mathcal{M} = \mathcal{M}_{g,1}, \mathcal{M}_{g,*}, \mathcal{M}_g$. However here we only give the final results. Detailed proof
for them will appear in [32] because they are technically rather complicated and long.

First we consider the case of $\mathcal{M}_{g,1}$. The representation ρ_2 lies over the previous
one

$$
\rho_1 : \mathcal{M}_{g,1} \longrightarrow \frac{1}{2}\mathcal{H}_1 \rtimes \mathrm{Sp}(2g; \mathbf{Z})
$$

where $\mathcal{H}_1 = \Lambda^3 H$. The restriction of ρ_1 to the Torelli group $\mathcal{I}_{g,1} = \mathcal{M}_{g,1}(1)$ is equal
to Johnson's homomorphism

$$
\tau_1 : \mathcal{I}_{g,1} \longrightarrow \mathcal{H}_1.
$$

It was proved by Johnson [13] that $\mathrm{Ker}\,\tau_1 = \mathcal{M}_{g,1}(2)$ coincides with the subgroup
$\mathcal{K}_{g,1}$ of $\mathcal{M}_{g,1}$ which is the one generated by Dehn twists along separating simple closed
curves on $\Sigma_g \backslash \mathrm{Int} D$. In our paper [26], we determined the image of the homomorphism

$$
\tau_2 : \mathcal{K}_{g,1} \longrightarrow \mathcal{H}_2^0
$$

where recall that $\mathcal{H}_2^0 = \mathrm{Ker}(H \otimes \mathcal{L}_3^0 \to \mathcal{L}_4^0)$. In that paper, we expressed $\mathcal{H}_2^0$ as follows.
Let $S^2(\Lambda^2 H)$ be the second symmetric power of $\Lambda^2 H$. We write simply $a \wedge b \leftrightarrow c \wedge d$
instead of $(a \wedge b) \otimes (c \wedge d) + (c \wedge d) \otimes (a \wedge b) \in S^2(\Lambda^2 H)(a, b, c, d \in H)$. We can define a
homomorphism

$$
p : S^2(\Lambda^2 H) \longrightarrow \mathcal{H}_2^0
$$

by setting

$$(4.1) \quad p(a \wedge b \leftrightarrow c \wedge d) = a \otimes [b, [c, d]] - b \otimes [a, [c, d]] + c \otimes [d, [a, b]] - d \otimes [c, [a, b]].$$

The right hand side is an element of $H \otimes \mathcal{L}_3^0$. But it is easy to see that it is in fact contained in $\mathcal{H}_2^0$. Now $\mathrm{Sp}(2g; \mathbf{Z})$ acts on $S^2(\Lambda^2 H)$ as well as $\mathcal{H}_2^0$ naturally and the homomorphism p is $\mathrm{Sp}(2g; \mathbf{Z})$-equivariant. Moreover the vector spaces $S^2(\Lambda^2 H_{\mathbf{Q}})$ and $\mathcal{H}_2^0 \otimes \mathbf{Q}$ are rational representations of $\mathrm{Sp}(2g; \mathbf{Q})$ so that they can be expressed as a direct sum of irreducible representations of $\mathrm{Sp}(2g; \mathbf{Q})$.

Here we use the terminologies of the book [7] to express irreducible representations of $\mathrm{Sp}(2g; \mathbf{Q})$. Thus let $\mathfrak{sp}(2g; \mathbf{C})$ be the Lie algebra of $\mathrm{Sp}(2g; \mathbf{C})$ and let $\mathfrak{h}$ be its Cartan subalgebra consisting of diagonal matrices. Choose a system of fundamental weights $L_i : \mathfrak{h} \to \mathbf{R} (i = 1, \cdots, g)$ as in [7] and for each g-tuple $(a_1, \cdots, a_g)$ of natural numbers, let $\Gamma_{a_1, \cdots, a_g}$ denote the irreducible representation with highest weight $(a_1 + \cdots + a_g)L_1 + (a_2 + \cdots + a_g)L_2 + \cdots + a_g L_g$. These representations are all rational representations defined over $\mathbf{Q}$ so that we can consider them as irreducible representations of $\mathrm{Sp}(2g; \mathbf{Q})$. For example $H_{\mathbf{Q}} = \Gamma_1$ and $U_{\mathbf{Q}} = \Gamma_{0,0,1}$.

Lemma 4.1. *We have the following irreducible decompositions.*

   (i) $S^2(\Lambda^2 H_{\mathbf{Q}}) \cong 2\mathbf{Q} + 2\Gamma_{0,1} + \Gamma_{0,2} + \Gamma_{0,0,0,1}$ *for* $g \geq 4$.

   (ii) $\mathcal{H}_2^0 \otimes \mathbf{Q} \cong \mathbf{Q} + \Gamma_{0,1} + \Gamma_{0,2}$ *for* $g \geq 2$.

Moreover the mapping $p : S^2(\Lambda^2 H_{\mathbf{Q}}) \longrightarrow \mathcal{H}_2^0 \otimes \mathbf{Q}$ *is surjective so that* $\mathrm{Ker}\, p \cong \mathbf{Q} + \Gamma_{0,1} + \Gamma_{0,0,0,1}$.

Remark 4.2. The irreducible decompositions for small values of g can be also given and it turns out that p is surjective for all $g \geq 2$. However here we omit them. This remark also applies to similar statements in the following sections.

Now we have two homomorphisms

$$\tau_1 : \mathcal{I}_{g,1} \to \mathcal{H}_1, \quad \tau_2 : \mathcal{K}_{g,1} \to \mathcal{H}_2^0$$

and it can be shown that the associated extension

$$1 \longrightarrow \mathcal{K}_{g,1} / \mathrm{Ker}\, \tau_2 \longrightarrow \mathcal{I}_{g,1} / \mathrm{Ker}\, \tau_2 \overset{\tau_1}{\longrightarrow} \mathcal{H}_1 \longrightarrow 1$$

is a central extension. Moreover the Euler class $\chi \in H^2(\mathcal{H}_1; \mathrm{Im}\tau_2)$ of this central extension was determined in [27] as

$$\begin{aligned}
(4.2) \quad \chi(\xi, \eta) = &-(a \cdot d)b \wedge c \leftrightarrow e \wedge f - (a \cdot e)b \wedge c \leftrightarrow f \wedge d - (a \cdot f)b \wedge c \leftrightarrow d \wedge e \\
&- (b \cdot d)c \wedge a \leftrightarrow e \wedge f - (b \cdot e)c \wedge a \leftrightarrow f \wedge d - (b \cdot f)c \wedge a \leftrightarrow d \wedge e \\
&- (c \cdot d)a \wedge b \leftrightarrow e \wedge f - (c \cdot e)a \wedge b \leftrightarrow f \wedge d - (c \cdot f)a \wedge b \leftrightarrow d \wedge e
\end{aligned}$$

where $\xi = a \wedge b \wedge c, \eta = d \wedge e \wedge f \in \mathcal{H}_1 = \Lambda^3 H$ $(a, b, c, d, e, f \in H)$. If we substitute (4.1) into (4.2), then we obtain the expression of χ as an element of $H^2(\mathcal{H}_1; \mathcal{H}_2^0)$ and we can conclude that

$$\chi(\xi, \eta) = -[\xi, \eta]$$

where $[\;,\;] : \mathcal{H}_1 \otimes \mathcal{H}_1 \to \mathcal{H}_2^0$ denotes the bracket of the graded Lie algebra $\oplus_{k \geq 1} \mathcal{H}_k^0$. We remark that the minus sign here comes from our sign convention in the definition of τ_2 in [26] so that it is not essential.

In view of the above facts, if we define a 2-cocycle of $\mathcal{H}_1$ with values in $\frac{1}{2}\mathcal{H}_2^0$ by

$$(4.3) \qquad \mathcal{H}_1 \times \mathcal{H}_1 \ni (\xi, \eta) \mapsto \frac{1}{2}[\xi, \eta] \in \frac{1}{2}\mathcal{H}_2^0,$$

then the restriction of ρ_2 to the Torelli group $\mathcal{I}_{g,1}$ can be explicitly described by a representation

$$(4.4) \qquad \rho_2 : \mathcal{I}_{g,1} \longrightarrow \frac{1}{2}\mathcal{H}_2^0 \widetilde{\times} \mathcal{H}_1$$

where the target group is the central extension of $\mathcal{H}_1$ by $\frac{1}{2}\mathcal{H}_2^0$ defined by the above 2-cocycle.

The cocycle (4.3) is clearly invariant under the natural action of $\mathrm{Sp}(2g; \mathbf{Z})$ so that we can make the semi-direct product of $\frac{1}{2}\mathcal{H}_2^0 \widetilde{\times} \mathcal{H}_1$ with $\mathrm{Sp}(2g; \mathbf{Z})$. Moreover in doing so we can allow any natural number for the denominator other than the number 2. Now we have the following result whose proof will be given in [32].

Theorem 4.3. *The representation* $\rho_2 : \mathcal{M}_{g,1} \to \mathrm{Aut} N_2^0$ *can be described explicitly by a homomorphism*

$$\rho_2 : \mathcal{M}_{g,1} \longrightarrow \left(\frac{1}{24}\mathcal{H}_2^0 \widetilde{\times} \frac{1}{2}\mathcal{H}_1\right) \rtimes \mathrm{Sp}(2g; \mathbf{Z})$$

whose restriction to the Torelli group $\mathcal{I}_{g,1}$ *is equal to the representation* $\rho_2 : \mathcal{I}_{g,1} \to \frac{1}{2}\mathcal{H}_2^0 \widetilde{\times} \mathcal{H}_1$ *given above.*

Next we consider the case of $\mathcal{M}_{g,*}$. The restriction of ρ_2 to the subgroup $\mathcal{K}_{g,1}$ is equal to the homomorphism

$$\tau_2 : \mathcal{K}_{g,1} \longrightarrow \mathcal{H}_2^0.$$

Let $\iota \in \mathbf{Z} \subset \mathcal{K}_{g,1}$ be the generator of $\mathrm{Ker}(\mathcal{M}_{g,1} \to \mathcal{M}_{g,*}) \cong \mathbf{Z}$. Then by the result of [26], we have

$$\tau_2(\iota) = -\sum_{i,j=1}^{g} \left\{ x_i \otimes [y_i, [x_j, y_j]] - y_i \otimes [x_i, [x_j, y_j]] \right\} \in \mathcal{H}_2^0.$$

Lemma 4.4. *We have the following irreducible decompositions.*

    (i) $I_3 \otimes \mathbf{Q} = \Gamma_1 = H_{\mathbf{Q}}$ *for* $g \geq 2$.
    (ii) $I_4 \otimes \mathbf{Q} = \Gamma_2 + \Gamma_{0,1}$ *for* $g \geq 2$.
    (iii) $H_{\mathbf{Q}} \otimes I_3 = \mathbf{Q} + \Gamma_2 + \Gamma_{0,1}$ *for* $g \geq 2$.

Since the natural mapping $\mathrm{Hom}(H, I_3) \cong H \otimes I_3 \to I_4$ is surjective (cf the commutative diagram (2.1)), in view of Lemma 4.4 above, we can conclude that

$$\mathrm{Ker}(\mathcal{H}_2^0 \longrightarrow \mathcal{H}_2) \cong \mathbf{Z}.$$

It follows that

$$(4.5) \qquad \mathcal{H}_2 \otimes \mathbf{Q} \cong \Gamma_{0,1} + \Gamma_{0,2}$$

by Lemma 4.1. Moreover it is easy to see that the element $\tau_2(\iota)$ is the generator of $\mathrm{Ker}(\mathcal{H}_2^0 \longrightarrow \mathcal{H}_2) = \mathbf{Z}$. Since ι is the generator of $\mathrm{Ker}(\mathcal{K}_{g,1} \to \mathcal{M}_{g,*}(2)) \cong \mathrm{Ker}(\mathcal{M}_{g,1} \to \mathcal{M}_{g,*}) \cong \mathbf{Z}$, we obtain the following corollary to Theorem 4.3.

Corollary 4.5. *The representation $\rho_2 : \mathcal{M}_{g,*} \to \mathrm{Aut}\, N_2$ can be described explicitly by a homomorphism*

$$\rho_2 : \mathcal{M}_{g,*} \longrightarrow \Big(\frac{1}{24}\mathcal{H}_2 \tilde{\times} \frac{1}{2}\mathcal{H}_1\Big) \rtimes \mathrm{Sp}(2g; \mathbf{Z}).$$

Finally we consider the case of $\mathcal{M}_g$. We know that $\pi_1 \Sigma_g \cap \mathcal{M}_{g,*}(2) = \Gamma_1(\pi_1 \Sigma_g)$ and also that $\Gamma_1(\pi_1 \Sigma_g)/\Gamma_2(\pi_1 \Sigma_g) = \mathcal{L}_2$ injects into $\mathcal{H}_2$ and we have $\overline{\mathcal{H}}_2 = \mathcal{H}_2/\mathcal{L}_2$ by the definition. Clearly we have $\mathcal{L}_2 \otimes \mathbf{Q} \cong \Gamma_{0,1}$ so that

$$\overline{\mathcal{H}}_2 \otimes \mathbf{Q} \cong \Gamma_{0,2}$$

by (4.5). On the other hand, it is easy to see that the bracket $\mathcal{H}_1 \otimes \mathcal{H}_1 \to \mathcal{H}_2$ induces $\overline{\mathcal{H}}_1 \otimes \overline{\mathcal{H}}_1 \to \overline{\mathcal{H}}_2$. In other words, the 2-cocycle (4.3) induces

$$\overline{\mathcal{H}}_1 \times \overline{\mathcal{H}}_1 \ni (\xi, \eta) \mapsto \frac{1}{2}[\xi, \eta] \in \frac{1}{2}\overline{\mathcal{H}}_2$$

which is a 2-cocycle of $\overline{\mathcal{H}}_1$ with values in $\frac{1}{2}\overline{\mathcal{H}}_2$. Hence we can make the central extension $\frac{1}{2}\overline{\mathcal{H}}_2 \tilde{\times} \overline{\mathcal{H}}_1$ and we have

Corollary 4.6. *The representation $\bar{\rho}_2 : \mathcal{M}_g \to \mathrm{Out}\, N_2$ can be described explicitly by a homomorphism*

$$\bar{\rho}_2 : \mathcal{M}_g \longrightarrow \Big(\frac{1}{24}\overline{\mathcal{H}}_2 \tilde{\times} \frac{1}{2}\overline{\mathcal{H}}_1\Big) \rtimes \mathrm{Sp}(2g; \mathbf{Z}).$$

5. The universal Euler class $e \in H^2(\mathcal{M}_{g,*}; \mathbf{Z})$

In this section we represent the universal Euler class $e \in H^2(\mathcal{M}_{g,*}; \mathbf{Z})$, which is the Euler class of the extension $1 \to \mathbf{Z} \to \mathcal{M}_{g,1} \to \mathcal{M}_{g,*} \to 1$, in terms of the representation

$$\rho_1 : \mathcal{M}_{g,*} \longrightarrow \frac{1}{2}\Lambda^3 H \rtimes \mathrm{Sp}(2g; \mathbf{Z})$$

mentioned in §2.

For that we define two 2-cocycles z_1, z_2 of the group $\frac{1}{2}\Lambda^3 H \rtimes \mathrm{Sp}(2g; \mathbf{Z})$ as follows. We can define two $\mathrm{Sp}(2g; \mathbf{Z})$-equivariant homomorphisms $C_i : \frac{1}{2}\Lambda^3 H \otimes \frac{1}{2}\Lambda^3 H \to \mathbf{Z}$ by making use of two different ways of contractions. Namely, for any homology classes $a_i, b_i \in H$ ($i = 1, 2, 3$), we set

$$C_1(a_1 \wedge a_2 \wedge a_3, b_1 \wedge b_2 \wedge b_3) = \sum_{\sigma, \tau \in \mathfrak{S}_3} \mathrm{sgn}\sigma \mathrm{sgn}\tau\, (a_{\sigma(1)} \cdot a_{\sigma(2)})(b_{\tau(1)} \cdot b_{\tau(2)})(a_{\sigma(3)} \cdot b_{\tau(3)})$$

$$C_2(a_1 \wedge a_2 \wedge a_3, b_1 \wedge b_2 \wedge b_3) = \sum_{\sigma, \tau \in \mathfrak{S}_3} \mathrm{sgn}\sigma \mathrm{sgn}\tau\, (a_{\sigma(1)} \cdot b_{\tau(1)})(a_{\sigma(2)} \cdot b_{\tau(2)})(a_{\sigma(3)} \cdot b_{\tau(3)})$$

where $\mathfrak{S}_3$ is the symmetric group of degree 3 and $a \cdot b \in \mathbf{Z}$ denotes the intersection number of a with b. For example, if $x_1, \cdots, x_g, y_1, \cdots, y_g$ is a symplectic basis of H, then $C_1(x_1 \wedge y_1 \wedge x_2, x_1 \wedge y_1 \wedge y_2) = 4$ and $C_2(x_1 \wedge y_1 \wedge x_2, x_1 \wedge y_1 \wedge y_2) = 6$. It will be proved below that the above two contractions in fact form a basis of $\mathrm{Sp}(2g;\mathbf{Z})$-equivariant homomorphisms $\Lambda^3 H \otimes \Lambda^3 H \to \mathbf{Z}$ (see Lemma 5.3). Now, for any two elements $(\xi, A), (\eta, B) \in \frac{1}{2}\Lambda^3 H \rtimes \mathrm{Sp}(2g;\mathbf{Z})$, we define

$$z_i((\xi, A), (\eta, B)) = C_i(\xi, A\eta) \quad (i = 1, 2).$$

It is easy to show that z_i is in fact a 2-cocycle of $\frac{1}{2}\Lambda^3 H \rtimes \mathrm{Sp}(2g;\mathbf{Z})$.

Theorem 5.1. *The Euler class $e \in H^2(\mathcal{M}_{g,*};\mathbf{Q})$ is represented by the 2-cocycle*

$$-\frac{1}{2g(2g+1)}\rho_1^*(z_1 + z_2)$$

of the group $\mathcal{M}_{g,*}$.

Observe here that the 2-cocycle $\rho_1^*(z_i) \in Z^2(\mathcal{M}_{g,*};\mathbf{Z})$ is described explicitly by

$$\rho_1^*(z_i)(\varphi, \psi) = C_i(\bar{k}(\varphi), \varphi_* \bar{k}(\psi)) \quad (\varphi, \psi \in \mathcal{M}_{g,*}).$$

The best way to give a conceptual proof of the above theorem is to use the explicit description of the representation ρ_2, which lies over ρ_1 and was sketched in §4 (details of which will be given in our future paper [32]).

Before proving Theorem 5.1, we prepare a few technical results. According to Lemma 4.1, we have two linearly independent $\mathrm{Sp}(2g;\mathbf{Z})$-equivariant homomorphisms $S^2(\Lambda^2 H) \to \mathbf{Z}$. These are explicitly given in our previous papers [26, 27]. One is $\bar{d}$ given in [26, Proposition 4.8] and the other is $\vec{d}$ given in [27, Proposition 5.1]. As was proved in [27], the latter one is trivial on $\mathrm{Ker}(S^2(\Lambda^2 H) \to \mathcal{H}_2^0)$ so that it defines an $\mathrm{Sp}(2g;\mathbf{Z})$-equivariant homomorphism $\vec{d} : \mathcal{H}_2^0 \to \mathbf{Z}$ (we use the same letter) which is uniquely defined up to non-zero scalars by Lemma 4.1 (ii). More precisely, we have

Lemma 5.2. *The $\mathrm{Sp}(2g;\mathbf{Z})$-equivariant homomorphism $\vec{d} : S^2(\Lambda^2 H) \to \mathbf{Z}$ defined by*

$$\vec{d}(a\wedge b \leftrightarrow c\wedge d) = -4(a \cdot b)(c \cdot d) - 2(a \cdot c)(b \cdot d) + 2(a \cdot d)(b \cdot c) \quad (a, b, c, d \in H)$$

descends to an $\mathrm{Sp}(2g;\mathbf{Z})$-equivariant homomorphism $\vec{d} : \mathcal{H}_2^0 \to \mathbf{Z}$ which is given explicitly by

$$\vec{d}(a \otimes [b, [c, d]]) = -(a \cdot b)(c \cdot d) - \frac{1}{2}(a \cdot c)(b \cdot d) + \frac{1}{2}(a \cdot d)(b \cdot c).$$

Proof. The former part is proved in [27]. Then the latter part follows immediately by applying the homomorphism $p : S^2(\Lambda^2 H) \to \mathcal{H}_2^0$ given in (4.1). $\square$

Lemma 5.3. *We have the following irreducible decomposition*

$$\Lambda^3 H_{\mathbf{Q}} = \Gamma_1 + \Gamma_{0,0,1}$$

so that

$$\dim \mathrm{Hom}(\Lambda^3 H_{\mathbf{Q}} \otimes \Lambda^3 H_{\mathbf{Q}}, \mathbf{Q})^{\mathrm{Sp}(2g;\mathbf{Q})} = 2$$

for all $g \geq 3$.

Proof. The former statement follows from a direct computation. Having this, we can deduce the latter statement because of the following two general facts about (rational) representations of $\mathrm{Sp}(2g; \mathbf{Q})$ (see [7]). One is that for any two irreducible representations V, W of $\mathrm{Sp}(2g; \mathbf{Q})$, the tensor product $V \otimes W$ contains a trivial representation if and only if $V \cong W$ and in this case the multiplicity of the trivial part is exactly equal to 1. The other is the fact that the dual of any irreducible representation V of $\mathrm{Sp}(2g; \mathbf{Q})$ is canonically isomorphic to V itself. $\square$

Lemma 5.4. *The two contractions* $C_i : \Lambda^3 H \otimes \Lambda^3 H \to \mathbf{Z}\,(i = 1, 2)$ *defined in the beginning of this section form a basis of*

$$\mathrm{Hom}(\Lambda^3 H_{\mathbf{Q}} \otimes \Lambda^3 H_{\mathbf{Q}}, \mathbf{Q})^{\mathrm{Sp}(2g;\mathbf{Q})}.$$

Proof. Clearly both of C_i define non trivial homomorphisms. Hence, by Lemma 5.3, it is enough to prove that C_1 and C_2 are linearly independent. To show this, we set $\xi_1 = x_1 \wedge y_1 \wedge x_2, \eta_1 = x_1 \wedge y_1 \wedge y_2$ and $\xi_2 = x_1 \wedge x_2 \wedge x_3, \eta_2 = y_1 \wedge y_2 \wedge y_3$. Then direct computation yields

$$C_1(\xi_1, \eta_1) = 4, \quad C_2(\xi_1, \eta_1) = 6$$
$$C_1(\xi_2, \eta_2) = 0, \quad C_2(\xi_2, \eta_2) = 6.$$

This completes the proof. $\square$

Proposition 5.5. *Let* $\xi, \eta \in \mathcal{H}_1$ *be any two elements and let* $[\xi, \eta] \in \mathcal{H}_2^0$ *be the bracket of them. Then we have*

$$(z_1 + z_2)(\xi, \eta) = -\vec{d}\,([\xi, \eta]).$$

Proof. Clearly both of $z_1 + z_2$ and $-\vec{d}\,([\,,\,])$ define $\mathrm{Sp}(2g; \mathbf{Z})$-equivariant homomorphisms

$$\mathcal{H}_1 \otimes \mathcal{H}_1 \longrightarrow \mathbf{Z}$$

which can be tensored with $\mathbf{Q}$. Hence, by Lemma 5.4, it is enough to prove that the above two homomorphisms take the same values on two vectors $(\xi_i, \eta_i)\,(i = 1, 2)$ which were defined in the proof of Lemma 5.4. By (4.2), we have

$$[\xi_1, \eta_1] = p(y_1 \wedge x_2 \leftrightarrow y_2 \wedge x_1 - x_2 \wedge x_1 \leftrightarrow y_1 \wedge y_2 + x_1 \wedge y_1 \leftrightarrow x_1 \wedge y_1)$$
$$[\xi_2, \eta_2] = p(x_2 \wedge x_3 \leftrightarrow y_2 \wedge y_3 + x_3 \wedge x_1 \leftrightarrow y_3 \wedge y_1 + x_1 \wedge x_2 \leftrightarrow y_1 \wedge y_2).$$

Hence

$$\vec{d}\,([\xi_1, \eta_1]) = -10, \quad \vec{d}\,([\xi_2, \eta_2]) = -6.$$

On the other hand, we have

$$z_1(\xi_1, \eta_1) = 4, \quad z_2(\xi_1, \eta_1) = 6$$
$$z_1(\xi_2, \eta_2) = 0, \quad z_2(\xi_2, \eta_2) = 6.$$

This completes the proof. $\square$

Proof of Theorem 5.1. We have the following commutative diagram of group extensions

$$
\begin{array}{ccc}
\mathbf{Z} & =\!=\!= & \mathbf{Z} \\
\downarrow & & \downarrow \\
\mathcal{M}_{g,1} \xrightarrow{\ \rho_2^0\ } & (\frac{1}{24}\mathcal{H}_2^0 \tilde{\times} \frac{1}{2}\mathcal{H}_1) \rtimes \mathrm{Sp}(2g;\mathbf{Z}) \\
\downarrow & & \downarrow \\
\mathcal{M}_{g,*} \xrightarrow{\ \rho_2\ } & (\frac{1}{24}\mathcal{H}_2 \tilde{\times} \frac{1}{2}\mathcal{H}_1) \rtimes \mathrm{Sp}(2g;\mathbf{Z})
\end{array}
$$

so that the central extension on the left vertical line is the pull back of that on the right vertical line by the homomorphism ρ_2. Hence the Euler class $e \in H^2(\mathcal{M}_{g,*};\mathbf{Q})$ is the pull back of the Euler class

$$
\chi \in H^2\Big((\frac{1}{24}\mathcal{H}_2 \tilde{\times} \frac{1}{2}\mathcal{H}_1) \rtimes \mathrm{Sp}(2g;\mathbf{Z});\mathbf{Q}\Big)
$$

of the latter central extension. We claim that χ is represented by the 2-cocycle

$$
-\frac{1}{2g(2g+1)}(z_1 + z_2) \in Z^2\big(\frac{1}{2}\Lambda^3 H \rtimes \mathrm{Sp}(2g;\mathbf{Z});\mathbf{Q}\big)
$$

which is now considered as the cocycle of the above group through the projection

$$
(\frac{1}{24}\mathcal{H}_2 \tilde{\times} \frac{1}{2}\mathcal{H}_1) \rtimes \mathrm{Sp}(2g;\mathbf{Z}) \longrightarrow \frac{1}{2}\Lambda^3 H \rtimes \mathrm{Sp}(2g;\mathbf{Z}).
$$

By the Gysin exact sequence of central extensions, we have only to check the following two conditions. One is that the pull back of χ to the group

$$
(5.1) \qquad\qquad (\frac{1}{24}\mathcal{H}_2^0 \tilde{\times} \frac{1}{2}\mathcal{H}_1) \rtimes \mathrm{Sp}(2g;\mathbf{Z})
$$

should be zero. The other is the fact that χ is the transgression of $1 \in H^1(\mathbf{Z};\mathbf{Z}) \cong \mathbf{Z}$.

First we consider the former condition. We prove that the pull back of the cocycle $z_1 + z_2$ to the group (5.1) is a coboudary. We define a 1-cochain

$$
f : (\frac{1}{24}\mathcal{H}_2^0 \tilde{\times} \frac{1}{2}\mathcal{H}_1) \rtimes \mathrm{Sp}(2g;\mathbf{Z}) \longrightarrow \mathbf{Q}
$$

by

$$
f(\kappa,\xi,A) = \vec{d}(\kappa) \quad (\kappa \in \frac{1}{24}\mathcal{H}_2^0, \xi \in \frac{1}{2}\mathcal{H}_1, A \in \mathrm{Sp}(2g;\mathbf{Z})).
$$

We claim that

$$
(5.2) \qquad\qquad z_1 + z_2 = -2\delta f \in C^2\Big((\frac{1}{24}\mathcal{H}_2^0 \tilde{\times} \frac{1}{2}\mathcal{H}_1) \rtimes \mathrm{Sp}(2g;\mathbf{Z});\mathbf{Q}\Big).
$$

To show this, let $(\kappa,\xi,A), (\lambda,\eta,B)$ be any two elements of the above group. Then we have

$$
(\kappa,\xi,A)(\lambda,\eta,B) = (\kappa + A\lambda + \frac{1}{2}[\xi, A\eta], \xi + A\eta, AB).
$$

Therefore

$$(5.3) \qquad \delta f((\kappa, \xi, A), (\lambda, \eta, B)) = \vec{d}(\lambda) - \vec{d}(\kappa + A\lambda + \tfrac{1}{2}[\xi, A\eta]) + \vec{d}(\kappa)$$

$$(5.4) \qquad = -\frac{1}{2}\vec{d}([\xi, A\eta]).$$

Here we have used the fact that the homomorphism $\vec{d} : \mathcal{H}_2^0 \to \mathbf{Z}$ is $\mathrm{Sp}(2g; \mathbf{Z})$-equivariant.

On the other hand, we have

$$(5.5) \qquad (z_1 + z_2)((\kappa, \xi, A), (\lambda, \eta, B)) = (z_1 + z_2)(\xi, A\eta).$$

Now if we apply Proposition 5.5 to (5.3),(5.4), we can conclude $z_1 + z_2 = 2\delta f$. Hence the pull back of the cohomology class χ to the group (5.1) is zero as required.

Next we check the second condition. For that, it is enough to evaluate the cocycles $\rho_1^*(z_i)$ $(i = 1, 2)$ on the fundamental cycle of $\pi_1\Sigma_g \subset \mathcal{M}_{g,*}$. It was shown in [30] that the restriction of $\tilde{k}$ to $\pi_1\Sigma_g \subset \mathcal{M}_{g,*}$ is a homomorphism and it is equal to the abelianization $\pi_1\Sigma_g \to H$ followed by the embedding $i : H \to \Lambda^3 H$ so that we have the following commutative diagram

$$
\begin{array}{ccc}
\pi_1\Sigma_g & \xrightarrow{\ \tilde{k}\ } & \tfrac{1}{2}\Lambda^3 H \\
\downarrow & & \| \\
H & \xrightarrow[\ i\ ]{} & \tfrac{1}{2}\Lambda^3 H.
\end{array}
$$

Now the fundamental cycle $\sigma_0 \in Z_2(\pi_1\Sigma_g; \mathbf{Z})$ of $\pi_1\Sigma_g$ goes to the 2-cycle

$$c_0 = \sum_{i=1}^{g} \left\{ (x_i, y_i) - (y_i, x_i) \right\} \in Z_2(H; \mathbf{Z})$$

of H under the abelianization. On the other hand, a direct computation shows that the composition

$$H \otimes H \xrightarrow{i \otimes i} \frac{1}{2}\Lambda^3 H \otimes \frac{1}{2}\Lambda^3 H \xrightarrow{C_i} \mathbf{Z}$$

is equal to $(2g - 2)^2\mu$, $6(g - 1)\mu$ for $i = 1, 2$, respectively. Here $\mu : H \otimes H \to \mathbf{Z}$ denotes the intersection pairing and is considered as a 2-cocycle of the abelian group H. Hence we have

$$\begin{aligned}
\rho_1^*(z_1 + z_2)(\sigma_0) &= \{(2g - 2)^2 + 6(g - 1)\}\mu(c_0) \\
&= 4g(g - 1)(2g + 1).
\end{aligned}$$

It follows that the cocyle given in the proposition takes the right value $2 - 2g$ on σ_0. This completes the proof. $\square$

In our paper [25, Theorem 1.7], we proved the following result.

Theorem 5.6. *The 2-cocycle $\rho_1^*(z_1)$ of the group $\mathcal{M}_{g,*}$ represents the cohomology class $-4g(g - 1)e - e_1 \in H^2(\mathcal{M}_{g,*}; \mathbf{Q})$.*

By combining Theorem 5.1 and Theorem 5.6, we can conclude that the first characteristic class $e_1 \in H^2(\mathcal{M}_{g,*}; \mathbf{Q})$ is represented by the 2-cocycle

$$-\frac{3}{2g+1}\rho_1^*(z_1) + \frac{2(g-1)}{2g+1}\rho_1^*(z_2)$$

of the group $\mathcal{M}_{g,*}$.

Lemma 5.7. *If we set*

$$C_0 = \frac{1}{2g+1}\big(-3C_1 + 2(g-1)C_2\big),$$

then $C_0 = 0$ on $H \otimes \Lambda^3 H + \Lambda^3 H \otimes H \subset \Lambda^3 H \otimes \Lambda^3 H$ so that it induces an $\mathrm{Sp}(2g; \mathbf{Z})$-equivariant homomorphism

$$\overline{C}_0 : \Lambda^3 H/H \otimes \Lambda^3 H/H \longrightarrow \mathbf{Z}.$$

Proof. For any element $a \in H$ and $\xi \in \Lambda^3 H$, we have

$$C_1(a \wedge \omega_0 \otimes \xi) = 2(g-1)a \cdot C\xi, \quad C_2(a \wedge \omega_0 \otimes \xi) = 3a \cdot C\xi$$

where $C : \Lambda^3 H \to H$ denotes the contraction. The required result follows from this. $\square$

Now we consider the representation

$$\bar{\rho}_1 : \mathcal{M}_g \longrightarrow \frac{1}{2}\Lambda^3 H/H \rtimes \mathrm{Sp}(2g; \mathbf{Z}).$$

We can define a 2-cocycle z of the group $\frac{1}{2}\Lambda^3 H/H \rtimes \mathrm{Sp}(2g; \mathbf{Z})$ by setting

$$z((\xi, A), (\eta, B)) = \overline{C}_0(\xi, A\eta) \quad (\xi, \eta \in \frac{1}{2}\Lambda^3 H/H).$$

Then we can summarize the above argument as follows.

Theorem 5.8. *The first characteristic class $e_1 \in H^2(\mathcal{M}_g; \mathbf{Q})$ is represented by the 2-cocycle $\bar{\rho}_1^*(z)$ of $\mathcal{M}_g$ which is described explicitly by*

$$\bar{\rho}_1^*(z)(\varphi, \psi) = \overline{C}_0(\bar{k}(\varphi), \varphi_* \bar{k}(\psi)) \quad (\varphi, \psi \in \mathcal{M}_g)$$

where $\bar{k} : \mathcal{M}_g \to \frac{1}{2}\Lambda^3 H/H$ is the crossed homomorphism associated to the representation $\bar{\rho}_1$.

6. Determination of the kernel of $\tau_1^* : H^2(\Lambda^3 H; \mathbf{Q}) \to H^2(\mathcal{I}_{g,*}; \mathbf{Q})$

The restriction of the representation $\rho_1 : \mathcal{M}_{g,*} \to \frac{1}{2}\Lambda^3 H \rtimes \mathrm{Sp}(2g; \mathbf{Z})$ to the Torelli group $\mathcal{I}_{g,*}$ is equal to Johnson's homomorphism

$$\tau_1 : \mathcal{I}_{g,*} \longrightarrow \Lambda^3 H.$$

In this section, we determine the kernel of the homomorphism $\tau_1^* : H^2(\Lambda^3 H; \mathbf{Q}) \to H^2(\mathcal{I}_{g,*}; \mathbf{Q})$ induced by τ_1. This is an extension of a result of Hain [9] in which he determines $\mathrm{Ker}\big(\tau_1^* : H^2(\Lambda^3 H/H; \mathbf{Q}) \to H^2(\mathcal{I}_g; \mathbf{Q})\big)$ as follows, where $\tau_1 : \mathcal{I}_g \to \Lambda^3 H/H = \overline{\mathcal{H}}_1$ is Johnson's homomorphism for the Torelli group $\mathcal{I}_g$. We use the same terminology as in §4,5 for irreducible representations of $\mathrm{Sp}(2g; \mathbf{Q})$.

Lemma 6.1. (Hain) *The representation $\overline{\mathcal{H}}_1 \otimes \mathbf{Q} = \Lambda^3 H_\mathbf{Q}/H_\mathbf{Q}$ of $\mathrm{Sp}(2g;\mathbf{Q})$ is irreducible and is isomorphic to $\Gamma_{0,0,1}$. Moreover we have the following irreducible decomposition*

$$H^2(\Lambda^3 H/H; \mathbf{Q}) \cong \Lambda^2 \Gamma_{0,0,1} = \mathbf{Q} + \Gamma_{0,2} + \Gamma_{0,1} + \Gamma_{0,0,0,1} + \Gamma_{0,1,0,1} + \Gamma_{0,0,0,0,0,1}$$

for all $g \geq 6$.

Theorem 6.2. (Hain) *Let $\tau_1 : \mathcal{I}_g \rightarrow \Lambda^3 H/H$ be Johnson's homomorphism. Then we have*

$$\mathrm{Ker}\big(\tau_1^* : H^2(\Lambda^3 H/H; \mathbf{Q}) \rightarrow H^2(\mathcal{I}_g; \mathbf{Q})\big) \cong \mathbf{Q} + \Gamma_{0,2}.$$

Let us explain "half" of the above result from our point of view. Namely we show that the summand $\mathbf{Q} + \Gamma_{0,2}$ is contained in $\mathrm{Ker}\,\tau_1^*$.

We consider the representation

$$\bar{\rho}_2 : \mathcal{M}_g \longrightarrow (\frac{1}{24}\overline{\mathcal{H}}_2 \tilde{\times} \frac{1}{2}\overline{\mathcal{H}}_1) \rtimes \mathrm{Sp}(2g;\mathbf{Z})$$

or rather its rational analogue

$$\bar{\rho}_2 : \mathcal{M}_g \longrightarrow (\overline{\mathcal{H}}_2^\mathbf{Q} \tilde{\times} \overline{\mathcal{H}}_1^\mathbf{Q}) \rtimes \mathrm{Sp}(2g;\mathbf{Z})$$

where $\overline{\mathcal{H}}_i^\mathbf{Q} = \overline{\mathcal{H}}_i \otimes \mathbf{Q}$. We know that $\overline{\mathcal{H}}_1^\mathbf{Q} = \Gamma_{0,0,1}$ by Lemma 6.1 and also that $\overline{\mathcal{H}}_2^\mathbf{Q} = \Gamma_{0,2}$ by the result of §4. Now the extension class of the nilpotent group $\overline{\mathcal{H}}_2^\mathbf{Q} \tilde{\times} \overline{\mathcal{H}}_1^\mathbf{Q}$ can be expressed as an $\mathrm{Sp}(2g;\mathbf{Q})$-equivariant homomorphism

$$(6.1) \qquad \Lambda^2 \overline{\mathcal{H}}_1^\mathbf{Q} \cong \Lambda^2 \Gamma_{0,0,1} \longrightarrow \overline{\mathcal{H}}_2^\mathbf{Q} \cong \Gamma_{0,2}.$$

We know that it is non-trivial by the results of §4. Hence in view of Lemma 6.1 above, we can conclude that (6.1) is a non-trivial projection onto the summand $\Gamma_{0,2}$. Now we consider the restriction of the representation $\bar{\rho}_2$ to the Torelli group

$$\bar{\rho}_2 : \mathcal{I}_g \longrightarrow \overline{\mathcal{H}}_2^\mathbf{Q} \tilde{\times} \overline{\mathcal{H}}_1^\mathbf{Q}.$$

This representation projects to the homomorphism $\tau_1 : \mathcal{I}_g \rightarrow \overline{\mathcal{H}}_1$ so that we can conclude that the summand $\Gamma_{0,2}$ is contain in $\mathrm{Ker}\,\tau_1^*$. Next we consider the trivial summand $\mathbf{Q}$. If we restrict the 2-cocycle $\bar{\rho}_1^*(z)$ of $\mathcal{M}_g$, given in Theorem 5.8, to the Torlli group $\mathcal{I}_g$, it takes the following form:

$$\bar{\rho}_1^*(z)(\varphi, \psi) = \overline{C}_0(\tau_1(\varphi), \tau_1(\psi)) \quad (\varphi, \psi \in \mathcal{I}_g)$$

where $\overline{C}_0 : \overline{\mathcal{H}}_1 \otimes \overline{\mathcal{H}}_1 \rightarrow \mathbf{Z}$ is the $\mathrm{Sp}(2g;\mathbf{Z})$-equivariant homomorphism given in Lemma 5.7. If we compare this with Lemma 6.1, then we can conclude that $\overline{C}_0$ represents the trivial summand $\mathbf{Q}$ appearing in $\Lambda^2 \Gamma_{0,0,1}$. On the other hand, it is known that the class e_1 vanishes in $H^2(\mathcal{I}_g; \mathbf{Q})$ (see [24]) so that the restriction of the above 2-cocycle $\bar{\rho}_1^*(z)$ to $\mathcal{I}_g$ is cohomologous to zero. Hence the trivial summand $\mathbf{Q}$ in $\Lambda^2 \Gamma_{0,0,1}$ is contained in $\mathrm{Ker}\,\tau_1^*$. This completes our topological proof of "half" of Theorem 6.2. Hain proved the other half by constructing explicit 2-cycles of the Torelli group $\mathcal{I}_g$ which have non-trivial values on each of the remaining summands $\Gamma_{0,1}$, $\Gamma_{0,0,0,1}$, $\Gamma_{0,1,0,1}$ and $\Gamma_{0,0,0,0,0,1}$ of $\Lambda^2 \Gamma_{0,0,1}$.

Now we extend Theorem 6.2 to the case of the Torelli group $\mathcal{I}_{g,*}$ with respect to the base point. We consider the following natural homomorphisms

$$C : \Lambda^3 H \longrightarrow H, \quad p : \Lambda^3 H \longrightarrow \Lambda^3 H/H$$

both of which are $\mathrm{Sp}(2g; \mathbf{Z})$-equivariant. It is easy to see that these two homomorphisms induce an isomorphism

$$(6.2) \qquad\qquad (C, p)_* : \Lambda^3 H_{\mathbf{Q}} \cong H_{\mathbf{Q}} \oplus U_{\mathbf{Q}}.$$

Lemma 6.3. *We have the following irreducible decomposition*

$$
\begin{aligned}
H^2(\Lambda^3 H_{\mathbf{Q}}; \mathbf{Q}) &= \Lambda^2(H_{\mathbf{Q}}^* \oplus U_{\mathbf{Q}}^*) \\
&= \Lambda^2 H_{\mathbf{Q}}^* \oplus \left(H_{\mathbf{Q}}^* \otimes U_{\mathbf{Q}}^*\right) \oplus \Lambda^2 U_{\mathbf{Q}}^* \\
&\cong \begin{cases} \mathbf{Q}' + \Gamma_{0,1}^f + \\ \Gamma_{0,1}^m + \Gamma_{1,0,1} + \Gamma_{0,0,0,1} + \\ \mathbf{Q} + \Gamma_{0,2} + \Gamma_{0,1}^b + \Gamma_{0,0,0,1} + \Gamma_{0,1,0,1} + \Gamma_{0,0,0,0,0,1} \end{cases}
\end{aligned}
$$

which holds for all $g \geq 6$. Here the trivial summand $\mathbf{Q}'$ appearing in $\Lambda^2 H_{\mathbf{Q}}^*$ is represented by the symplectic form ω_0^* and we put the symbols f, m, b to the summands $\Gamma_{0,1}$ to distinguish them (f for the fibre, m for mixed and b for the base).

Lemma 6.4. *The unique irreducible component $\Gamma_{0,1}$ in $\mathcal{H}_2^0(\subset H \otimes \mathcal{L}_3^0)$ is detected by the $\mathrm{Sp}(2g; \mathbf{Z})$-equivariant homomorphism*

$$q : H \otimes \mathcal{L}_3^0 \longrightarrow \Lambda^2 H$$

defined by

$$q(a \otimes [b, [c, d]]) = 2(a \cdot b)c \wedge d + (a \cdot c)b \wedge d - (a \cdot d)b \wedge c \quad (a, b, c, d \in H).$$

Proof. A direct computation shows that the above homomorphism q followed by the projection $\Lambda^2 H \rightarrow \Lambda^2 H/\mathbf{Z}$ is non-trivial where $\mathbf{Z}$ is the submodule of $\Lambda^2 H$ generated by the symplectic class ω_0. $\square$

We are now ready to state the main result of this section.

Theorem 6.5. *Let $\tau_1 : \mathcal{I}_{g,*} \rightarrow \Lambda^3 H$ be Johnson's homomorphism. Then we have*

$$\mathrm{Ker}\left(\tau_1^* : H^2(\Lambda^3 H; \mathbf{Q}) \rightarrow H^2(\mathcal{I}_{g,*}; \mathbf{Q})\right) \cong \mathbf{Q} + \Gamma_{0,2} + \Gamma_{0,1}$$

where the summand $\Gamma_{0,1}$ stands for the "diagonal" in $\Gamma_{0,1}^f + \Gamma_{0,1}^m + \Gamma_{0,1}^b$ (except for the cases $g = 2, 3$ where there is no $\Gamma_{0,1}^b$ summand).

Proof. By a result of [24], the Hochschild-Serre spectral sequence of the rational cohomology of the extension

$$1 \longrightarrow \pi_1 \Sigma_g \longrightarrow \mathcal{I}_{g,*} \longrightarrow \mathcal{I}_g \longrightarrow 1$$

degenerates at the E_2 term so that we have an isomorphism

$$H^2(\mathcal{I}_{g,*}; \mathbf{Q}) \cong H^2(\pi_1\Sigma_g) \oplus \left(H^1(\pi_1\Sigma_g; \mathbf{Q}) \otimes H^1(\mathcal{I}_g; \mathbf{Q}) \right) \oplus H^2(\mathcal{I}_g; \mathbf{Q})$$

$$= \mathbf{Q} \oplus \left(H^{\bullet}_{\mathbf{Q}} \otimes U^{\bullet}_{\mathbf{Q}} \right) \oplus H^2(\mathcal{I}_g; \mathbf{Q}).$$

Hence, in view of Theorem 6.2, we have only to prove that the summand $\Gamma_{0,1}$ is really contained in $\operatorname{Ker}\tau_1^*$. We consider the homomorphism

$$\rho_2 : \mathcal{I}_{g,*} \longrightarrow \frac{1}{2}\mathcal{H}_2 \tilde{\times} \mathcal{H}_1$$

given in (4.4). The Euler class of the target group of the above representation is given by an $\mathrm{Sp}(2g; \mathbf{Z})$-equivariant homomorphism

$$\frac{1}{2}[\ ,\ ] : \Lambda^2\mathcal{H}_1 \longrightarrow \frac{1}{2}\mathcal{H}_2$$

which is induced from (4.3). In view of Lemma 6.3 together with Lemma 4.1 (ii) and Lemma 6.4, it is enough to prove the following fact. Namely under the following composition of $\mathrm{Sp}(2g; \mathbf{Z})$-equivariant homomorphisms

$$\Lambda^3 H \otimes \Lambda^3 H \xrightarrow{\chi} S^2(\Lambda^2 H) \xrightarrow{p} \mathcal{H}_2^0 \xrightarrow{q} \Lambda^2 H \longrightarrow \Lambda^2 H/\mathbf{Z}$$

each of the three submodules

$$H \otimes H,\ \ H \otimes \operatorname{Ker} C,\ \ \operatorname{Ker} C \otimes \operatorname{Ker} C \subset \Lambda^3 H \otimes \Lambda^3 H$$

maps to $\Lambda^2 H/\mathbf{Z}$ *surjectively*. Here, as before, H is considered to be a submodule of $\Lambda^3 H$ by the embedding $H \ni u \mapsto u \wedge \omega_0 \in \Lambda^3 H$ and $C : \Lambda^3 H \to H$ is the contraction. Recall that the composition $\operatorname{Ker} C \subset \Lambda^3 H \to \Lambda^3 H/H$ is an isomorphism after tensoring with $\mathbf{Q}$. Now let us check each of the three cases.

(i) We choose two elements $x_1, y_1 \in H$ from a symplectic basis $x_1, \cdots, x_g, y_1, \cdots, y_g$ of H. Then we have

$$\chi(x_1 \wedge \omega_0, y_1 \wedge \omega_0)$$
$$= -\sum_{i,j>1} x_i \wedge y_i \leftrightarrow x_j \wedge y_j + \sum_{i>1}\left(-y_i \wedge x_1 \leftrightarrow y_1 \wedge x_i + x_1 \wedge x_i \leftrightarrow y_i \wedge y_1 \right)$$

so that

$$p \circ \chi(x_1 \wedge \omega_0, y_1 \wedge \omega_0) =$$
$$- \sum_{i,j>1}\left(x_i \otimes [y_i, [x_j, y_j]] - y_i \otimes [x_i, [x_j, y_j]] + x_j \otimes [y_j, [x_i, y_i]] - y_j \otimes [x_j, [x_i, y_i]] \right)$$
$$+ \sum_{i>1}\left\{ \begin{array}{l} -y_i \otimes [x_1, [y_1, x_i]] + x_1 \otimes [y_i, [y_1, x_i]] - y_1 \otimes [x_i, [y_i, x_1]] + x_i \otimes [y_1, [y_i, x_1]] \\ +x_1 \otimes [x_i, [y_i, y_1]] - x_i \otimes [x_1, [y_i, y_1]] + y_i \otimes [y_1, [x_1, x_i]] - y_1 \otimes [y_i, [x_1, x_i]] \end{array} \right\}.$$

Hence we have

$$q \circ p \circ \chi(x_1 \wedge \omega_0, y_1 \wedge \omega_0) = -4(g-1)x_1 \wedge y_1 - 8g \sum_{i>1} x_i \wedge y_i$$

which projects to a non-zero element of $\Lambda^2 H/\mathbf{Z}$.

(ii) We choose $x_1 \in H$ and $y_1 \wedge x_2 \wedge x_3 \in \operatorname{Ker} C$. Then we have

$$\chi(x_1 \wedge \omega_0, y_1 \wedge x_2 \wedge x_3) =$$
$$- \sum_{i>1} x_i \wedge y_i \leftrightarrow x_2 \wedge x_3 + x_1 \wedge x_2 \leftrightarrow x_3 \wedge y_1 + x_1 \wedge x_3 \leftrightarrow y_1 \wedge x_2.$$

Hence

$$p \circ \chi(x_1 \wedge \omega_0, y_1 \wedge x_2 \wedge x_3) =$$
$$- \sum_{i>1} \Big(x_i \otimes [y_i, [x_2, x_3]] - y_i \otimes [x_i, [x_2, x_3]] + x_2 \otimes [x_3, [x_i, y_i]] - x_3 \otimes [x_2, [x_i, y_i]] \Big)$$
$$+ x_1 \otimes [x_2, [x_3, y_1]] - x_2 \otimes [x_1, [x_3, y_1]] + x_3 \otimes [y_1, [x_1, x_2]] - y_1 \otimes [x_3, [x_1, x_2]]$$
$$+ x_1 \otimes [x_3, [y_1, x_2]] - x_3 \otimes [x_1, [y_1, x_2]] + y_1 \otimes [x_2, [x_1, x_3]] - x_2 \otimes [y_1, [x_1, x_3]].$$

Then we have

$$q \circ p \circ \chi(x_1 \wedge \omega_0, y_1 \wedge x_2 \wedge x_3) = -4(g+1) x_2 \wedge x_3$$

which projects to a non-zero element of $\Lambda^2 H/\mathbf{Z}$.

(iii) Finally we choose two elements $x_1 \wedge x_2 \wedge x_3, y_1 \wedge y_2 \wedge y_3 \in \operatorname{Ker} C$ and compute

$$\chi(x_1 \wedge x_2 \wedge x_3, y_1 \wedge y_2 \wedge y_3) =$$
$$- x_2 \wedge x_3 \leftrightarrow y_2 \wedge y_3 - x_3 \wedge x_1 \leftrightarrow y_3 \wedge y_1 - x_1 \wedge x_2 \leftrightarrow y_1 \wedge y_2.$$

From this, we have

$$p \circ \chi(x_1 \wedge x_2 \wedge x_3, y_1 \wedge y_2 \wedge y_3) =$$
$$-x_2 \otimes [x_3, [y_2, y_3]] + x_3 \otimes [x_2, [y_2, y_3]] - y_2 \otimes [y_3, [x_2, x_3]] + y_3 \otimes [y_2, [x_2, x_3]]$$
$$-x_3 \otimes [x_1, [y_3, y_1]] + x_1 \otimes [x_3, [y_3, y_1]] - y_3 \otimes [y_1, [x_3, x_1]] + y_1 \otimes [y_3, [x_3, x_1]]$$
$$-x_1 \otimes [x_2, [y_1, y_2]] + x_2 \otimes [x_1, [y_1, y_2]] - y_1 \otimes [y_2, [x_1, x_2]] + y_2 \otimes [y_1, [x_1, x_2]].$$

Therefore

$$q \circ p \circ \chi(x_1 \wedge x_2 \wedge x_3, y_1 \wedge y_2 \wedge y_3) = -4(x_1 \wedge y_1 + x_2 \wedge y_2 + x_3 \wedge y_3)$$

which projects to a non-zero element of $\Lambda^2 H/\mathbf{Z}$ for all $g > 3$.

This completes the proof. $\square$

7. Proof of the main theorem

In this section, we give a sketch of the proof of our main theorem (Theorem 3.4 of §3). For that we consider the representation

$$\rho_2 : \mathcal{M}_{g,*} \longrightarrow \Big(\frac{1}{24} \mathcal{H}_2 \tilde{\times} \frac{1}{2} \mathcal{H}_1 \Big) \rtimes \operatorname{Sp}(2g; \mathbf{Z})$$

which was given in Corollary 4.5. For simplicity, let us write

$$N = \frac{1}{24} \mathcal{H}_2 \tilde{\times} \frac{1}{2} \mathcal{H}_1$$

so that we can write

$$(7.1) \qquad\qquad \rho_2 : \mathcal{M}_{g,*} \to N \rtimes \operatorname{Sp}(2g; \mathbf{Z}).$$

Since N is a torsion free nilpotent group, it can be considered as a discrete subgroup (more precisely, a lattice) of the real nilpotent Lie group $N_{\mathbf{R}} = N \otimes \mathbf{R}$ and the quotient space $M = N \backslash N_{\mathbf{R}}$ is a compact C^∞ manifold which is a $K(N,1)$ space. It is easy to see that $\mathrm{Sp}(2g; \mathbf{R})$ acts naturally on $N_{\mathbf{R}}$ as an automorphism group and the restriction of this action to the subgroup $\mathrm{Sp}(2g; \mathbf{Z}) \subset \mathrm{Sp}(2g; \mathbf{R})$ clearly preserves the lattice N. Hence $\mathrm{Sp}(2g; \mathbf{Z})$ acts on the nilmanifold $M = N \backslash N_{\mathbf{R}}$ by diffeomorphisms.

We realize the classifying space of the semi-direct product $N \rtimes \mathrm{Sp}(2g; \mathbf{Z})$ as a *foliated M-bundle*

$$M \longrightarrow E \longrightarrow B\mathrm{Sp}(2g; \mathbf{Z})$$

over the classifying space $B\mathrm{Sp}(2g; \mathbf{Z})$ of $\mathrm{Sp}(2g; \mathbf{Z})$. Here E is defined to be the quotient of $M \times E\mathrm{Sp}(2g; \mathbf{Z})$ with respect to the diagonal action of $\mathrm{Sp}(2g; \mathbf{Z})$ where $E\mathrm{Sp}(2g; \mathbf{Z})$ denotes the total space of the universal $\mathrm{Sp}(2g; \mathbf{Z})$-bundle over $B\mathrm{Sp}(2g; \mathbf{Z})$. We consider the mapping

$$(7.2) \qquad\qquad B\mathcal{M}_{g,*} \longrightarrow E$$

which is the geometric realization of the homomorphism (7.1). Since we can approximate any skeleton of the classifying spaces $B\mathcal{M}_{g,*}$ and $B\mathrm{Sp}(2g; \mathbf{Z})$ by finite dimensional C^∞ manifolds, in the following arguments we may assume that $B\mathcal{M}_{g,*}$ and E are in fact C^∞ manifolds and the mapping (7.2) is smooth.

Now let $\mathfrak{n}_{\mathbf{R}}$ be the Lie algebra of $N_{\mathbf{R}}$. Then we can consider $\Lambda^* \mathfrak{n}_{\mathbf{R}}^*$ as the set of all left invariant differential forms on $N_{\mathbf{R}}$ and hence also as a subcomplex of the de Rham complex $A^*(M)$ of the manifold M. $\mathrm{Sp}(2g; \mathbf{R})$ acts on $N_{\mathbf{R}}$ as an automorphism group so that $\Lambda^* \mathfrak{n}_{\mathbf{R}}^*$ is a representation of $\mathrm{Sp}(2g; \mathbf{R})$. Let

$$(\Lambda^* \mathfrak{n}_{\mathbf{R}}^*)^{\mathrm{Sp}}$$

denote the $\mathrm{Sp}(2g; \mathbf{R})$-invariant subspace of $\Lambda^* \mathfrak{n}_{\mathbf{R}}^*$. Any element $\omega \in (\Lambda^* \mathfrak{n}_{\mathbf{R}}^*)^{\mathrm{Sp}}$ of this subspace can be considered as a differential from on M which is invariant under the action of $\mathrm{Sp}(2g; \mathbf{Z})$ so that we have a homomorphism

$$(\Lambda^* \mathfrak{n}_{\mathbf{R}}^*)^{\mathrm{Sp}} \longrightarrow A^*(E).$$

The projection $N_{\mathbf{R}} \to \mathcal{H}_1 \otimes \mathbf{R}$ induces a homomorphism $(\Lambda^*(\mathcal{H}_1 \otimes \mathbf{R})^*)^{\mathrm{Sp}} \to (\Lambda^* \mathfrak{n}_{\mathbf{R}}^*)^{\mathrm{Sp}}$ and the composition

$$(\Lambda^*(\mathcal{H}_1 \otimes \mathbf{R})^*)^{\mathrm{Sp}} \to (\Lambda^* \mathfrak{n}_{\mathbf{R}}^*)^{\mathrm{Sp}} \to A^*(E) \to A^*(B\mathcal{M}_{g,*})$$

is nothing but the "geometric realization" of the homomorphism

$$(7.3) \qquad\qquad \rho_1^* : \left(\Lambda^*(\Lambda^3 H_{\mathbf{Q}})^*\right)^{\mathrm{Sp}} \longrightarrow H^*(\mathcal{M}_{g,*}; \mathbf{Q})$$

given in Proposition 3.3. Now by a result of Nomizu [34] about the cohomology of nilpotent groups or alternatively by Sullivan's general theory [35], we have a natural isomorphism

$$H^*(\mathfrak{n}_{\mathbf{R}}) \cong H^*(N; \mathbf{R}).$$

Moreover this isomorphism clearly respects the action of $\mathrm{Sp}(2g; \mathbf{Z})$ so that we have

$$H^*(\mathfrak{n}_{\mathbf{R}})^{\mathrm{Sp}} \cong H^*(N; \mathbf{R})^{\mathrm{Sp}}.$$

184 SHIGEYUKI MORITA

Hence the homomorphism (7.3) can be written as the composition

$$\left(\Lambda^*(\Lambda^3 H_{\mathbf{Q}})^*\right)^{\mathrm{Sp}} \xrightarrow{p^*} H^*(N;\mathbf{Q})^{\mathrm{Sp}} \longrightarrow H^*(\mathcal{M}_{g,*};\mathbf{Q})$$

where the first homomorphism is induced from the natural projection $p : N \to \frac{1}{2}\mathcal{H}_1$.

Now by (6.2), we have an isomorphism

$$(\Lambda^3 H_{\mathbf{Q}})^* \cong H_{\mathbf{Q}}^* \oplus U_{\mathbf{Q}}^*$$

and this induces an isomorphism

$$(7.4) \quad \Lambda^i(\Lambda^3 H_{\mathbf{Q}})^* \cong La^i H_{\mathbf{Q}}^* \oplus \left(\Lambda^{i-1} H_{\mathbf{Q}}^* \otimes U_{\mathbf{Q}}^*\right) \oplus \left(\Lambda^{i-2} H_{\mathbf{Q}}^* \otimes \Lambda^2 U_{\mathbf{Q}}^*\right)$$
$$\oplus \cdots \oplus \left(\Lambda^{i-j} H_{\mathbf{Q}}^* \otimes \Lambda^j U_{\mathbf{Q}}^*\right) \oplus \cdots$$
$$\oplus \left(\Lambda^2 H_{\mathbf{Q}}^* \otimes \Lambda^{i-2} U_{\mathbf{Q}}^*\right) \oplus \left(H_{\mathbf{Q}}^* \otimes \Lambda^{i-1} U_{\mathbf{Q}}^*\right) \oplus \Lambda^i U_{\mathbf{Q}}^*$$

for any i. In particular, as was already described in Lemma 6.3, we have

$$\Lambda^2(\Lambda^3 H_{\mathbf{Q}})^* \cong \Lambda^2 H_{\mathbf{Q}}^* \oplus \left(H_{\mathbf{Q}}^* \otimes U_{\mathbf{Q}}^*\right) \oplus \Lambda^2 U_{\mathbf{Q}}^*$$

and

$$\left(\Lambda^2(\Lambda^3 H_{\mathbf{Q}})^*\right)^{\mathrm{Sp}} \cong \mathbf{Q}' \oplus \mathbf{Q}$$

where the first trivial summand $\mathbf{Q}'$ is generated by the symplectic form $\omega_0^* \in \Lambda^2 H_{\mathbf{Q}}^*$. On the other hand, it is easy to deduce from Theorem 5.1 that the Euler class $e \in H^2(\mathcal{M}_{g,*};\mathbf{Q})$ is represented by the pull back (under the homomorphism ρ_1) of certain element

$$\varepsilon \in \left(\Lambda^2(\Lambda^3 H_{\mathbf{Q}})^*\right)^{\mathrm{Sp}} = \mathbf{Q}' \oplus \mathbf{Q}$$

which has non-zero values on both factors.

Now let $\Gamma_{0,1}$ be the "diagonal" in $\Gamma_{0,1}^f + \Gamma_{0,1}^m + \Gamma_{0,1}^b \subset \Lambda^2(\Lambda^3 H_{\mathbf{Q}})^*$ which was described in Theorem 6.5. Then the proof of Theorem 6.5 actually shows that

$$p^*(\Gamma_{0,1}) = \{0\} \subset H^2(N;\mathbf{Q}).$$

If we combine this fact with $\Lambda^2 H_{\mathbf{Q}}^* = \mathbf{Q}' \oplus \Gamma_{0,1}^f$, then we can conclude as follows. Namely for any element $\alpha \in \Lambda^i(\Lambda^3 H_{\mathbf{Q}})^*$, we can use the above $\Gamma_{0,1}$ to reduce the degrees of α with respect to the $H_{\mathbf{Q}}^*$ coordinates (the "fibre" direction) within the same cohomology class so that eventually we obtain certain element $\alpha' \in \Lambda^i(\Lambda^3 H_{\mathbf{Q}})^*$ such that α' has non-zero components only in the following three summands

$$\omega_0^* \wedge \Lambda^{i-2} U_{\mathbf{Q}}^* \ (\subset \Lambda^2 H_{\mathbf{Q}}^* \otimes \Lambda^{i-2} U_{\mathbf{Q}}^*), \quad H_{\mathbf{Q}}^* \otimes \Lambda^{i-1} U_{\mathbf{Q}}^*, \quad \Lambda^i U_{\mathbf{Q}}^*$$

in the decomposition of $\Lambda^i(\Lambda^3 H_{\mathbf{Q}})^*$ given in (7.4) and

$$p^*(\alpha) = p^*(\alpha') \in H^i(N;\mathbf{Q}).$$

Now let us apply the above argument to the element $\varepsilon^{i+1} \in \left(\Lambda^{2i+2}(\Lambda^3 H_{\mathbf{Q}})^*\right)^{\mathrm{Sp}}$ and project the result to the Sp-invariant part. Then we can conclude that there exists certain element

$$\varepsilon_i \in \left(\Lambda^{2i+2}(\Lambda^3 H_{\mathbf{Q}})^*\right)^{\mathrm{Sp}}$$

which can be written as

$$(7.5) \qquad \varepsilon_i = \omega_0^\bullet \wedge \varepsilon_i^f + \varepsilon_i^m + \varepsilon_i^b$$

where

$$\varepsilon_i^f \in (\Lambda^{2i} U_{\mathbf{Q}}^\bullet)^{\mathrm{Sp}}, \quad \varepsilon_i^m \in \left(H_{\mathbf{Q}}^\bullet \otimes \Lambda^{2i+1} U_{\mathbf{Q}}^\bullet\right)^{\mathrm{Sp}}, \quad \varepsilon_i^b \in \left(\Lambda^{2i+2} U_{\mathbf{Q}}^\bullet\right)^{\mathrm{Sp}}$$

such that

$$p^*(\varepsilon^{i+1}) = p^*(\varepsilon_i) \in H^{2i+2}(N; \mathbf{Q})^{\mathrm{Sp}}.$$

Now, by the definition, the characteristic class $e_i \in H^{2i}(\mathcal{M}_g; \mathbf{Q})$ is defined to be the integral of $\rho_1^*(\varepsilon^{i+1})$ along the fibre of $\mathcal{M}_{g,*} \to \mathcal{M}_g$ (see [24]). By the above result, it can be also represented by the integral of $\rho_1^*(\varepsilon_i)$ along the fibre. But by the form of the element ε_i described in (7.5), we have

$$\pi_*(\rho_1^*(\varepsilon_i)) = 2g\,\overline{\rho}_1^*(\varepsilon_i^f) \in H^{2i}(\mathcal{M}_g; \mathbf{Q})$$

because clearly we have $\pi_*(\rho_1^*(\omega_0^\bullet)) = 2g$ and $\pi_*(\rho_1^*(\varepsilon_i^m)) = \pi_*(\rho_1^*(\varepsilon_i^b)) = 0$ where π_* denotes the integral along the fibre of $\pi : \mathcal{M}_{g,*} \to \mathcal{M}_g$. In view of the result of §3, we can now conclude that there exists certain element $\beta_i \in \mathbf{Q}[\beta_\Gamma; \Gamma \in \mathcal{G}^0]$ which has degree $2i$ such that

$$\Phi(\beta_i) = e_i.$$

This completes the proof of Theorem 3.4.

References

1. M. Asada, *Two properties of the filtration of the outer automorphism groups of certain groups*, Math. Z. **218** (1995), 123-133.

2. M. Asada and M. Kaneko, *On the automorphism group of some pro-ℓ fundamental groups*, Adv. Studies in Pure Math. **12** (1987), 137-159.

3. M. Asada and H. Nakamura, *On the graded quotient modules of mapping class groups of surfaces*, Israel J. Math. **90** (1995), 93-113.

4. A. Borel, *Stable real cohomology of arithmetic groups*, Ann. Sci. Ecole Norm. Sup. **7** (1974), 235-272.

5. A. Borel, *Stable real cohomology of arithmetic groups II*, in *Manifolds and Groups, Papers in Honor of Yozo Matsushima*, Progress in Math. **14**, Birkhäuser, Boston, 1981, 21-55.

6. C.J. Earle and J. Eells, *A fiber bundle description of Teichmüller theory*, J. Differential Geometry **3** (1969), 19–43.

7. W. Fulton and J. Harris, *Representation Theory*, Graduate Texts in Math. **129**, Springer Verlag, 1991.

8. R. Hain, *Completions of mapping class groups and the cycles $C - C^-$*, in *Mapping Class Groups and Moduli Spaces of Riemann Surfaces*, Contemporary Math. **150** (1993), 75-105.

9. R. Hain, *Infinitesimal presentations of the Torelli groups*, preprint.

10. J. Harer, *Stability of the homology of the mapping class group of an orientable surface*, Ann. of Math. **121** (1985), 215-249.

11. D. Johnson, *An abelian quotient of the mapping class group $\mathcal{I}_g$*, Math. Ann. **249** (1980), 225-242.

12. D. Johnson, *A survey of the Torelli group*, Contemporary Math. **20** (1983), 165-179.

13. D. Johnson, *The structure of the Torelli group II and III*, Topology **24** (1985), 113-144.

14. N. Kawazumi, *A generalization of the Morita-Mumford classes to extended mapping class groups for surfaces*, preprint.

15. N. Kawazumi, *On the stable cohomology algebra of extended mapping class groups for surfaces*, preprint.

16. N. Kawazumi, *An infinitesimal approach to the stable cohomology of the moduli of Riemann surfaces*, These Proceedings, 79–100.

17. N. Kawazumi and S. Morita, in preparation.

18. M. Kontsevich, *Formal (non)-commutative symplectic geometry*, in *Gelfand Mathematical Seminars*, 1990-92, Birkhäuser, Boston, 1993, 173-188.

19. M. Kontsevich, *Feynman diagrams and low-dimensional topology*, in *Proceedings of the first European Congress of Mathematicians*, vol. 2, Progress in Math. **120**, Birkhäuser, Boston, 1994, 97-121.

20. J. Labute, *On the descending central series of groups with a single defining relation*, J. Algebra **14** (1970), 16-23.

21. W. Magnus, A. Karrass and D. Solitar, *Combinatorial Group Theory*, Interscience Publ., New York, 1966.

22. E.Y. Miller, *The homology of the mapping class group*, J. Differential Geometry **24** (1986), 1-14.

23. S. Morita, *Characteristic classes of surface bundles*, Bull. Amer. Math. Soc. **11** (1984), 386-388.

24. S. Morita, *Characteristic classes of surface bundles*, Invent. Math. **90** (1987), 551-577.

25. S. Morita, *Families of Jacobian manifolds and characteristic classes of surface bundles II*, Math. Proc. Camb. Phil. Soc. **105** (1989), 79-101.

26. S. Morita, *Casson's invariant for homology 3-spheres and characteristic classes of surface bundles I*, Topology **28** (1989), 305-323.

27. S. Morita, *On the structure of the Torelli group and the Casson invariant*, Topology **30** (1991), 603-621.

28. S. Morita, *Mapping class groups of surfaces and three dimensional manifolds*, Proceedings of the International Congress of Mathematicians, Kyoto 1990, Springer, 1991, 665-674.

29. S. Morita, *The structure of the mapping class group and characteristic classes of surface bundles*, in *Mapping Class Groups and Moduli Spaces of Riemann Surfaces*, Contemporary Math. **150**, Amer. Math. Soc., 1993, 303-315.

30. S. Morita, *The extension of Johnson's homomorphism from the Torelli group to the mapping class group*, Invent. Math. **111** (1993), 197-224.

31. S. Morita, *Abelian quotients of subgroups of the mapping class group of surfaces*, Duke Math. J. **70** (1993), 699-726.

32. S. Morita, in preparation.

33. D. Mumford, *Towards an enumerative geometry of the moduli space of curves*, in *Arithmetic and Geometry*, Progress in Math. **36**, Birkhäuser, Boston, 1983, 271-328.

34. K. Nomizu, *On the cohomology of compact homogeneous spaces of nilpotent Lie groups*, Ann. of Math. **59** (1954), 531-538.

35. D. Sullivan, *Infinitesimal computations in topology*, Publ. Math. I.H.E.S. **47** (1977), 269-331.

DEPARTMENT OF MATHEMATICAL SCIENCES,, UNIVERSITY OF TOKYO,, KOMABA, MEGURO-KU,, TOKYO 153 JAPAN

E-mail address: morita@ms.u-tokyo.ac.jp

Proceedings of
THE 37TH TANIGUCHI SYMPOSIUM ON
TOPOLOGY AND TEICHMÜLLER SPACES
held in Finland, July 1995
ed. by Sadayoshi KOJIMA *et al.*
©1996 World Scientific Publishing Co.
pp. 187–220

MATHEMATICS IN AND OUT OF STRING THEORY

SUBHASHIS NAG

Dedicated to the memory of my father and mother

(Received December 14, 1995)

The upsurge of excitement amongst theoretical physicists, over the subject of string theory, has filtered through to an appreciable extent into the mathematical community. Whereas the basic reason for the excitement amongst the physicists has been the vision of unification for the fundamental forces of nature via string theory, in mathematics its interest has been the wide range of deep ideas that have been involved, and the fascinating interconnections that have emerged.

"Stringy" ideas in mathematics include :

(1) The investigation of path integrals over the spaces of Riemann surfaces, leading to a natural modular-invariant measure ("the Polyakov volume form") on the Teichmüller spaces. The description of this measure by complex geometry of the Teichmüller spaces, involving the Mumford isomorphisms.

(2) A search for infinite-dimensional "Universal Teichmüller spaces" of Riemann surfaces that parametrize simultaneously complex structures on surfaces of all topologies, – and the canonical relationships between various natural candidates for such a moduli space. This is important for a non-perturbative formulation of string theory.

(3) The study of the unitary (and projective unitary) representations of the diffeomorphism group of the circle (the closed string!); at the infinitesimal level, this is the representation theory of the Virasoro algebra. Indeed, there is an intimate relationship ([NV]) between the group $\text{Diff}(S^1)$ and the Teichmüller spaces – demonstrating that (3) is deeply related to (2).

It goes without saying that the above topics by no means exhaust the mathematical challenges raised by string theory. Owing to restrictions of space and time, and, more importantly, of the author's knowledge, we shall deal in these notes only with some matters pertaining to items (1) and (2) above. For more directions, see the references

---

Expanded version of the opening lecture in the 37th International Taniguchi Symposium: *"Topology and Teichmüller spaces"*, July 1995.

cited. We will provide here an exposition of the Polyakov-Mumford construction on the Teichmüller space $\mathcal{T}_h$ of Riemann surfaces of fixed genus $h \geq 2$; we will then explain our recent work (with Indranil Biswas and Dennis Sullivan) coherently fitting together this construction over the Universal Commensurability Teichmüller space, $\mathcal{T}_\infty$. In fact, $\mathcal{T}_\infty$ qualifies as a parameter space of the type desired in (2), because it comprises compact Riemann surfaces of all genus.

The connecting thread intertwining all the mathematics we discuss is the natural appearance of the Teichmüller/moduli spaces of Riemann surfaces. The most fundamental point is that, mathematically speaking, the quantum theory of the closed bosonic string is the theory of a sum over random surfaces ("world-sheets") that are swept out by strings propagating in spacetime. Owing to conformal invariance properties of the "Polyakov action", that sum finally reduces to an integral over the parameter space of Riemann surfaces. *Namely, the quantum string theory, as a sum over random surfaces, precipitates a natural measure – the Polyakov measure – on each moduli space $\mathcal{M}_h$.*

We adopt the attitude that we are addressing mathematicians with no prior exposure to (the pulling of) strings. We have taken particular pains (Sections 1 and 2) to explain to a mathematical audience the reduction of the Polyakov functional integral from an ill-defined and infinite dimensional situation to a finite dimensional one.

Acknowledgements. It is a pleasure to record my thanks to the *Taniguchi Foundation of Japan* for inviting me to speak at the 37th Taniguchi Mathematics Symposium: "Topology and Teichmüller Spaces" in Finland, and for inviting this exposition for the Proceedings. The generous financial support given by the Foundation provides me very pleasant memories of my travels in Finland during July-August, 1995. I thank sincerely all the many mathematicians who have listened to my talks over the years and have helped me to understand the material being presented here.

The paper is organized as follows:
Section 1: Polyakov action and the string path integral.
Section 2: Polyakov volume form on the moduli space.
Section 3: Geometry of Teichmüller space and Polyakov volume.
Section 4: The Universal Teichmüller space of compact surfaces.
Section 5: Universal Polyakov-Mumford on $\mathcal{T}_\infty$.

1. Polyakov action and the string path integral

1.1. Mechanics

The uninitiated mathematician may not object to being reminded of how path integrals arise in the first place. Mechanics, whether classical or quantum can be formulated as arising from a Lagrangian that allows one to assign a weight, called the "action", to any choice of an admissible path in configuration space connecting given initial and final boundary conditions.

For instance, suppose a particle (or a mechanical system) is moving in configuration space $\mathbf{R}^d$, with position at time t being $x(t) = (x_1(t), \ldots, x_d(t))$ from $x(t_o) = x_o$ to $x(t_1) = x_1$. Then the Lagrangian, $L(x(t), \dot{x}(t), t)$, is a functional of the path and the *action* for that choice of path is defined by:

$$(1.1) \qquad S(x(t)) = \int_{t_0}^{t_1} L(x(t), \dot{x}(t), t) dt$$

Example. Particle moving in a potential $V : \mathbf{R}^d \to \mathbf{R}$. L may be taken as "kinetic energy minus potential energy", namely, $L = \frac{m}{2} \|\dot{x}(t)\|^2 - V(x(t))$.

Amongst all admissible paths interpolating between $x(t_0) = x_0$ and $x(t_1) = x_1$, the actual classical path followed by the particle is the "best" path - namely the value of the action should be extremal (minimal). Thus the classical equations of motion are the Euler-Lagrange equations for the variational problem of minimising the action (1.1). Applied to the case of the example above, the reader can easily check that the Euler-Lagrange equations are simply Newton's equations of motion.

In quantum mechanics [Feynman's path integral formulation] there is no determined "preferred" path from (x_0, t_0) to (x_1, t_1). Rather "all" joining paths are possible histories of transition, and the proper question therefore is not which path the particle follows, but what is the probability amplitude that a particle at x_0 (at time t_0) will be at x_1 (at time t_1). That probability is taken to be a certain weighted average over all interpolating paths, where a path is weighted by $exp(-Action(path))$. Thus notice that the classical path (with the minimum action) is given the highest weight, and paths near to the classical one would get relatively high weighting. The *path integral* answering the basic quantum query is

$$(1.2) \qquad Z = \int_{\{x(t)\}} e^{-S(x(t))} Dx$$

where $\{x(t)\}$ is the family of all admissible (say continuous) paths joining the given initial and final conditions, and Dx represents some Wiener-type measure on this path-family.

The analog of this infinite-dimensional path integral is what we will now describe for Polyakov's bosonic string. The crucial discovery is that in a particularly happy situation (namely when the spacetime dimension $d = 26$), the integral reduces from the infinite-dimensional space of possible paths (=world-sheets) to the finite-dimensional moduli spaces of Riemann surfaces.

1.2. String Theory

String theory is the theory of fundamental particles considered as being one dimensional, "strings", rather than as zero dimensional ("point-like") objects. Thus a closed string, (therefore a *circle* - there being only one closed 1-manifold), propagating in a spacetime $\mathbf{R}^d$ sweeps out a 2-dimensional surface called its *world-sheet*. Several strings can interact, and more may be created or annihilated, without the world-sheet (which is a 2-manifold) becoming singular.

The figure illustrates that a configuration of two strigns at an initial time can become (say) three at a later time while sweeping out a *non-singular* history Σ. Compare this with the case for point-like particles!

In Polyakov's string theory [Pol1] the action assigned to any particular world-sheet Σ depends on its location (embedding) in $\mathbf{R}^d$ as well as on an arbitrarily chosen

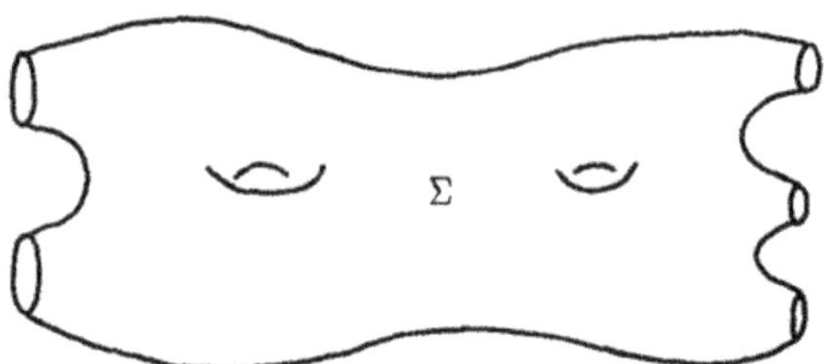

FIGURE 1.1. String world-sheet

smooth Riemannian metric on Σ. Both these freedoms, in the choice of embedding and metric, have to be integrated out (i.e., averaged over) in setting up the path integral.

The fundamental path integral that one needs to evaluate is the "vacuum to vacuum" amplitude – meaning that both the initial and final configurations are taken to be empty. Consequently, the world-sheets are without boundary – namely *closed surfaces* embedded in Euclidean space of d dimension, (which is taken to be the background spacetime). We therefore assume henceforth that the world-sheets are closed and orientable surfaces, and attempt to work out the contribution to the path integral for each fixed genus.

Fix for reference a closed oriented smooth surface Σ of genus h (number of handles), and consider an arbitrary smooth embeeding of Σ in $\mathbf{R}^d$:

$$(1.3) \qquad X \equiv (X^1, \ldots, X^d) : \Sigma \longrightarrow \mathbf{R}^d$$

and simultaneously consider an arbitrary Riemannian metric g on Σ:

$$(1.4) \qquad ds^2 = g_{ij} d\sigma^i d\sigma^j \quad , \quad i, j = 1, 2$$

Here (σ^1, σ^2) are local smooth coordinates on Σ. The image $X(\Sigma)$ is to be considered as a typical vacuum-to-vacuum world-sheet (path) and the random metric $((g_{ij}))$ (having nothing to do with the metric induced on Σ via the embedding into Euclidean space) is an extra dynamical variable which also has to be summed over in the Polyakov string theory.

The fundamental definition is the **Polyakov action** for that world-sheet and that assigned metric:

$$(1.5) \qquad S(X, g) = \iint_\Sigma \left[g^{ab} \frac{\partial X^\mu}{\partial \sigma^a} \frac{\partial X^\mu}{\partial \sigma^b} \right] \sqrt{g}\, d\sigma^1 d\sigma^2$$

Clearly, the part of the integrand (i.e. Lagrangian) in square brackets is a real-valued function on Σ, and it is being integrated with respect to the area-element, $d_g(\text{vol}) = \sqrt{g} d\sigma^1 d\sigma^2$, induced on Σ by the metric g.

Notation. Summation conventions over all repeated indices are in use in (1.4), (1.5), and subsequently. In (1.5), indices a and b are summed over $1 \le a, b \le 2$, and μ is summed over $1 \le \mu \le d$. Furthermore, $\sqrt{g}$ signifies the density $\sqrt{\det(g_{ij})}$, and $((g^{ab}))$ denotes the inverse matrix to $((g_{ij}))$, as usual.

The basic problem therefore is to analyse the functional integral :

$$(1.6) \qquad Z = \int_{\{g_{ij}\}} \int_{\{X\}} e^{-S(X,g)} DX.Dg$$

That will represent the basic "partition function", or vacuum-to-vacuum amplitude, for string propagations over world-sheets with h handles.

The remarkable thing is that, after taking care of certain symmetries in the Polyakov action, the above integral does have a sensible reduction to a *finite dimensional integration* over the moduli space $\mathcal{M}_h$ of complex structures on a genus h surface, provided the spacetime dimension d equals 26. Our first purpose, therefore, is to explain concisely to a mathematical audience how the action prescription (1.5) leads to a canonical measure – called naturally the Polyakov measure – on each moduli space $\mathcal{M}_h$.

Remark on the classical theory for (1.5). For a fixed choice of embedding X, one may enquire as to what is the "best" (extremal) metric for the action (1.5). It is easily derived that the action is extremised precisely for the metric g on Σ induced from the (Euclidean) target space $\mathbf{R}^d$ via the embedding X. Namely, $g_{ij} = \sum_{\mu=1}^d \frac{\partial X^\mu}{\partial \sigma^i} \frac{\partial X^\mu}{\partial \sigma^j}$, is the "classical" metric. So Polyakov action tells us to give this induced metric the highest weight but average over *all* metrics in the path-integral (1.6). Note that this fact, about the classical metric being the one induced from the target $\mathbf{R}^d$, remains true even if we choose an arbitrary Riemannian metric $((G_{\mu\nu}))$ in the background spacetime $\mathbf{R}^d$; here, of course, we replace the Lagrangian integrand in (1.5) by the more general: $[g^{ab} \frac{\partial X^\mu}{\partial \sigma^a} \frac{\partial X^\nu}{\partial \sigma^b} G_{\mu\nu}]$. As we will see, however, for the corresponding partition function (1.6), the integral over the embedding variables (for fixed $((g_{ij}))$) is *not* any more an (infinite dimensional) "Gaussian" if $((G_{\mu\nu}))$ is non-flat. For our purposes, therefore, we restrict to the case $((G_{\mu\nu})) =$ Euclidean, just as in (1.5).

1.3. Symmetries of the Polyakov action

To analyze (1.6), we first need to note that Polyakov's action has certain *symmetries*:

$$(1.7) \qquad S(X^\mu + c^\mu, g) = S(X^\mu, g), \quad \text{any } c^\mu \in \mathbf{R}^d.$$

$$(1.8) \qquad S(f^* X^\mu, f^* g) = S(X^\mu, g), \quad \text{any } f \in \mathrm{Diff}^+(\Sigma).$$

$$(1.9) \qquad S(X^\mu, e^\phi g) = S(X^\mu, g), \quad \text{any } \phi \in C^\infty(\Sigma, \mathbf{R}).$$

(1.7) corresponds to the fact that the action remains unchanged if the embedded surface is simply translated in $\mathbf{R}^d$. (1.8) says that, since (1.5) is invariantly defined, independent of choice of coordinates on Σ, if we use any (orientation preserving) diffeomorphism of Σ to pullback both the metric and the embedding, the action remains unperturbed. [Explicitly, $f^* X^\mu = X^\mu \circ f$, and $f^* g$ is the metric on Σ which assigns to any curve the length that g assigns to the f-image of the curve.] Finally, *(1.9) is the truly non-trivial symmetry*, and says that, for a fixed embedding, the value *of the action depends only on the conformal class of the metric g.* [Conformal scaling, $g \longmapsto (scaling function)g$ is called a "Weyl-scaling" by physicists.] Verification of (1.9) is immediate since the Weyl factor cancels off between the square-bracketed Lagrangian and the area-element term.

Clearly then, if the path integral (1.6) were actually computed over *all* embeddings X and *all* Riemannian metrics g, then one would be getting infinite answers simply because one is *overcounting* by (1) the "volume of $\mathbf{R}^d$", corresponding to the arbitrary c^μ of (1.7); (2) the "volume of $\mathrm{Diff}(\Sigma)$" because of (1.8); (3) the "volume of positive functions (conformal factors) on Σ" because of (1.9). In other words, we can only expect to make sense of (1.6) by quotienting out these symmetries – namely by integrating on the quotient space :

$$(1.10) \qquad \{\text{Embeddings}\} \times \{\text{Metrics}\}/\{\mathbf{R}^d \times \mathrm{Diff}^+(\Sigma) \times \mathrm{Conf}(\Sigma)\}$$

Notation. Write $\mathrm{Emb}(\Sigma) = \{X^\mu\}$ for the space of all (smooth) embeddings of Σ in $\mathbf{R}^d$, and $\mathrm{Met}(\Sigma)$ for the space of all (smooth) Riemannian metrics on Σ.

1.4. The integral over $\{X^\mu\}$

For every *fixed* $g \in \mathrm{Met}(\Sigma)$ we will show below that, in analogy with Gaussian (multivariate normal distribution) integrals, the integral over $\{\text{Embeddings}\}/\mathbf{R}^d$ can be carried out to produce a reasonable answer. One should consider this section as providing heuristic motivation for the following:

Proposition/Definition 1.1. *For every fixed g in* $\mathrm{Met}(\Sigma)$*, the inner integral in (1.6) is assigned the value:*

$$(1.11) \qquad \int_{\mathrm{Emb}(\Sigma)} e^{-S(X,g)} DX = \left[\frac{\det'(-\Delta_g)}{\int_\Sigma d_g(\mathrm{vol})}\right]^{-d/2}$$

where Δ_g is the Laplace-Beltrami operator on functions on Σ, and $\det'(A)$ denotes the determinant of an operator A after discarding any zero eigenvalues.

The Polyakov integral (1.6) therefore becomes:

$$(1.11^*) \qquad Z = \int_{\mathrm{Met}(\Sigma)} \left[\frac{\det'(-\Delta_g)}{\int_\Sigma d_g(\mathrm{vol})}\right]^{-d/2} Dg$$

with respect to a suitable volume element $[Dg]$ on the space of Riemannian metrics on Σ.

Note. Determinants will be computed by heat-kernel (zeta-function) regularization.

Since (Σ, g) is a Riemannian manifold, we introduce the standard L^2 inner-product on functions on Σ by

$$(1.12) \qquad \langle f_1, f_2 \rangle = \iint_\Sigma (f_1 f_2) d_g\, \mathrm{vol}$$

and obtain the Hilbert space $L^2(\Sigma)$ of square-integrable real-valued functions with respect to this scalar product. The Laplace-Beltrami operator, Δ_g, is the formally self-adjoint operator defined on sufficiently smooth functions on Σ by :

$$(1.13) \qquad \Delta_g(f) = \frac{1}{\sqrt{g}} \left[\frac{\partial}{\partial \sigma^a} \left(\sqrt{g} g^{ab} \frac{\partial f}{\partial \sigma^b} \right) \right]$$

Lemma 1.2. *The Polyakov action (1.5) can be rewritten as*

$$(1.14) \qquad S(X, g) = \sum_{\mu=1}^{d} \langle X^\mu, (-\Delta_g) X^\mu \rangle$$

Notice that $(-\Delta_g)$ is a positive operator.

Proof. Triangulate Σ so that each triangle falls in a typical (σ^1, σ^2) coordinate patch. Then integrating by parts in (1.5) with respect to σ^a, on any triangle, gives an integral over the boundary of the triangle plus the term

$$- \iint_{\text{triangle}} X^\mu \frac{1}{\sqrt{g}} \left[\frac{\partial}{\partial \sigma^a} \left(\sqrt{g} g^{ab} \frac{\partial X^\mu}{\partial \sigma^b} \right) \right] \sqrt{g} \, d\sigma^1 d\sigma^2$$

Summing up over all the triangles, the boundary terms cancel off and we are left with (1.14). $\square$

Now we are ready to give some heuristic arguments for (1.11). Since all the d target coordinates are on an equal footing, the equation (1.14) shows that $\int e^{-S(X,g)} DX$ will be the product of d identical integrals :

$$\int e^{\langle X^\mu, \Delta_g X^\mu \rangle} DX^\mu, \ \mu = 1, 2, \ldots, d.$$

Thus we are reduced to motivating the equation:

$$(1.11') \qquad \int_{\{Y: \Sigma \to \mathbf{R}\}} e^{-\langle Y, -\Delta_g Y \rangle} DY = \left[\frac{\det'(-\Delta_g)}{\text{Area}(\Sigma, g)} \right]^{-1/2}$$

This is clear. Indeed, suppose $\{e_o, e_1, e_2, \ldots\}$ is an orthonormal basis for $L^2(\Sigma)$ consisting of eigenfunctions of $-\Delta_g$. Since the Δ_g-harmonic functions are the constants, we have a 1-dimensional space of eigenfunctions with eigenvalue $\lambda_o = 0$ generated by the constant function e_o. We set $(-\Delta_g)e_k = \lambda_k e_k$, with $\lambda_k > 0$ for all $k = 1, 2, \ldots$.

Expanding in Fourier series an arbitrary $Y : \Sigma \to \mathbf{R}$ in terms of the basis, we set

$$Y = \sum_{k=0}^{\infty} y_k e_k$$

with $y_k = \langle Y, e_k \rangle$.

Then one has

$$(1.15) \qquad -\langle Y, -\Delta_g Y \rangle = - \sum_{k=0}^{\infty} |y_k|^2 \lambda_k$$

If we take the functional measure "DY" in (1.11') to mean $dy_o dy_1 dy_2 dy_3 \ldots$ with Lebesgue measure in each factor, we see that we have on our hands a *Gaussian integral*. Ignoring the dy_o integral for the time being, this should produce $(\lambda_1 \lambda_2 \lambda_3 \ldots)^{-1/2}$

by comparison with the standard multivariate normal integrals. Thus we make the reasonable definition that:

$$(1.16) \qquad \int e^{-\sum_{k=1}^{\infty} y_k^2 \lambda_k} \left[\Pi_{k=1}^{\infty} dy_k\right] = \left[\det'(-\Delta_g)\right]^{-1/2}$$

Now, the integral over y_o, which should simply be "the volume of y_o-space", clearly contributes an infinity which we wish to understand and cancel against the infinity produced by the translation symmetry of embeddings expressed in equation (1.7). We will explain this.

Since the eigenvector e_o (corresponding to $\lambda_o = 0$) is a constant, the normalization $\|e_o\|^2 = 1$ shows that the value of e_o is $(\text{Area}(\Sigma, g))^{-1/2}$. In order to perform the y_o-integral uniformly over all choices of the underlying metric g, we stipulate the normalization $\int e^{-\|y_o \cdot 1\|^2} dy_o = 1$. $[\| \cdot \|$ always denotes the L^2 norm from (1.12). The "1" inside the norm means the (unnormalized) eigenvector given by the constant (harmonic) function 1. Note also that we ignore the $\sqrt{2\pi}$ factors that appear in finite dimensional Gaussian integrals.]

We were interested in analyzing $\int e^{-\|y_o \cdot e_o\|^2 \lambda_o} dy_o$, where we think of λ_o as a *small* (but positive) eigenvalue that we will ultimately send to zero. By the normalized integral above, and the value of e_o, we are forced to the conclusion that the integral above must behave like $(\lambda_o e_o^2)^{-1/2} = \lambda_o^{-1/2}(\text{Area}(\Sigma, g))^{1/2}$.

So putting all the integrals over $(y_o, y_1, y_2 \ldots)$ together produces:

$$(1.17) \qquad \lambda_o^{-1/2}(\text{Area}(\Sigma, g))^{1/2}(\det'(-\Delta_g))^{-1/2}$$

Now remember the symmetry exhibited in (1.7); in fact, $Y \mapsto Y + c$, affects only the y_o-variable and produces the overcounting infinity mentioned before. We think of this infinity as "cancelling off" the $\lambda_o^{-1/2}$ infinity appearing in (1.17) (since $\lambda_o \to 0$). Thus (1.17) becomes (1.11'), which in turn motivates (1.11) itself.

In any event, we are only claiming that (1.11) is a well-motivated *definition*, and a little thought about the above arguments shows that it is the definition that is mathematically natural.

1.5. Met(Σ) and $\mathcal{M}_h$

Taking care of the symmetries enjoyed by the action law, the Polyakov integral (1.11*) has now metamorphosed to the following (still rather enigmatic) shape :

$$Z = \frac{1}{\text{vol}(\text{Diff}^+(\Sigma)) \times \text{vol}(\text{Conf}(\Sigma))} \int_{\text{Met}(\Sigma)} \left[\frac{\det'(-\Delta_g)}{\text{Area}(\Sigma, g)}\right]^{-d/2} Dg$$

$$(1.18) \qquad = \int_{\text{Met}(\Sigma)/\,\text{Diff}^+(\Sigma) \times \text{Conf}(\Sigma)} \left[\frac{\det'(-\Delta_g)}{\text{Area}(\Sigma, g)}\right]^{-d/2} Dg$$

The final space over which one is now supposed to be integrating is nothing other than the moduli space $\mathcal{M}_h$ parametrizing all complex analytic structures on Σ. In fact, two metrics g_1 and g_2 in Met(Σ) are equivalent in $\mathcal{M}_h$ provided $g_2 = f^*(e^\phi g_1)$, for some $e^\phi \in \text{Conf}(\Sigma)$ and some $f \in \text{Diff}^+(\Sigma)$. Thus the complex structures, τ_1 and τ_2, obtained via isothermal parameters from the metrics g_1 and g_2, respectively, are then biholomorphically equivalent via the biholomorphism $f : \Sigma_{\tau_2} \longrightarrow \Sigma_{\tau_1}$. So we define:

$$(1.19) \qquad \mathcal{M}_h = \mathrm{Met}(\Sigma)/\{\mathrm{Diff}^+(\Sigma) \times \mathrm{Conf}(\Sigma)\}$$

The fundamental question, therefore, is whether the integrand in (1.18) produces a well-defined measure on $\mathcal{M}_h$ for some choice of the free parameter d (the spacetime dimension). We will sketch in the next sections how that pleasant state of affairs transpires exactly when $d = 26$.

2. Polyakov volume form on the moduli space

2.1. The Riemannian structure of $\mathrm{Met}(\Sigma)$

Consider the infinite dimensional manifold of Riemannian metrics, $\mathrm{Met}(\Sigma)$, on the fixed smooth surface Σ. At the metric $g \in \mathrm{Met}(\Sigma)$ we can assign the following natural inner product to the tangent space of $\mathrm{Met}(\Sigma)$ thereat:

$$(2.1) \qquad \langle \delta g^1_{ab}, \delta g^2_{cd} \rangle = \int \int_\Sigma (g^{ac} g^{bd} \delta g^1_{ab} \delta g^2_{cd}) \, d_g(\mathrm{vol})$$

where $\delta g^1_{ab}, \delta g^2_{cd}$ – two symmetric 2nd-order covariant tensors on Σ – are small perturbations of g, representing two tangent vectors to $\mathrm{Met}(\Sigma)$. Thus $\mathrm{Met}(\Sigma)$ itself qualifies as a Riemannian manifold (of infinite dimension). *The Polyakov path integral, (1.11*) or (1.18), is to be interpreted as an integration over $\mathrm{Met}(\Sigma)$ with respect to the volume form induced on $\mathrm{Met}(\Sigma)$ by this Riemannian structure (2.1).*

Remark. It is possible to make the mathematical considerations regarding (2.1) completely rigorous by working with the Hilbert manifold $\mathrm{Met}(\Sigma)$ that one obtains by considering all metrics belonging to a suitable Sobolev class H^s. Since the finite-dimensional mathematics on the Teichmüller space that we finally arrive upon is independent of these technical subtleties, we will not say more about this matter in this exposition. In what follows it is possible to restrict oneself to metrics, conformal scalings and diffeomorphisms that are all of class C^∞ on Σ.

As we know, the infinite dimensional manifold, $\mathrm{Met}(\Sigma)$, is acted on by the infinite dimensional group $\mathcal{G} = \mathrm{Diff}^+(\Sigma) \times \mathrm{Conf}(\Sigma)$, (this is actually a semi-direct product), producing the quotient space $\mathcal{M}_h$ – an orbifold of finite dimension $(6h - 6)$.

Remark. If one assigns ∞ as the 'dimension' of the space of (local) functions on Σ, then the choice of a Riemannian metric tensor involves essentially three arbitrary functions – so $\mathrm{Met}(\Sigma)$ is 3∞ dimensional. In that sense $\mathrm{Conf}(\Sigma)$ is ∞ dimensional, and $\mathrm{Diff}^+(\Sigma)$ is 2∞ dimensional. Thus, the trading off of the parameters in $\mathrm{Met}(\Sigma)$ for those in the gauge group leaves a residual finite number of "moduli parameters" – leading to the interesting equation "$3\infty - 3\infty = 6h - 6$"!

We shall work with the universal covering space of $\mathcal{M}_h$ in the orbifold covering sense; that is the *Teichmüller space*, $\mathcal{T}(\Sigma) = \mathcal{T}_h$. To define it, replace the group

Diff$^+(\Sigma)$ in the above action by its identity component $\text{Diff}_0(\Sigma)$ (comprising diffeomorphisms homotopic to the identity). We obtain:

$$(2.2) \qquad \mathcal{T}_h = Met(\Sigma)/\{\text{Diff}_0(\Sigma) \times \text{Conf}(\Sigma)\}$$

Concomitantly, we shall denote the quotient projection from $\text{Met}(\Sigma)$ to $\mathcal{T}_h$ by:

$$(2.3) \qquad \mathcal{P} : \text{Met}(\Sigma) \to \mathcal{T}_h$$

Notice that the quotient $\text{Met}(\Sigma)/\text{Conf}(\Sigma)$ is precisely the space of (smooth) Beltrami coefficients on Σ (see [N1]).

Representing $\mathcal{T}_h$ as a slice within $\text{Met}(\Sigma)$. We henceforth assume that the genus h is at least two; (the situation for spheres is trivial, and for tori the case is special and easily treated.) Then $\mathcal{T}_h$ is a smooth manifold of real dimension $(6h - 6)$ (with a natural complex manifold structure); the discrete *"mapping class group"* (or *"modular group"*), $MCG(\Sigma) = MC_h = \text{Diff}^+(\Sigma)/\text{Diff}_0(\Sigma)$, acts biholomorphically and proper discontinuously on $\mathcal{T}_h$ producing the quotient $\mathcal{M}_h$ as a (complex analytic) orbifold.

The space $\mathcal{T}_h$ can be concretely pictured as the space of conjugacy classes of "marked" Fuchsian groups $\{\Gamma\}$ which are cocompact and which produce quotient surfaces of the given genus. A Fuchsian group Γ is marked by the choice of an isomorphism of the fundamental group of Σ onto it. The moduli space $\mathcal{M}_h$ is just the set of conjugacy classes of these Fuchsian groups (without markings). See [N1] for this basic material.

Any smooth section (=right-inverse) of the quotient map $\mathcal{P}$ will be called a *slice* in $\text{Met}(\Sigma)$.

> *Juliet: "What's in a name? that which we call a rose,*
> *By any other name would smell as sweet;"*

Thus a slice is an embedded copy of Teichmüller space, $\mathcal{T}_h$, in $\text{Met}(\Sigma)$. Slices are $(6h - 6)$ dimensional submanifolds of $\text{Met}(\Sigma)$, transverse to the orbits of the "gauge group" $\mathcal{G}$. Mathematically speaking, a slice represents the variation of the conformal moduli as variation of Riemannian metric; in physics one says that choosing a slice "fixes the gauge freedom".

The Poincare slices and Weil-Petersson. The uniformization theorem guarantees that every Riemann surface structure on Σ arises from a Poincare (hyperbolic) metric of constant negative curvature (-1), that metric being uniquely determined up to an arbitrary diffeomorphism. Define therefore the following subset (infinite dimensional submanifold) of $\text{Met}(\Sigma)$:

$$(2.4) \qquad Hyp(\Sigma) = \{g \in \text{Met}(\Sigma) : curvature(g) \equiv -1\}$$

The quotient $\text{Met}(\Sigma)/\text{Conf}(\Sigma)$ is therefore in natural bijection with the above submanifold of hyperbolic metrics.

Using the uniformization theorem with parameters, (see [N1]), we can choose a smoothly varying family of hyperbolic metrics $\gamma(t)$ in $Hyp(\Sigma)$ representing the Teichmüller space. Any such slice we call a "Poincare slice". Thus $Hyp(\Sigma)/\mathrm{Diff}_0(\Sigma)$ is the Teichmüller space, and we can choose special Poincare slices that are *orthogonal with respect to (2.1)* to the orbits of the gauge group $\mathrm{Diff}_0(\Sigma)$. A convenient name we will adopt for such a slice is *"horizontal Poincare slice"*. The fundamental metric (2.1), restricted to any horizontal Poincare slice, gives a Riemannian structure to $\mathcal{T}_h$ that is known classically as the *Weil-Petersson metric* on the Teichmüller space. See Section 3.1 for more in this direction.

Remark. Having fixed a Poincare slice $\gamma(t)$, any other Poincare slice is then given by $f_t^*(\gamma(t))$, where the $f_t \in \mathrm{Diff}_0(\Sigma)$ are an arbitrary family of diffeomorphisms that are chosen to depend smoothly on $t \in \mathcal{T}_h$. It is therefore clear that the metric (2.1) *cannot* induce the same metric on $\mathcal{T}_h$ via an arbitrarily chosen Poincare slice; however, (2.1) does induce Weil-Petersson on the *horizontal* Poincare slices defined above.

A natural way to select a Poincare slice is to utilize the unique *harmonic diffeomorphism* (Eells-Sampson), that exists in the homotopy class of the identity, between any two hyperbolic metrics on Σ. One may pullback the target metric via this diffeomorphism to obtain a specific choice of hyperbolic metric on Σ representing the Teichmüller class of the target metric. This Poincare slice is horizontal (because harmonic Beltrami coefficients are orthogonal to the Diff_0 directions). See [J],[N3],[W] and the references therein.

Facts about (2.1):

(a) *The surface diffeomorphisms, $\mathrm{Diff}^+(\Sigma)$, act on $\mathrm{Met}(\Sigma)$ as isometries of the Riemannian structure (2.1).*

(b) *However, the action of the Weyl-rescalings on $\mathrm{Met}(\Sigma)$ do not enjoy this compatibility with the metric (2.1).*

Remark. These are easily established. Fact(b) above, namely that conformal rescalings fail to preserve the metric on $\mathrm{Met}(\Sigma)$, is at the root of the "conformal anomaly" that we will have to grapple with in the material below.

2.2. Change of coordinates in $\mathrm{Met}(\Sigma)$

In order to analyse the integral (1.18) restricted to any slice, we must clearly understand the nature of the metric (2.1) on $\mathrm{Met}(\Sigma)$ in coordinates that are along the gauge orbits, and complementary coordinates in the Teichmüller directions along the slice. Indeed, we want to factor out of the Polyakov integral the integrals over the orbits of $\mathrm{Diff}^+(\Sigma)$ and $\mathrm{Conf}(\Sigma)$, as explained in (1.18). Therefore it is natural to want to express the integral 1.18 as an iterated integral over these gauge orbits and over the slice.

So fix once and for all some real analytic coordinates $((t)) = ((t_i))$, i running from 1 to $N = 6h - 6$, on $\mathcal{T}_h$; (the $((t_i))$ can be chosen as the real and imaginary parts of *holomorphic* coordinates, and this can even be done globally over Teichmüller space). We shall identify $\mathcal{T}_h$ as this domain in $((t))$-space, whenever convenient.

Let us *fix* a slice K, by choosing a right-inverse of $\mathcal{P}$:

$$(2.5) \qquad\qquad \gamma : \mathcal{T}_h \to \mathrm{Met}(\Sigma), \ \gamma(\mathcal{T}_h) = K$$

and note that an arbitrary metric $\rho \in \mathrm{Met}(\Sigma)$ has a unique expression:

$$(2.6) \qquad\qquad \rho = f^*[e^\phi \gamma((t_i))]$$

Here $((t_i))$ represents the Teichmüller point to which ρ projects by $\mathcal{P}$, e^ϕ is a conformal rescaling, and $f \in \mathrm{Diff}_0(\Sigma)$. Thus the new coordinates for ρ are $(f, \phi, ((t_i))) \in \mathrm{Diff}_0(\Sigma) \times \mathrm{Conf}(\Sigma) \times \mathcal{T}_h$. Rescale the entire slice of metrics by the fixed e^ϕ, and set:

$$g = g((t)) = e^\phi \gamma((t))$$

$g((t))$ is an associated gauge-fixing slice, and note that f becomes an isometry from the metric ρ to the the metric g because $\rho = f^*(g)$. We remark that since $\mathrm{Diff}_0(\Sigma)$ is the arc-component of the identity in $\mathrm{Diff}^+(\Sigma)$, it is possible to express f as $exp(\xi)$ where ξ is a smooth vector-field on Σ.

We need to compute the Riemannian structure (2.1) of $\mathrm{Met}(\Sigma)$ in these new coordinates. At any given point $\rho \in \mathrm{Met}(\Sigma)$, we must understand small metrical variations $\delta\rho_{ab}$ in terms of changes in these new $Gauge \times \mathcal{T}_h$ coordinates. Instead of working with a small change of ρ we will work with a corresponding small change of g; remembering (Fact (a) above), that the pullback action by the fixed diffeomorphism f is an isometric automorphism of $\mathrm{Met}(\Sigma)$, we lose nothing by this.

To this end we first write the arbitrary small change in the metric in the form:

$$(2.7) \qquad\qquad \delta\rho = f^*(\delta g)$$

utilizing the *same* diffeomorphism f as in (2.6). The definition of δg implies that it involves: a diffeomorphism close to the identity (which we write as $exp(\xi)$), a small change in the Weyl rescaling, $\delta\phi$, and small changes $((\delta t_i))$ in the Teichmüller coordinates; $\delta g = (exp\xi)^*[e^{\phi+\delta\phi}\gamma((t+\delta t))] - g$

We will work in the tangent space to $\mathrm{Met}(\Sigma)$ at the point g. Recall that the *trace* (with respect to g_{ab}) of a symmetric tensor σ_{ab} is by definition the contraction $g^{ab}\sigma_{ab}$. Observe that the traceless symmetric second-order covariant tensors, dh_{ij}, and the pure-trace tensors of the form $\delta\phi g_{ab}$, constitute *orthogonal* spaces with respect to the fundamental inner product (2.1). It is therefore convenient to express the general metrical deformation δg as a sum of traceless and pure-trace parts. That will be done below.

Notation. We shall sum over repeated indices in the formulae of this article. The roman letters a, b etc will usually vary over the surface coordinates (i.e., $1 \le a, b \le 2$), whereas the indices i, j, m, n will usually run through Teichmüller coordinates and thus range from 1 to $N = (6h - 6)$. Indices will be raised and lowered with respect to g. $D(\xi)$ will denote covariant derivative of the vector field ξ, with respect to g.

Lemma 2.1. *The infinitesimal change δg defined above decomposes as:*

$$(2.8) \qquad\qquad \delta g_{ab} = \delta\phi g_{ab} + (P\xi)_{ab} + \delta t^i T^i_{ab}$$

with the last two terms constituting the tracefree part. Here

$$P : \{\,Vector\ fields\ on\ \Sigma\,\} \to \{\,Symmetric\ traceless\ (2,0)\ tensors\ on\ \Sigma\,\}$$

is the 1st order elliptic operator (depending on g), given by

$$(P\xi)_{ab} = D_a(\xi_b) + D_b(\xi_a) - g_{ab}D_c(\xi^c)$$

and,

$$T^i_{ab} = \frac{\partial}{\partial t_i}[g_{ab}((t))] - \frac{1}{2}g^{cd}\frac{\partial}{\partial t_i}[g_{cd}((t))]$$

The $T^i((t))$ are the (trace free parts of) the tangent vectors in the Teichmüller directions to the gauge slice $g((t))$.

Proof. Differential geometry on the surface gives: $(exp\xi)^*g_{ab} = g_{ab} + D_a(\xi_b) + D_b(\xi_a)$; and by Taylor expansion one sees: $e^{\delta\phi}g_{ab} = g_{ab} + \delta\phi g_{ab} + o(\delta\phi)$. Computing with the help of these gives (2.8). In (2.8), by suitably redefining $\delta\phi$, all the pure-trace component of the δg has been absorbed in the first term on the right hand side. □

Now, the various spaces of tensor-fields on the surface (Σ, g) carry natural inner products induced by the metric g. For the space of vector-fields the pairing is explicitly:

$$(2.9a) \qquad (\xi, \eta)_g = \int\int_\Sigma [\xi^a\eta^b g_{ab}]d_g\,\text{vol}$$

and for the covariant 2nd-order traceless tensors the pairing is:

$$(2.9b) \qquad (R, S)_g = \int\int_\Sigma [R_{ab}S_{cd}g^{ac}g^{bd}]d_g\,\text{vol}$$

(Whenever the metric g is clear from the context we will take the liberty of suppressing that subscript.)

We can think of the operator P as an unbounded closed operator defined on the appropriate dense domain of the Hilbert space of vector fields. Then, by basic functional analysis, the target Hilbert space of symmetric 2nd order tracefree tensors decomposes into the orthogonal direct sum: {2nd order symmetric tracefree tensors} = {Range$P$} ⊕ {Ker P^*}.

The range of P constitutes the piece of the tangent space to Met(Σ) arising from pullbacks of the metric g by diffeomorphisms close to the identity. The Weyl (conformal) scalings of g are already absorbed in the pure-trace part of the infinitesimal deformation δg. Consequently, the residual piece given by $Ker P^*$ is the finite-dimensional part comprising tangent vectors in the Teichmüller (slice) directions.

The final upshot is that the tangent space at g to the infinite dimensional space Met(Σ) decomposes as an orthogonal direct sum:

$$(2.10) \qquad T_g(\text{Met}(\Sigma)) = \{pure\ trace\} \oplus \{Range\ P\} \oplus \{Ker\ P^*\}$$

In order to understand the Teichmüller deformations piece, which is of central interest to us, we need therefore to analyse the adjoint of P (computed with respect to the inner products (2.9a) and (2.9b)). One easily verifies that P^* is nothing other than the d-bar operator on the space of (smooth) quadratic differentials on (Σ, g)

(thought of as a Riemann surface). *Consequently, the kernel, $Ker P^*$, is the $6h - 6$ dimensional (real) subspace of the tracefree 2nd order symmetric tensors consisting of the (real and imaginary parts of) holomorphic quadratic differentials.*

Teichmüller's Lemma: $T_t^*(\mathcal{T}_h) \equiv H^0(X_t, K(X_t)^2)$. The upshot of the discussion above will be recognized by those familiar with Teichmüller theory as a variant of "Teichmüller's lemma"; that lemma describes the complex cotangent space (at the conformal structure g) to the Teichmüller space $\mathcal{T}_h$, as precisely (namely, canonically isomorphic to) the vector space of holomorphic quadratic differentials on (Σ, g). See [N1]. Indeed, the tangent space at $t \in \mathcal{T}_h$ to Teichmüller space, is, by Kodaira-Spencer, canonically isomorphic to the first cohomology $H^1(X_t, K^{-1})$, which, by Serre duality, is canonically isomorphic to the dual space of global holomorphic sections of K^2. By the Riemann-Roch theorem, this last space is a complex vector space of complex dimension $(3h - 3)$ (for each $t \in \mathcal{T}_h$). (Note: $K = K(X_t)$ denotes the holomorphic cotangent bundle of the Riemann surface X_t.)

Bases for $H^0(X_t, K^2)$ as X_t varies in $\mathcal{T}_h$. To facilitate our work, we therefore introduce an arbitrary auxiliary choice of basis for the holomorphic quadratic differentials on each of the Riemann surfaces $X_t = (\Sigma, g((t)))$ – i.e., all over the Teichmüller space $\mathcal{T}_h$. Thus let $Q^m((t))$, $1 \le m \le (6h - 6)$, be a basis of the $6h - 6$ dimensional (over reals) vector space of holomorphic quadratic differentials on the Riemann surface $(\Sigma, g((t)))$, where the Q^m are assumed to vary smoothly with $((t)) \in \mathcal{T}_h$. (That is easily accomplished by, for instance, utilizing Poincare theta-series to define the Q^m with respect to real-analytically varying Fuchsian groups $\Gamma_{((t))}$.) Notice the important fact that once we have chosen bases as above on each surface along a particular slice, the job is accomplished for *all* slices, – because *the Q^m depend only on the conformal class* of the metric under consideration.

Clearly, Teichmüller's lemma, asserts that each $Q^m((t))$ can be considered as a nowhere vanishing 1-form over $\mathcal{T}_h$. We will need this interpreation in Lemma 2.4 below.

Substituting (2.8) into the Riemannian structure formula (2.1) we obtain immediately the desired re-expression for the Riemannian norm on the tangent space to $\text{Met}(\Sigma)$ at g:

$$(2.11) \qquad \|\delta g\|^2 = \|\delta\phi\|^2 + (P\xi, P\xi) + (T^i, Q^m)((Q^m, Q^n))^{-1}(Q^n, T^j)\delta t_i \delta t_j$$

We explain the notations: The L^2 norm for functions on the surface (Σ, g) was already shown in (1.12), – and that defines the first term $\|\delta\phi\|^2$. The $T^i((t))$ are, as shown above, the (trace free parts of) the tangent vectors in the Teichmüller directions. Each T^i is, of course, a second-order symmetric traceless covariant 2nd-order tensor on (Σ, g). All the pairings appearing in the second and third terms on the right of (2.11) are the inner products exhibited in (2.9b). Thus, setting $R = S = P\xi$ in (2.9b) gives the second term, and similarly for the pieces in the third term. (Note that the inversion involved in the last term is matrix inversion.)

Proposition 2.2. *The volume measure induced by the Riemannian structure (2.1) on the infinite dimensional manifold* $\mathrm{Met}(\Sigma)$ *expressed in the coordinates* $(\phi, \xi, ((t_i)))$ *is:*

(2.12)

$$[Dg] = [(\det P^* P)^{1/2}][\det((Q^m, Q^n))]^{-1/2}[\det(T^i, Q^m)][D\phi][D\xi]dt_1 \wedge \cdots \wedge dt_{6h-6}$$

By $[D\phi]$ *we mean the volume measure on* $\mathrm{Conf}(\Sigma)$ *induced by the* L^2*-norm (1.12) on the space of rescaling functions* ϕ*. Similarly,* $[D\xi]$ *is the volume element on* $\mathrm{Diff}_0(\Sigma)$ *arising from the inner product (2.9a) on vector fields.*

Proof. The expression of the Riemannian structure (2.1) in the form (2.11) is adapted precisely to the slice K. Thus the usual formula $d_g \mathrm{vol} = \sqrt{\det((g_{ij}))}dx_1 \wedge \cdots \wedge dx_M$ for the Riemannian volume element on a M-dimensional manifold, when applied formally to this infinite dimensional $\mathrm{Met}(\Sigma)$, gives (2.12). Indeed, in the infinite dimensional pieces corresponding to the first two terms on the right of (2.11), one needs to compute the determinants of the operators $I = Identity$ and P, respectively. For later convenience we have substituted $|\det P| = [(\det P^* P)^{1/2}]$. $\square$

Recall that the integrations over the gauge variables, (ϕ, ξ), is expected to produce the volumes of the gauge groups that we have been desiring to factor out (see (1.18)). Therefore, ignoring the $[D\phi][D\xi]$ in the above volume element of $\mathrm{Met}(\Sigma)$ is clearly a reasonable method of accomplishing that aim, – and it is the physicist's way of reducing the Polyakov integral to a finite dimensional integral over the given slice. Dropping $[D\phi][D\xi]$ from (2.12) we obtain therefore *a volume element on the slice* K:

$$(2.13) \qquad d\mu_K = \left[\frac{(\det P_g^* P_g)}{\det((Q^m, Q^n)_g)} \right]^{1/2} [\det((T^i, Q^m)_g)]dt_1 \wedge \cdots \wedge dt_{6h-6}$$

We emphasize that as the metric $g = g((t))$ varies over K, the operator $P = P_g$ varies, and thus the infinite determinant $[(\det P^* P)^{1/2}]$ above is a non-trivial function of $((t))$.

Now we have reached the goal of expressing the Polyakov integral as a well-defined finite dimensional integration over any chosen slice:

The Polyakov prescription. *In the light of Proposition 2.2 and the discussion above, we interpret the Polyakov integral (1.18) as the following integral over the (arbitrarily chosen) slice:*

$$(2.14) \qquad \int_{Slice\ K} \left[\frac{\det'(-\Delta_g)}{\mathrm{Area}(\Sigma, g)} \right]^{-d/2} d\mu_K$$

The integrand above is (for any choice of d) a well-defined volume form on the slice K, and since K is an embedded copy of $\mathcal{T}_h$ the volume form can be considered as living on $\mathcal{T}_h$.

Remark. We have been working over the Teichmüller space, $\mathcal{T}_h$, rather than over the moduli space, $\mathcal{M}_h$ – as was prescribed by (1.18). So, strictly speaking, the Polyakov integral (2.14) should be divided by the "cardinality of the mapping class group MC_h". Alternatively, we should think of the integral in (2.14) only over a fundamental domain for the action of MCG. In any event, we are only interested in the *volume form = integrand* on $\mathcal{T}_h$ or $\mathcal{M}_h$, rather than in the integral itself. The basic fact is that the Polyakov measure on $\mathcal{T}_h$ is MCG-invariant, but its total integral over $\mathcal{M}_h$ (i.e., over any fundamental domain) is infinite (see (3.3) below).

2.3. Finally! The Polyakov measure on $\mathcal{T}_h$

Now that we have reduced ourselves from the original Polyakov path integral (1.6) to the integral (2.14), the fundamental question is whether (2.14) is *independent of the choice of the slice*. The main result is:

Theorem 2.3. *When $d/2 = 13$, the integrand in (2.14), as a volume form on the Teichmüller space $\mathcal{T}_h$, is* **independent** *of the choice of the slice K. This is the Polyakov volume form, d(Poly), on $\mathcal{T}_h$. It is mapping class group invariant, is absolutely continuous with respect to the Lebesgue measure class on $\mathcal{T}_h$.*

Addendum to Theorem. *d(Poly) has the following expression in terms of the Weil-Petersson volume form on $\mathcal{T}_h$:*

$$(2.15) \qquad d(\text{Poly}) = [(Z'_G(1))^{-13} Z_G(2)]d(Weil - Pet)$$

Here $Z_G(s)$ is the Selberg zeta function (an entire function) associated to the (variable) Fuchsian group G.

Remark. Throughout the above considerations, we are ignoring overall constant factors in expressions for the Polyakov form. It is our attitude that the Polyakov volume form on each genus moduli is an interesting mathematical object defined only up to an arbitrary scaling constant. See (3.3) below also.

The notations in (2.15) will be explained as we go along. Combining (2.13) and (2.14) we realize that we have to deal with the following volume form on the slice K:

$$(2.16)$$

$$dP_K = \left[\frac{\det'(-\Delta_g)}{\text{Area}(\Sigma, g)}\right]^{-d/2} \left[\frac{(\det P_g^* P_g)}{\det((Q^m, Q^n)_g)}\right]^{1/2} [\det((T^i, Q^m)_g)]dt_1 \wedge \cdots \wedge dt_{6h-6}$$

(the metric g varies over K.)

The aim is to understand how far the above volume density depends on the choice of the Teichmüller slice. Let us break up the work into natural pieces:

Lemma 2.4. *The volume form on $\mathcal{T}_h$ given by:*

$$(2.17) \qquad d\lambda = [\det((T^i, Q^m)_g)]dt_1 \wedge \cdots \wedge dt_{6h-6}$$

is independent of the choice of the slice. (It depends only on the choice of the bases of quadratic differentials $Q^m((t))$, which we have fixed, once and for all, over the entire Teichmüller space $\mathcal{T}_h$.)

Indeed, recalling that the $Q^m((t))$ can be considered as 1-forms over $\mathcal{T}_h$, this volume form is the following top dimensional form:

$$(2.18) \qquad d\lambda = Q^1 \wedge \cdots \wedge Q^{6h-6}$$

Proof. A new slice $K^\sharp$ is obtained from the slice K by scaling the metrics comprising K in a $((t))$-dependent fashion: $g^\sharp((t)) = e^{\phi((t))}g((t))$. The trace-free parts of the tangent vectors to the slices are related by: $T^{i\sharp} = e^\phi T^i$. Computing the relevant pairings now, by the law (2.9b) (with respect to the metrics g and $g^\sharp$, respectively), we obtain

$$(2.19) \qquad (T^{i\sharp}, Q^m)_{g^\sharp} = (T^i, Q^m)_g$$

Note that the above equality simply represents the fact that the T^i are tangent vectors to the Teichmüller space, whereas the Q^m are co-tangent vectors to the same space, and that the pairing is just the *duality pairing* of tangents with cotangents. Hence, it is not surprising that (2.19) holds independent of scalings in the metric.

Consequently, (2.17) is independent of the choice of the slice, as claimed. The rest is straightforward. $\square$

Consider therefore the fundamental function F defined below (for each value of the unspecified "space-time" dimension d). Each $F = F_d$ is a real-valued function on the entire space of metrics, Met(Σ):

$$(2.20) \qquad F_d(g) = \left[\frac{\det'(-\Delta_g)}{\text{Area}(\Sigma, g)}\right]^{-d/2} \left[\frac{(\det P_g^* P_g)}{\det((Q^m, Q^n)_g)}\right]^{1/2}$$

We shall see that F_{26} is singled out!:

Proposition 2.5. *The function F_d is invariant along the orbits of the gauge group $\{\text{Diff}^+(\Sigma) \times \text{Conf}(\Sigma)\}$ if and only if $d = 26$; F_{26} thus descends as a function to the moduli space $\mathcal{M}_h$. Namely one has the remarkable invariance:*

$$(2.21) \qquad F_{26}(f^*(e^\phi g)) = F_{26}(g)$$

for all $g \in \text{Met}(\Sigma)$, under all conformal rescalings and pullbacks by diffeomorphisms.

Proof. Let us first note the significance of the two infinite dimensional determinant factors that constitute the function F_d. Define:

$$(2.22) \qquad D_0(g) = \left[\frac{\det'(-\Delta_g)}{\text{Area}(\Sigma, g)}\right]$$

$$(2.23) \qquad D_1(g) = \left[\frac{(\det P_g^* P_g)}{\det((Q^m, Q^n)_g)}\right]$$

Remember that the operator Δ_g is the Laplacian acting on smooth functions on Σ, and that $P_g^* P_g$ is the Laplacian acting on smooth vector-fields.

The problem is to understand the variation of these two determinantal functions on Met(Σ) under a conformal rescaling of the metric. This is a standard type of problem for the application of heat kernel techniques – the results are clearly exposed in Alvarez[Alv]. See also [AN]. Recall the heat-kernel (i.e., zeta-function) regularized determinant for operators in infinite dimensions:

$$(2.24) \qquad \log(\det' D) = -\lim_{\epsilon \to 0} \int_\epsilon^\infty [Trace'(e^{-tD})] \frac{dt}{t}$$

The notation $\det'$ and $Trace'$ mean that the determinant or trace is being calculated after discarding the zero eigenvalues – namely on the orthogonal complement of the kernel of D.

Consider an arbitrary infinitesimal conformal scaling of the metric g to $\gamma = e^{\delta\phi}g$. The corresponding perturbation in the non-zero eigenvalues of the relevant operators appearing in (2.22) and (2.23) can be calculated. The first variations turn out as below, (R_g denotes the scalar curvature for metric g):

$$(2.25) \qquad \delta(\log D_0) = \frac{1}{12\pi}\int_\Sigma R_g\delta\phi d_g(\text{vol})$$

$$(2.26) \qquad \delta(\log D_1) = \frac{13}{12\pi}\int_\Sigma R_g\delta\phi d_g(\text{vol})$$

Therefore, δF_d is $\frac{1}{12\pi}(13-\frac{d}{2})$ times the integral appearing on the right of the formulae (2.25) and (2.26). The invariance property (2.21) for F_{26} follows immediately. $\square$

Proof of Theorem 2.3. Combining the gauge-invariant function F_{26} on Met(Σ) with the slice-independent volume form $d\lambda$ of Lemma 2.4, we see that we have proved that *the Polyakov volume form on T_h has the expression:*

$$(2.27) \qquad d(\text{Poly}) = F_{26}d\lambda$$

Since the bases for quadratic differentials were introduced simply as an artifice to help simplify our formulae, it is obvious that the volume form $d(\text{Poly})$ above is also independent of the choice of these bases $\{Q^m((t))\}$. [Note: Glancing at (2.16), we may see directly that the combination of Q-dependent pieces: $[\det((T^i, Q^m)_g)][\det((Q^m, Q^n)_g)]^{-1/2}$, evidently remains invariant when the Q^m are subjected to an arbitrary (in general t-dependent) change of basis.]

We have therefore established the existence of the Polyakov volume form on T_h, independent of all arbitrary choices, as stated in Theorem 2.3. $\square$

As general references for the matter we presented see [Alv],[AN],[Nel]. The expression for $d(\text{Poly})$ in terms of $d(Weil - Pet)$ will be dealt with in the next section.

3. Geometry of Teichmüller space and Polyakov volume

3.1. Weil-Petersson and Polyakov

Weil-Petersson Riemannian structure on T_h. Since the cotangent space to T_h at any point t is canonically isomorphic to the vector space of holomorphic quadratic differentials on the Riemann surface $X_t = (\Sigma, \gamma(t))$, a hermitian structure is induced on T_h by utilizing the Petersson pairing of holomorphic quadratic differentials on X_t. In fact, let $\gamma(t)$ be a hyperbolic metric representing the conformal structure $t \in T_h$; if the metric is expressed as $\gamma = \lambda(z)|dz|$ in terms of local holomorphic (=isothermal) coordinates, then set

$$(3.1) \qquad (Q^1, Q^2)_{WP} = \int_\Sigma Q^1(z)\overline{Q^2(z)}\lambda(z)^{-2}dz \wedge d\bar{z}$$

This is Weil-Petersson hermitian metric of T_h. See [IT],[J],[N1]. The corresponding volume form on T_h we shall denote by $d(Weil - Pet)$.

Lemma 3.1.

(1) *The Riemannian structure (2.1) of* $\mathrm{Met}(\Sigma)$*, restricted to a horizontal Poincare slice, is Weil-Petersson on* $\mathcal{T}_h$*.*

(2) *Choose the bases* $\{Q^m((t))\}$ *for the spaces of holomorphic quadratic differentials to be orthonormal (pairing (2.9b)) with respect to the corresponding hyperbolic metrics* $\gamma((t))$*. Then the (slice-independent) volume form* $d\lambda = Q^1 \wedge \cdots \wedge Q^{6h-6}$ *described in Lemma 2.4 is* $d(Weil - Pet)$*.*

Proof. (1):   A holomorphic quadratic differential Q on (Σ, γ) is a cotangent vector to $\mathcal{T}_h$ there; the Weil-Petersson pairing sets up an isomorphism of the cotangent space with the tangent space – thus Q corresponds to a certain tangent vector v_Q to $\mathcal{T}_h$. Identifying $\mathcal{T}_h$ with the slice, the tangent v_Q is given by an infinitesimal change in the hyperbolic metric γ. But that perturbation of metric, say $\delta\gamma_Q$, is, in its turn, represented by a certain Beltrami differential on (Σ, γ). Indeed, one sees that the co-vector Q corresponds to the harmonic (Bers') Beltrami differential $\mu_Q = \overline{Q(z)}y^2$, $(z = x + iy)$.

Above, we are utilizing uniformization to represent the Riemann surface (Σ, γ) as H/Γ, H being the Poincare upper half-plane, and Γ a Fuchsian group operating thereon. The hyperbolic metric $\gamma = y^{-1}|dz|$ therefore gets deformed to $\gamma + \delta\gamma = y^{-1}|dz + \varepsilon\mu(z)d\bar{z}|$ with $\mu = \mu_Q$ as shown. See [N1] for details.

Expanding out the right side we may compute the $\delta\gamma = \delta\gamma_Q$; then substituting in the formula (2.1) we get $(\delta\gamma_{Q^1}, \delta\gamma_{Q^2}) = (Q^1, Q^2)_{WP}$, as desired.

(2):   Clearly, at every point $((t)) \in \mathcal{T}_h$, the co-vectors $\{Q^m((t))\}$ now constitute an orthonormal basis for $T^*_{((t))}(\mathcal{T}_h)$ with respect to the Weil-Petersson inner-product. Therefore, $Q^1 \wedge \cdots \wedge Q^{6h-6} = d(Weil - Pet)$ $\square$

Remark.   The bases of quadratic differentials above can be constructed using Poincare theta series with respect to varying Fuchsian group $\Gamma((t))$ and then applying the Gram-Schmidt orthonormalization procedure. Since the bases are arbitrary up to any smooth choice $\mathcal{T}_h \to GL(6h - 6, \mathbf{R})$, we note that the corresponding measure $Q^1 \wedge \cdots \wedge Q^{6h-6} = d\lambda$ can be, a priori, chosen quite arbitrarily.

Selberg zeta and determinants for Laplacians.   Associated to the hyperbolic Riemann surface $(\Sigma, \gamma) = H/\Gamma$, Γ the uniformizing (purely hyperbolic, cocompact) Fuchsian group, Selberg defines

$$(3.2) \qquad Z_\Gamma(s) = \prod_\alpha \prod_{n \geq 0} [1 - exp\{-l(g_\alpha)(s + n)\}]$$

where $\{g_\alpha\}$ are a complete set of representatives for the conjugacy classes of the primitive elements of Γ (i.e., those allowing no non-trivial n-th roots in Γ); $l(g)$ denotes the length of the geodesic in the free homotopy class determined by g. This function is defined by the above product for $Re(s) > 1$, and can be shown to have an analytic continuation as an entire function in the whole s-plane.

The Selberg trace formula, that goes hand in hand with the above zeta function, allows one to express the heat-kernel regularized determinants of the Laplace operators on (Σ, γ) – on functions, vector-fields and higher order tensor fields – as certain values of the above holomorphic function. As explained, for example, in [DP], one gets $\det'(-\Delta) = Z'(1)$ (up to a constant independent of the Fuchsian group), and similarly, $\det(P^*P) = Z(2)$. (All the numbers $Z(k)$, integral k, have interpretations as determinants of Laplacians acting on suitable spaces of tensors on (Σ, γ).)

Proof of Addendum to Theorem 2.3. Glancing now at (2.22), (2.23), and armed with (2) of Lemma 3.1, we have clearly obtained the expression for $d(\text{Poly})$ claimed in the Addendum to Theorem 2.3. (Note: Over hyperbolic metrics the Gauss-Bonnet shows that $\text{Area}(\Sigma, \gamma) = 4\pi(h-1)$; moreover, the determinant appearing in the denominator of (2.23) we had already normalized to unity by our choice of orthonormal bases for quadratic differentials.) $\square$

Remark. The asymptotics of the Selberg zeta functions, when the Fuchsian groups approach the boundary (Deligne-Mumford boundary) of $\mathcal{M}_h$, can be used to show that the Polyakov volume form blows up near the boundary so fast that the fundamental amplitude integral is divergent:

$$(3.3) \qquad \int_{\mathcal{M}_h} d(\text{Poly}) = \infty$$

(Recall that the Weil-Petersson volume of $\mathcal{M}_h$, on the other hand, is finite.) (3.3) demonstrates that no finite answer is forthcoming via the Polyakov prescription for the "vacuum-to-vacuum" amplitude of the bosonic string.

3.2. Complex geometry and Polyakov

We are now at the point where we can explain our deepest reason for excitement about the Polyakov measure. *In fact, $d(\text{Poly})$ has a natural and simple construction (over each $\mathcal{T}_h$) by simply the complex analytic geometry of $\mathcal{T}_h$ – with no inputs whatsoever from physics.*

The universal family and allied bundles over $\mathcal{T}_h$. The Teichmüller spaces are fine moduli spaces. Namely, the total space $\Sigma \times \mathcal{T}_h$ admits a natural complex structure such that the projection to the second factor

$$(3.4) \qquad \psi_h : \mathcal{C}_h := \Sigma \times \mathcal{T}_h \to \mathcal{T}_h$$

gives the universal Riemann surface over $\mathcal{T}_h$. This means that for any $\eta \in \mathcal{T}_h$, the submanifold $\Sigma \times \eta$ is a complex submanifold of $\mathcal{C}_h$, and the complex structure on Σ induced by this embedding is represented by η. As is well-known, (Chapter 5 in [N1]), the family $\mathcal{C}_h \to \mathcal{T}_h$ is the *universal* object in the category of holomorphic families of genus h marked Riemann surfaces.

We recall now the *determinant of cohomology* construction for obtaining line bundles over any parameter space of Riemann surfaces. Let X be a compact Riemann surface and L be a holomorphic line bundle on X. The determinant of cohomology of L is defined to be the 1-dimensional complex vector space $\det H^*(L) = \det(L)$:

$$(3.5) \qquad \det(L) = (\overset{top}{\wedge} H^0(X, L)) \bigotimes (\overset{top}{\wedge} H^1(X, L)^*)$$

We are ready to pass to families. Let $\pi : \mathcal{X} \longrightarrow S$ be a holomorphic family of compact Riemann surfaces parametrized by a base S which is complex analytic space. The prototypical instance is the universal family described over $\mathcal{T}_h$. The definition demands that π be a holomorphic submersion with compact and connected fibers of complex dimension one – each fiber being of genus h.

For any point $s \in S$, the construction (3.5) gives a complex line $\det(L_s)$. The basic fact is that these lines fit together to give a holomorphic line bundle on S (see [BGS]), which is called the *determinant of cohomology bundle* of L_S, and is denoted by $\det(L_S)$. (Note: This is not to be confused with the top exterior power of L_S – which is a line bundle over the total space $\mathcal{X}$.) The general definition arises from the direct image functors R_* of algebraic geometry. See [KM], [D].

Over each genus Teichmüller space we thus have a sequence of natural determinant of cohomology bundles arising from the powers of the relative tangent bundles along the fibers of the universal curve. Indeed, let $\omega_h \longrightarrow \mathcal{C}_h$ be the relative cotangent bundle for the projection ψ_h in (3.4). The determinant line bundle over $\mathcal{T}_h$ arising from its n-th tensor power is fundamental for us, and we shall denote it by:

$$(3.6) \qquad \mathrm{DET}_{n,h} := \det(\omega_h{}^n) \longrightarrow \mathcal{T}_h, \ n \in \mathbf{Z}$$

Applying Serre duality shows that there is a canonical isomorphism $\mathrm{DET}_{n,h} = \mathrm{DET}_{1-n,h}$, for all n. $\mathrm{DET}_{0,h} = \mathrm{DET}_{1,h}$ is called the *Hodge* line bundle over $\mathcal{T}_g$.

Remark.

(i) The determinant of cohomology bundle, $\det(L_S)$, is functorial with respect to base change. Given any morphism $\gamma : S' \longrightarrow S$ this means, in particular, that there is a canonical isomorphism of the determinant of cohomology associated to the pulled back family over S' with the pullback by γ of the original determinant bundle over S. See [BNS] for details.

(ii) The determinant of cohomology construction, $\det(L_S)$, produces a bundle over the parameter space S induced by the bundle over the total space $\mathcal{X}$; now, the Grothendieck-Riemann-Roch (GRR) theorem (see [BGS], [D]) gives a canonical isomorphism of $\det(L_S)$ with a combination of certain bundles (on S) obtained from the direct images of the bundle L_S and the relative tangent bundle $T_{\mathcal{X}/S}$. The GRR theorem is important for our work below.

(iii) Whenever the bundles L_S and the relative tangent bundle above are assigned smooth hermitian structures, the determinant of cohomology inherits a smooth hermitian structure, due to Quillen [Q], in a functorial way. See [BGS], [D], [BNS]. The canonical isomorphisms arising from base change (see (i)) then become *isometric isomorphisms*.

The vertical (=relative) tangent bundle for the universal family, namely the bundle $T_{\mathcal{C}_h/\mathcal{T}_h} \to \mathcal{C}$ comprising the tangent bundles of the fibering Riemann surfaces, carries a canonical hermitian structure by assigning the Poincaré hyperbolic metric on each surface (utilizing the uniformization theorem with moduli parameters). As a consequence, applying remark (i) above, the holomorphic line bundles $\mathrm{DET}_{n,h}$ carry natural *Quillen hermitian metric* arising from the Poincaré metrics on the fibers of $\mathcal{C}_h$.

Observe that by the naturality of the above constructions it follows that the action of MC_h on $\mathcal{T}_h$ has a natural lifting as automorphisms of these DET bundles. These automorphisms are unitary with respect to the Quillen structure.

The fibers of $\mathrm{DET}_{n,h}$**:.** Notice that the fiber of Hodge, over $X \in \mathcal{T}_h$, is the top exterior power of the vector space of holomorphic Abelian differentials on X. Similarly, the fiber of $\mathrm{DET}_{2,h}$ is the $(3h - 3)$ exterior power of the vector space of holomorphic quadratic differentials on X, and so on. By some complex geometry (e.g., Poincare theta series with respect to holomorphically varying quasi-Fuchsian groups), one can create the natural *MCG equivariant holomorphic vector bundle*, $V_{n,h}$ over $\mathcal{T}_h$ by attaching over X the fiber $H^0(X, K_X^n)$ $(n = 1, 2, \cdots)$. Therefore, (3.6) shows that the top exterior powers of these vector bundles over $\mathcal{T}_h$ are nothing other than the determinant of cohomology bundles $\mathrm{DET}_{n,h}$ we are describing.

In particular, by Teichmüller's lemma we know $V_{2,h}$ is canonically isomorphic to the holomorphic cotangent bundle of Teichmüller space. Thus, there are canonical isomorphisms:

$$(3.7) \qquad \overset{(3h-3)}{\wedge} V_{2,h} = \overset{top}{\wedge} T^*\mathcal{T}_h = \mathrm{DET}_{2,h}$$

each of the above being a description of the canonical line bundle over $\mathcal{T}_h$.

We also recall that holomorphic 1-forms can be paired naturally, in the L^2 sense, on any Riemann surface X via the "Hodge pairing"

$$(3.8) \qquad (\alpha, \beta) = -i \int \int_X \alpha \wedge \bar{\beta}$$

This gives a *MCG* invariant hermitian metric on $V_{1,h}$, and hence induces a *MCG* invariant hermitian structure (called the Hodge metric) on the Hodge bundle. (The Hodge metric has to be modified suitably by a factor given by the determinant of the Laplacian to obtain the Quillen metric on that bundle.)

Mumford isomorphisms and Polyakov volume. These natural line bundles over $\mathcal{T}_h$ will be considered as *MCG*-equivariant line bundles, and the isomorphisms we talk about will be *MCG*-equivariant isomorphisms. By applying the Grothendieck-Riemann-Roch theorem indicated above, Mumford [Mum] had shown that $\mathrm{DET}_{n,h}$ is canonically isomorphic to a certain fixed (*genus-independent!*) tensor power of the Hodge bundle.

Proposition 3.2.

$$(3.9) \qquad \mathrm{DET}_{n,h} \cong \mathrm{DET}_{1,h}{}^{\otimes(6n^2-6n+1)}$$

The above isomorphism is isometric with respect to the Quillen metrics. An isomorphism (3.9) is ambiguous only up to a non-zero scalar.

Remark.

(i) It can be shown that the Picard group of (isomorphism classes of) all MC_h-equivariant holomorphic line bundles over $\mathcal{T}_h$, (which can be identified as the group of holomorphic line bundles over the moduli space $\mathcal{M}_h$), is a cyclic group generated by the Hodge bundle. This can be shown from work of Harer, Mumford and Arbarello-Cornalba. That cyclic group is of order 10 for $h = 2$, and is infinite cyclic for $h > 2$. The above Proposition shows that the natural sequence of determinant of cohomology line bundles pick out a certain explicit subsequence from the Picard group.

(ii) The $MCG(\equiv MC_h)$ invariant holomorphic functions on $\mathcal{T}_h$ – namely those that descend to $\mathcal{M}_h$ – are it constant, for all $h > 2$. This follows from the Satake compactification theory for $\mathcal{M}_h$. Thus, although $\mathcal{M}_h$ is non-compact, in certain aspects it behaves like a compact analytic space. Consequently, (at least when $h > 2$), any two isomorphisms between bundles over $\mathcal{M}_h$ can only be ambiguous up to a scale factor, as asserted.

There is a remarkable connection, discovered by Belavin and Knizhnik [BK], between the Mumford isomorphism above for the case $n = 2$, [i.e., that DET_2 is the 13-th tensor power of Hodge], and the existence of the Polyakov string measure on the moduli space $\mathcal{M}_g$. In fact this 13 is the same lucky number as the value of $d/2$ in Proposition 2.5. See [BK],[Bos],[Nel].

First, it is elementary to see that assigning a hermitian metric on the canonical bundle of any complex space gives a measure on that space. Indeed, fixing a volume density on a space simply amounts to fixing a fiber metric on the canonical line bundle, K, – because then we know the unit vectors in K – and absolute square gives volume. Now, K for the Teichmüller space is nothing other than $\mathrm{DET}_{2,h}$ – recall (3.7) above. Thus, any natural hermitian structure on $\mathrm{DET}_{2,h}$ becomes a choice of a natural volume form on $\mathcal{T}_h$.

Theorem 3.3. *The Hodge bundle, $Hodge \equiv \mathrm{DET}_{1,h}$, has its natural Hodge metric, arising from (3.8). We may transport the corresponding metric on $Hodge^{13}$ to $\mathrm{DET}_{2,h}$ by Mumford's isomorphism, (the choice of isomorphism being unique up to scalar) – thereby obtaining a MCG invariant volume form, say $d(\mathrm{Mum})$, on $\mathcal{T}_h$ (also unambiguous up to the choice of a scalar). This $d(\mathrm{Mum})$ is none other than the Polyakov volume form $d(\mathrm{Poly})$ of Theorem 2.3.*

Sketch of Proof. Choose a local holomorphic frame, $(\omega_1(t), \cdots , \omega_h(t))$ for the rank h vector bundle $V_{1,h}$ over a local holomorphic t-patch in $\mathcal{T}_h$. Suppose that $(\psi_1(t), \cdots , \psi_{3h-3}(t))$ is a local holomorphic frame for $V_{2,h}$ over the same neighbourhood such that under the Mumford isomorphism:

$$(3.10) \qquad [\omega_1 \wedge \cdots \wedge \omega_h]^{13} \mapsto (\psi_1 \wedge \cdots \wedge \psi_{3h-3})$$

Then the Mumford volume form on the t-coordinate patch is immediately seen to have the expression:

$$(3.11) \qquad d(\mathrm{Mum}) = [\det((\omega_i(t), \omega_j(t)))]^{-13} d\lambda$$

210 SUBHASHIS NAG

with the matrix of pairings being the Hodge pairings of (3.8), and where the measure $d\lambda$ is given by Lemma 2.4, equation (2.18): $d\lambda = \psi_1 \wedge \cdots \wedge \psi_{3h-3} \wedge \overline{\psi_1} \wedge \cdots \wedge \overline{\psi_{3h-3}}$

Utilizing therefore the bases Q^m for quadratic differentials given by precisely these ψ_k and their conjugates, we may compare $d(\text{Mum})$ of (3.11) with $d(\text{Poly})$ as shewn in equation (2.20); we derive that the *ratio* of these two volume forms on $\mathcal{T}_h$ is the following function:

$$(3.12) \quad \frac{d(\text{Poly})}{d(\text{Mum})} = F_{26}(g(t))[\det((\omega_i(t), \omega_j(t)))]^{13}$$

$$= \left[\frac{\det'(-\Delta_g)}{\text{Area}(\Sigma, g)}\right]^{-13} \left[\frac{(\det P_g^* P_g)}{\det((\psi_m, \psi_n)_g)}\right]^{1/2} [\det((\omega_i(t), \omega_j(t)))]^{13}$$

Above, $g = g(t)$ is any Riemannian metric representing the conformal structure $t \in \mathcal{T}_h$.

Since both volume forms under consideration are, by construction, modular invariant, we note that the above ratio, say $G(t)$, is a real valued MCG invariant function on the Teichmüller space.

Claim. $\frac{d(\text{Poly})}{d(\text{Mum})} = G(t)$ is the absolute value square of a global *holomorphic* function f on $\mathcal{T}_h$.

Basically one computes $\partial\bar{\partial}[\log G(t)]$ and shows that this is identically zero. This local computation is explained in [BK],[Nel] and other references. Thus *holomorphic* $f = f_U$ exists on any local (t)-neighbourhood U satisfying $G(t) = |f(t)|^2$. But $\mathcal{T}_h$ is simply connected, and it is clear that these local f_U can be patched up to analytically continue along all paths in $\mathcal{T}_h$. It follows directly that there is a holomorphic function f defined on the entire $\mathcal{T}_h$ such that $G(t) \equiv |f(t)|^2$.

Finally, we show that this global f is necessarily modular invariant. In fact, under any modular transformation f can only be multiplied by factor of absolute value one – thus f gives rise to a character for the modular group. But, since that group, MCG, is generated by elements of finite order (!), which all have fixed points in their action on $\mathcal{T}_h$, it follows that the character must be trivial. That is as desired.

But such f, and hence G, must be constant (recall the Satake compactification remark) – and we are through. $\square$

Remark.

    (i) The expression for $d(\text{Mum})$ in (3.11) also allows one to study the blow up of the volume form as one approaches the Deligne-Mumford boundary of the moduli space. This can be used to reprove the divergence of the string amplitude stated in (3.3).

    (ii) As we said above Theorem 3.3, any natural hermitian structure on $\text{DET}_{2,h}$ is the assignment of a natural volume form on $\mathcal{T}_h$. Therefore the Quillen hermitian structure on this DET bundle assigns also a volume element to Teichmüller/moduli space. This is being studied in relation to the Polyakov volume. For example, is the Quillen volume of $\mathcal{M}_h$ finite?

4. The universal Teichmüller space of compact surfaces

The moral of the story above is that the presence of the *Mumford isomorphisms* over the moduli space of genus g Riemann surfaces describes the Polyakov measure structure thereon! It is consequently a natural problem to try to create *genus-independent* versions of these constructions by working over some universal parameter space that parametrizes Riemann surfaces of varying genus. Indeed, the string may sweep over world-sheets of arbitrary genus – so that the problem raised is fundamental to non-perturbative string theory.

In this and the following section we give a concise report of our work [BNS] where we succeed in finding a genus-independent description of the Mumford isomorphisms over a certain universal parameter space $\mathcal{T}_\infty$. We would like to emphasize one point: all our constructions are equivariant under the action on $\mathcal{T}_\infty$ of an intriguing new mapping class group MC_∞.

4.1. The direct limit, $\mathcal{T}_\infty$, of classical Teichmüller spaces

We start with a fundamental topological situation. Let

$$(4.1) \qquad\qquad \pi : \tilde{X} \longrightarrow X$$

be an *unramified covering map*, orientation preserving, between two compact connected oriented two manifolds $\tilde{X}$ and X of genera $\tilde{g}$ and g, respectively. Assume $g \geq 2$. The degree of the covering π, which will play an important role, is the ratio of the respective Euler characteristics; namely, $\deg(\pi) = (\tilde{g} - 1)/(g - 1)$.

Given any complex structure on X, we may pull back this structure via π to a complex structure on $\tilde{X}$. Now, the homotopy lifting property guarantees that there is a unique diffeomorphism $\tilde{f} \in \mathrm{Diff}_0(\tilde{X})$ which is a lift of any given $f \in \mathrm{Diff}_0(X)$. Mapping f to $\tilde{f}$ defines an injective homomorphism of $\mathrm{Diff}_0^+(X)$ into $\mathrm{Diff}_0^+(\tilde{X})$. Consequently, π induces an injection of the smaller Teichmüller space into the larger one:

$$(4.2) \qquad\qquad \mathcal{T}(\pi) : \mathcal{T}_g \longrightarrow \mathcal{T}_{\tilde{g}}$$

It is known that this map $\mathcal{T}(\pi)$ is a *proper holomorphic embedding* between these finite dimensional complex manifolds; $\mathcal{T}(\pi)$ respects the quasiconformal-distortion (=Teichmüller) metrics. From the definitions it is evident that this embedding between the Teichmüller spaces depends only on the (unbased) isotopy class of the covering π.

At the level of Fuchsian groups, one should note that any covering space π corresponds to the choice of a subgroup H of finite index ($= \deg(\pi)$) in the uniformizing group G for X, and the embedding (4.2) is then the standard inclusion mapping $\mathcal{T}(G) \to \mathcal{T}(H)$; (see Chapter 2, [N1]).

Remark. One notices that $\mathcal{T}$ is a contravariant functor from the category of closed oriented topological surfaces, morphisms being covering maps, to the category of finite dimensional complex manifolds and holomorphic embeddings. We shall have more to say along these lines below.

We construct a category $\mathcal{A}$ of certain topological objects and morphisms: the objects, $Ob(\mathcal{A})$, are a set of compact oriented topological surfaces each equipped with a base point $(\star)$, there being exactly one surface of each genus $g \geq 0$; let the object of genus g be denoted by X_g. The morphisms are based isotopy classes of pointed covering mappings

$$\pi : (X_{\tilde{g}}, \star) \to (X_g, \star)$$

there being one arrow for each such isotopy class. *Note that the monomorphism of fundamental groups induced by (any representative of) the based isotopy class π, is unambiguously defined.*

The direct system of classical Teichmüller spaces: Fix a genus g and let $X = X_g$. Observe that all the morphisms with the fixed target X_g:

$$(4.3) \qquad M_g = \{\alpha \in Mor(\mathcal{A}) : Range(\alpha) = X_g\}$$

constitute a *directed set* under the partial ordering given by factorisation of covering maps. Thus if α and β are two morphisms from the above set, then $\beta \succ \alpha$ if and only if the image of the monomorphism $\pi_1(\beta)$ is contained within the image of $\pi_1(\alpha)$; that happens if and only if there is a commuting triangle of morphisms: $\beta = \alpha \circ \theta$. It is important to note that the factoring morphism θ is *uniquely determined* because we are working with base points. [Remark: Notice that the object of genus 1 in $\mathcal{A}$ only has morphisms to itself – so that this object together with all its morphisms (to and from) form a subcategory.]

By (4.2), each morphism of $\mathcal{A}$ induces a proper, holomorphic, Teichmüller-metric preserving embedding between the corresponding finite-dimensional Teichmüller spaces. We can thus create the natural *direct system of Teichmüller spaces* over the above directed set M_g, by associating to each $\alpha \in M_g$ the Teichmüller space $\mathcal{T}(X_{g(\alpha)})$, where $X_{g(\alpha)} \in Ob(\mathcal{A})$ denotes the domain surface for the covering α. To each $\beta \succ \alpha$ one associates the corresponding holomorphic embedding $\mathcal{T}(\theta)$ (with θ as in the paragraph above). From this direct system we form the *direct limit Teichmüller space over* $X = X_g$:

$$(4.4) \qquad \mathcal{T}_\infty(X_g) = \mathcal{T}_\infty(X) := ind.lim.\mathcal{T}(X_{g(\alpha)})$$

$\mathcal{T}_\infty(X)$ is a metric space with the Teichmüller metric, and it also has a natural Weil-Petersson Riemannian structure obtained from scaling the Weil-Petersson pairing on each finite dimensional stratum, $\mathcal{T}_g$, by the factor $(g-1)^{-1}$. See [NS].

$\mathcal{T}_\infty$ is our *commensurability Teichmüller space* – which will serve as the base space for universal Mumford isomorphisms.

The Teichmüller space, $\mathcal{T}(H_\infty)$, of the hyperbolic solenoid:. Over the same directed set M_g we may also define a natural *inverse system of surfaces*, by associating to $\alpha \in M_g$ a certain copy, S_α of the pointed surface $X_{g(\alpha)}$. [Note: Fix a universal covering over $X = X_g$. S_α can be taken to be the universal covering quotiented by the action of the subgroup $Im(\pi_1(\alpha)) \subset \pi_1(X, \star)$.] If $g \geq 2$, then the inverse limit of this system is the *universal solenoidal surface* $H_\infty(X) = proj.lim.X_{g(\alpha)}$ that was studied in [S],[NS].

In fact, H_∞ is a compact topological space that fibers over X with the fibers being Cantor sets. The path components of H_∞, (with leaf-topology), are simply connected two-manifolds restricted to each of which the projection $\pi_\infty : H_\infty \to X$ becomes a universal covering. There are uncountably many of these path components ("leaves") in H_∞, and each is a dense subset in H_∞. Each leaf is thus, morally speaking, a hyperbolic plane. That is why we call H_∞ the universal hyperbolic solenoid. The facts above follow from a careful study of this inverse system of surfaces, the main tool being the lifting of paths in X to its coverings.

As explained in [S],[NS], the solenoid H_∞ has a natural Teichmüller space comprising equivalence classes of complex structures on the leaves – the leaf complex structures being required to vary continuously in the fiber (Cantor) directions. In particular, any complex structure assigned to any of the surfaces $X_{g(\alpha)}$ appearing in the inverse tower can be pulled back to all the surfaces above it – and therefore assigns a complex structure of the sort demanded on H_∞ itself. These complex structures that arise from some finite stage can be characterized as the "transversely locally constant" (TLC) ones (see [NS]), and they comprise precisely the *dense* subset $\mathcal{T}_\infty(X)$ sitting within the separable Banach manifold $\mathcal{T}(H_\infty(X))$. We collect some of these thoughts in the:

Proposition 4.1. *The "ind-space" $\mathcal{T}_\infty(X)$, (see [Sha]), is the inductive limit of finite dimensional complex manifolds, and hence carries a complex structure defined strata-wise. The completion of $\mathcal{T}_\infty(X)$ with respect to the Teichmüller metric is the separable complex Banach manifold $\mathcal{T}(H_\infty(X))$.*

Alternatively, $\mathcal{T}_\infty(X)$ can be embedded in Bers' universal Teichmüller space, $\mathcal{T}(\Delta)$, as a directed union of the Teichmüller spaces of Fuchsian groups. (The Fuchsian groups vary over the finite index subgroups of a fixed cocompact Fuchsian group G, $X = \Delta/G$.) The closure in $\mathcal{T}(\Delta)$ of $\mathcal{T}_\infty(X)$ is a Bers-embedded copy of $\mathcal{T}(H_\infty(X))$.

Remark. It is evident, but important to note, that the spaces $\mathcal{T}_\infty(X)$ and $\mathcal{T}(H_\infty(X))$ we are dealing with do *not* really depend on the choice of X. If we were to start with a surface X' of different genus (also greater than one), then we could pass to a common covering surface (always available!), and hence the limit spaces we construct would be isomorphic.

4.2. The commensurability mapping class group $MC_\infty = Aut(\mathcal{T}_\infty)$

A remarkable but obvious fact about the above construction is that every morphism $\pi : Y \to X$ of $\mathcal{A}$ induces a natural *Teichmüller metric preserving homeomorphism*

$$(4.5) \qquad \mathcal{T}_\infty(\pi) : \mathcal{T}_\infty(Y) \longrightarrow \mathcal{T}_\infty(X)$$

$\mathcal{T}_\infty(\pi)$ is invertible simply because the morphisms of $\mathcal{A}$ with target Y are cofinal with those having target X (thus all finite ambiguities are forgotten in passing to the inductive limits!). It is also clear that $\mathcal{T}_\infty(\pi)$ is a biholomorphic identification (with respect to the strata-wise complex structures). [Note that $\mathcal{T}_\infty$ is covariant – whereas the Teichmüller functor $\mathcal{T}$ itself was contravariant.]

It follows that each $\mathcal{T}_\infty(X)$, and so also its metric completion $\mathcal{T}(H_\infty(X))$, is equipped with a large *automorphism group* – one from each undirected cycle of morphisms of $\mathcal{A}$ starting from X and returning to X. By repeatedly using pull-back diagrams (i.e., by choosing appropriate connected component of the fiber product of covering maps), it is fairly easy to see that the automorphism of $\mathcal{T}_\infty(X)$ arising from any (many arrows) cycle can be obtained simply from a two-arrow cycle $\tilde{X} \underset{\rightarrow}{\overset{\rightarrow}{}} X$.

Namely, whenever we have (the isotopy class of) a "self-correspondence" of X given by any two non-isotopic coverings, say α and β,

$$(4.6) \qquad\qquad \tilde{X} \underset{\rightarrow}{\overset{\rightarrow}{}} X$$

we can create a corresponding automorphism $R \in Aut(\mathcal{T}_\infty(X))$ defined as the composition: $R = \mathcal{T}_\infty(\beta) \circ (\mathcal{T}_\infty(\alpha))^{-1}$.

These automorphisms constitute a group that we shall call the *commensurability modular group*:

$$(4.7) \qquad\qquad MC_\infty(X) = Aut(\mathcal{T}_\infty(X))$$

acting on $\mathcal{T}_\infty(X)$ and on $\mathcal{T}(H_\infty(X))$.

To clarify matters further, consider the abstract graph (1-complex), $\Gamma(\mathcal{A})$, obtained from the topological category $\mathcal{A}$ by looking at the objects as vertices and the (undirected) arrows as edges. It is clear from the definition above that the fundamental group of this graph, viz. $\pi_1(\Gamma(\mathcal{A}), X)$, is acting on $\mathcal{T}_\infty(X)$ as these automorphisms. (We may fill in triangular 2-cells in this abstract graph whenever two morphisms (edges) compose to give a third edge; the thereby-reduced fundamental group of this 2-complex also produces on $\mathcal{T}_\infty(X)$ the action of $MC_\infty(X)$.)

Making explicit the genus one situation. For the genus one object X_1 in $\mathcal{A}$, we know that the Teichmüller space for any unramified covering is but a copy of the upper half-plane H. The maps $\mathcal{T}(\pi)$ are Möbius identifications of copies of the half-plane with itself, and we easily see that the pair $(\mathcal{T}_\infty(X_1), CM_\infty(X_1))$ is identifiable as $(H, PGL(2, \mathbb{Q}))$. Notice that the action has dense orbits in this case. Anticipating for a moment the definition of $Vaut$ given below, we remark that $GL(2, \mathbb{Q}) \cong Vaut(\mathbb{Z} \oplus \mathbb{Z})$, and $Vaut^+$ is precisely the subgroup of index 2 therein, as expected.

On the other hand, if $X \in Ob(\mathcal{A})$ is of any genus $g \geq 2$, then we get an infinite dimensional "ind-space" as $\mathcal{T}_\infty(X)$ with the action of $\mathcal{G}(X)$ on it as described. Since the tower of coverings over X and Y (both of genus higher than 1) eventually become cofinal, it is clear that *for any choice of genus higher than one we get **one** isomorphism class of pairs* $(\mathcal{T}_\infty, MC_\infty)$.

Virtual automorphism group of $\pi_1(X)$ and MC_∞:. In the classical situation, the action of the mapping class group $MCG(X)$ on $\mathcal{T}(X)$ was induced by the action of (isotopy classes of) self-homeomorphisms of X; in the direct limit set up we now have the more general (isotopy classes of) self-correspondences of X inducing the new mapping class automorphisms on $\mathcal{T}_\infty(X)$. In fact, we will see that our group MC_∞ corresponds to "virtual automorphisms" of the fundamental group $\pi_1(X)$, –

generalizing exactly the classical situation where the usual $Aut(\pi_1(X))$ appears as the action via modular automorphisms on $\mathcal{T}(X)$.

Given any group G, one may look at its "partial" or "virtual" automorphisms; as opposed to usual automorphisms that are defined on all of G, for virtual automorphisms we demand only that they be defined on some finite index subgroup. To be precise, consider isomorphisms $\rho : H \to K$ where H and K are subgroups of finite index in G. Two such isomorphisms (say ρ_1 and ρ_2) are considered equivalent if there is a finite index subgroup (sitting in the intersection of the two domain groups) on which they coincide. The equivalence class $[\rho]$ – which is like the *germ* of the isomorphism ρ – is called a *virtual automorphism* of G; clearly the virtual automorphisms of G constitute a group, christened $Vaut(G)$, under the obvious law of composition, (i.e., compose after passing to deeper finite index subgroups, if necessary).

Clearly $Vaut(G)$ is trivial unless G is infinite (though there do exist infinite groups – see [MT] – such that $Vaut$ is trivial). Also evident is the fact that $Vaut(group) = Vaut(any\ finite\ index\ subgroup)$. Since we shall apply this concept to the fundamental group of a surface of genus g, $(g > 1)$, the last remark shows that our $Vaut(\pi_1(X_g))$ *is genus independent!*

In fact, $Vaut$ presents us a neat way of formalizing the "two-arrow cycles" (4.6) which we introduced to represent elements of MC_∞. Letting $G = \pi_1(X)$, (recall that X is already equipped with a base point), we see that the diagram (4.6) corresponds exactly to the following virtual automorphism of G:

$$(4.8) \qquad [\rho] = [\beta_* \circ \alpha_*^{-1} : \alpha_*(\pi_1(\tilde{X})) \to \beta_*(\pi_1(\tilde{X}))]$$

Here α_* denotes the monomorphism of the fundamental group $\pi_1(\tilde{X})$ into $\pi_1(X) = G$, and similarly β_* etc.. We let $Vaut^+(\pi_1(X))$ denote the subgroup of $Vaut$ arising from pairs of orientation preserving coverings.

As we said, all the automorphisms arising from arbitrarily complicated cycles of coverings (i.e., any finite sequence of morphisms starting and ending at X), are each obtained from these simple two-arrow pictures (4.6). The reduction of any many-arrow cycle in $\Gamma(\mathcal{A})$ to a two-arrow cycle utilizes successive fiber product diagrams; there is some amount of choice in this reduction process, and one may obtain different two-arrow cycles starting from the same cycle – but the virtual automorphism that is defined is unambiguous. The final upshot is:

Proposition 4.2. *One has natural surjective group homomorphisms:*

$$\pi_1(\Gamma(\mathcal{A}), X) \to Vaut^+(\pi_1(X)) \to MC_\infty(X)$$

The second arrow is an isomorphism.

Remark.

(i) The concept of $Vaut$ has already arisen in group theory papers – for example [Ma],[MT]. I am grateful to Chris Odden for pointing out these references to me, and for interesting discussions.

(ii) There is a natural representation of $Vaut(\pi_1(X))$ in the homeomorphism group of the unit circle S^1, by the standard theory of *boundary homeomorphisms* (see, for example, [N1]). This leads to many obvious questions whose answers are not obvious. In recent work, Robert Penner and the present author have proved an isomorphism between basically the group MC_∞ and a direct limit of Penner's Ptolemy groups that operate on his "Tessellations" version of universal Teichmüller space. That isomorphism is connected with the following conjecture, and we hope to report on these matters in forthcoming publications.

Topological transitivity of MC_∞ on $\mathcal{T}_\infty$ and allied issues:.   Does MC_∞ act with dense orbits in $\mathcal{T}_\infty$? That is a basic query. This question is directly seen to be equivalent to the following old conjecture which, we understand, is due to L.Ehrenpreis and C.L.Siegel:

Conjecture 4.3. *Given any two compact Riemann surfaces, X_1 (of genus $g_1 \geq 2$) and X_2 (of genus $g_2 \geq 2$), and given any $\epsilon > 0$, can one find finite unbranched coverings π_1 and π_2 (respectively) of the two surfaces such that the corresponding covering Riemann surfaces $\tilde{X}_1$ and $\tilde{X}_2$ are of the same genus and there exists a $(1+\epsilon)$ quasiconformal homeomorphism between them. (Namely, $\tilde{X}_1$ and $\tilde{X}_2$ come ϵ-close in the Teichmüller metric.)*

Remark. Since the uniformization theorem guarantees that the *universal* coverings of X_1 and X_2 are *exactly* conformally equivalent, the conjecture asks whether we can obtain high *finite* coverings that are *approximately* conformally equivalent.

5. Universal Polyakov-Mumford on $\mathcal{T}_\infty$

The goal is to construct natural geometrical fiber bundles over the commensurability Teichmüller space, which when restricted to the finite dimensional strata, $\mathcal{T}_g$, become the Hodge (or higher n) DET bundles thereon. The bundles over $\mathcal{T}_\infty$ should be related by universal Mumford isomorphisms, which restrict to the strata as the finite dimensional Mumford isomorphisms explained in (3.9). We are able to carry out this entire program in a MC_∞ equivariant fashion ([BNS]).

5.1. The Main Lemma

We invoke into play the arbitrary unramified finite covering $\pi : \tilde{X} \to X$, and recall that $\mathcal{T}(\pi)$ (4.2) is the associated holomorphic embedding of Teichmüller spaces.

The fundamental question is whether there exists any natural relationship between the line bundle $\mathrm{DET}_{n,g}$ over the Teichmüller space $\mathcal{T}_g = \mathcal{T}(X)$ and the bundle $\mathcal{T}(\pi)^* \mathrm{DET}_{n,\tilde{g}}$ obtained as the pullback of the corresponding determinant of cohomology bundle over the larger Teichmüller space $\mathcal{T}_{\tilde{g}} = \mathcal{T}(\tilde{X})$. For example, we are asking, is there any natural relationship between the two Hodge bundles?

We have an elegant answer to this question that forms the foundation for our genus-independent description of Mumford isomorphisms. In effect, $\mathrm{DET}_{n,g}$ raised to the tensor power $\deg(\pi)$, simply extends naturally over the larger Teichmüller space $\mathcal{T}_{\tilde{g}}$ as the $\text{DET}_{n,\tilde{g}}$ bundle thereon! We prove this by utilizing the Grothendieck-Riemann-Roch (GRR) theorem (of [D]) in crucial ways.

Lemma 5.1. *The two holomorphic line bundles (with Quillen hermitian structures),* $(\text{DET}_{n,g})^{\deg(\pi)}$ *and* $\mathcal{T}(\pi)^{\ast}\text{DET}_{n,\tilde{g}}$, *on* $\mathcal{T}_g$ *are canonically isometrically isomorphic for every integer* n. *(The isomorphism is canonical up to the choice of a 12th root of unity.) In other words, there is a canonical isometrical line bundle morphism* $\Gamma(\pi)$ *lifting* $\mathcal{T}(\pi)$ *and making the following diagram commute:*

$$
\begin{array}{ccc}
\text{DET}_{n,g}{}^{\deg(\pi)} & \xrightarrow{\ \Gamma(\pi)\ } & \text{DET}_{n,\tilde{g}} \\[2pt]
\Big\downarrow & & \Big\downarrow \\[8pt]
\mathcal{T}_g & \xrightarrow{\ \mathcal{T}(\pi)\ } & \mathcal{T}_{\tilde{g}}
\end{array}
$$

The maps $\Gamma(\pi)$ *are functorial, so that for any commuting triangle of morphisms in the category* $\mathcal{A}$, *the corresponding* Γ-*lifts also commute.*

Curvature forms of DET **bundles and Lemma 5.1.** The existence of the canonical relating morphism between the above determinant bundles (fixed n) in the fixed covering situation was first conjectured and deduced by us (see [BN]) utilizing the differential geometry of the Quillen metrics. Recall that the Teichmüller spaces $\mathcal{T}_g$ and $\mathcal{T}_{\tilde{g}}$ carry natural symplectic forms – the Weil-Petersson Kähler forms – which are in fact the *curvature forms of the Quillen metrics* of these DET bundles ([Wol], [ZT], [BGS]). If the covering π is unbranched of degree d, a direct calculation shows that this natural WP form on $\mathcal{T}_{\tilde{g}}$ (appropriately renormalized by the degree d) pulls back to the WP form of $\mathcal{T}_g$ by $\mathcal{T}(\pi)$. Equality of the curvature forms leads one to expect the isomorphism of $\text{DET}_{n,g}{}^{d}$ with $\mathcal{T}(\pi)^{\ast}\text{DET}_{n,\tilde{g}}$. This intuition is what is behind the more sophisticated GRR proof of the Lemma above.

5.2. Power-law principal bundle morphisms over Teichmüller spaces

We desire to obtain certain canonical geometric objects over the inductive limit of the finite dimensional Teichmüller spaces by coherently fitting together the determinant line bundles $\text{DET}_{n,g}$ thereon. To this end it is necessary to find a canonical mapping relating $\text{DET}_{n,g}$ itself to $\text{DET}_{n,\tilde{g}}$ utilizing the Lemma above.

Now, given any complex line bundle $\lambda \to T$ over any base T, there is a certain *canonical* mapping of λ to any positive integral (d-th) tensor power of itself, given by:

$$(5.1) \qquad\qquad \omega_d : \lambda \longrightarrow \lambda^{\otimes d}$$

where ω_d on any fiber of λ is the map $z \mapsto z^d$. Observe that ω_d maps λ minus its zero section to $\lambda^{\otimes d}$ minus its zero section by a map which is of degree d on the $\mathbb{C}^{\ast}$ fibers. ω_d is a homomorphism of the associated principal $\mathbb{C}^{\ast}$ bundles. When T is a complex manifold, and λ is a line bundle in that category, then the map ω_d is a holomorphic morphism between the total spaces of the source and target bundles.

In the situation of Lemma 5.1, therefore, we may define a canonical principal bundle morphism relating the relevant bundles:

$$(5.2) \qquad\qquad \Omega(\pi) := \Gamma(\pi) \circ \omega_{deg(\pi)} : \text{DET}_{n,g} \to \text{DET}_{n,\tilde{g}}$$

where $\Gamma(\pi)$ is the canonical (GRR) line bundle morphism provided by the Lemma.

The canonical and functorial choice of these connecting maps, $\Omega(\pi)$, provides us with a direct system of line/principal bundles over the direct systems of Teichmüller spaces.

Given a direct system T_α of complex manifolds, and line bundles ξ_α over these, whenever there are connecting maps as the $\Omega(\pi)$ above, we may pass to the direct limit of the bundles themselves, simultaneously with passing to the limit of the base spaces. That precipitates our main "non-perturbative" result:

Theorem 5.2. *Fix integer* n. *Starting from any "base" surface* $X \in Ob(\mathcal{A})$, *we obtain a direct system of principal* $\mathbb{C}^*$ *bundles* $\mathcal{L}_n(\tilde{X}) := \mathrm{DET}_{n,g(\tilde{X})}$ *over the Teichmüller spaces* $\mathcal{T}(\tilde{X})$ *with connecting holomorphic homomorphisms* $\Omega(\pi)$ *between their total spaces.*

Passing to the direct limit, one therefore obtains over the universal commensurability Teichmüller space, $\mathcal{T}_\infty(X)$, *a principal* $\mathbb{C}^* \otimes \mathbb{Q}$ *bundle:*

$$\mathcal{L}_{n,\infty}(X) = ind.lim.\mathcal{L}_n(Y)$$

Since the maps $\Omega(\pi)$ *preserved the Quillen unit circles, the limit object also inherits such a Quillen "hermitian" structure. The commensurability modular group action* $CM_\infty(X)$ *on* $\mathcal{T}_\infty(X)$ *has a natural lifting to* $\mathcal{L}_{n,\infty}(X)$ - *acting by isometrical automorphisms.*

Finally, the Mumford isomorphisms persist as isometrical MC_∞ *equivariant isomorphisms:*

$$\mathcal{L}_{n,\infty}(X) = \mathcal{L}_{1,\infty}(X)^{\otimes(6n^2-6n+1)}$$

Remark.

(i) To be explicit, the Mumford isomorphism in the above theorem means that $\mathcal{L}_{n,\infty}$ and $\mathcal{L}_{1,\infty}$ are equivariantly isomorphic relative to the automorphism of $\mathbb{C}^* \otimes \mathbb{Q}$ induced by the homomorphism of $\mathbb{C}^*$ that raises to the power exhibited.

(ii) We could have used the Quillen hermitian structure to reduce the structure group from $\mathbb{C}^*$ to $U(1)$, and thus obtain direct systems of $U(1)$ bundles over the Teichmüller spaces. Passing to the direct limit would then produce $U(1) \otimes \mathbb{Q} :=$ "tiny circle" bundles over $\mathcal{T}_\infty$.

(iii) Another interpretation of this construction over $\mathcal{T}_\infty$ is to produce "rational line bundles" over it, with relating Mumford isomorphisms, utilizing Lemma 5.1 but not involving the power law mappings. See [BN], [BNS] for details.

(iv) We have also glued together the universal bundles over direct limits of the moduli spaces $\mathcal{M}_g$, by considering the tower of characteristic coverings over X. See [BN].

Above we have succeeded in fitting together the Hodge and higher DET_n bundles over the ind-space $\mathcal{T}_\infty$, together with the relating Mumford isomorphisms – our entire construction being MC_∞ equivariant. We thus have a structure on $\mathcal{T}_\infty$ that suggests a *genus-independent and universal* version of the finite dimensional Polyakov structure that we delineated in Sections 1 to 3 of this paper.

References

[Alv]   O. Alvarez, *Theory of strings with boundaries: fluctuations, topology and quantum geometry*, Nucl. Phys. **B216** (1983), 125–184.

[AN]   L. Alvarez-Gaume and P. Nelson, *Riemann surfaces and string theories*, in Trieste Spring School "Supersymmetry, supergravity,superstrings '86", World Scient., Singapore, (1986), 419–510.

[BGS]   J.Bismut, H.Gillet and C.Soulé, *Analytic torsion and holomorphic determinant bundles, I,II,III*, Comm. Math Phys. **115** (1988), 49–78, 79–126, 301–351.

[BK]   A. Belavin, V. Knizhnik, *Complex geometry and quantum string theory*, Phys. Lett., **168 B** (1986), 201–206.

[BN]   I. Biswas and S. Nag, *Weil-Petersson geometry and determinant bundles over inductive limits of moduli spaces*, preprint.

[BNS]   I. Biswas, S. Nag and D. Sullivan, *Determinant bundles, Quillen metrics and Mumford isomorphisms over the Universal Commensurability Teichmüller Space*, Acta Mathematica, **176** (1996), 145–169.

[Bos]   J.B. Bost, *Fibres determinants, determinants regularises et mesures sur les espaces de modules des courbes complexes*, Semin. Bourbaki **152–153** (1987), 113–149.

[D]   P. Deligne, *Le déterminant de la cohomologie*, Contemporary Math. **67** (1987), 93–177.

[DP]   E. D'Hoker and D. Phong, *Multiloop amplitudes for the bosonic Polyakov string*, Nucl. Phys. B. **269** (1986), 205–234.

[IT]   Y. Imayoshi and M. Taniguchi, *An introduction to Teichmüller spaces*, Springer-Verlag, Tokyo, (1992).

[J]   J. Jost, *Harmonic maps and curvature computations in Teichmüller theory*, Ann. Acad. Scient Fenn. **16** (1991), 13–46.

[KM]   F. Knudsen and D. Mumford, *The projectivity of the moduli space of stable curves I, preliminaries on "det" and "div"*, Math. Scand. **39** (1976), 19–55.

[Ma]   A. Mann, *Problem about automorphisms of infinite groups*, Second Intn'l Conf Group Theory, Debrecen, (1987).

[MT]   F. Menegazzo and M. Tomkinson, *Groups with trivial virtual automorphism group*, Israel Jour. Math. **71** (1990), 297–308.

[Mum]   D. Mumford, *Stability of projective varieties*, L'Enseign. Math. **23** (1977), 39–100.

[N1]   S. Nag, *The Complex Analytic Theory of Teichmüller Spaces*, Wiley-Interscience, New York, (1988).

[N2]   S. Nag, *Canonical measures on the moduli spaces of compact Riemann surfaces*, Proc. Indian Acad. Sci. (Math Sci.) **99** (1989), 103–111.

[N3]   S. Nag, *Self-dual Connections, hyperbolic metrics and harmonic Mappings on Riemann surfaces*, Proc. Indian Acad. Sci. (Math Sci.) **101** (1991), 215–218.

[N4]   S. Nag, *A period mapping in Universal Teichmüller space*, Bull. Amer. Math. Soc. **26, no.2** (1992), 280–287.

[NS]   S. Nag, and D. Sullivan, *Teichmüller theory and the universal period mapping via quantum calculus and the $H^{1/2}$ space on the circle*, Osaka J. Math. **32, no.1** (1995), 1–34. [Max-Planck-Inst. (Bonn) preprint MPI-94-54.]

[NV]   S. Nag and A. Verjovsky, *Diff(S^1) and the Teichmüller Spaces, (Parts I and II)*, Commun. Math. Phys. **130** (1990), 123–138. Part I with A.V.

[Nel]   P.Nelson, *Lectures on strings and moduli space*, Physics Reports **149, no. 6** (1987), 337–375.

[P]   R.Penner, *Universal constructions in Teichmüller theory*, Adv. Math. **98** (1993), 143–215.

[Pol1] A.M. Polyakov, *Quantum geometry of bosonic strings*, Physics Letters **103B, no. 3** (1981), 207–210.

[Pol2] A.M. Polyakov, *Quantum geometry of fermionic strings*, Physics Letters **103B, no. 3** (1981), 211–213.

[Q]   D. Quillen, *Determinants of Cauchy-Riemann operators over a Riemann surface*, Func. Anal. Appl. **19** (1985), 31–34.

[S]   D. Sullivan, *Relating the universalities of Milnor-Thurston, Feigenbaum and Ahlfors-Bers*, in *Milnor Festschrift*, Topological Methods in Modern mathematics, (*ed.* L. Goldberg, A. Phillips), Publish or Perish, (1993), 543–563.

[Sha]   I.R. Shafarevich, *On some infinite-dimensional groups II*, Math. USSR Izvest. **18** (1982), 185–194.

[W]   M. Wolf, *The Teichmüller theory of harmonic maps*, Jour. Diff. Geom. **29** (1989), 449–479.

[Wol]   S. Wolpert, *Chern forms and the Riemann tensor for the moduli space of curves*, Invent. Math. **85** (1986), 119–145.

[ZT]   P.G. Zograf and L.A. Takhtadzhyan, *A local index theorem for families of $\bar{\partial}$-operators on Riemann surfaces*, Russian Math Surveys **42** (1987), 169–190.

THE INSTITUTE OF MATHEMATICAL SCIENCES, MADRAS 600113, INDIA
E-mail address: nag@imsc.ernet.in

Proceedings of
THE 37TH TANIGUCHI SYMPOSIUM ON
TOPOLOGY AND TEICHMÜLLER SPACES
held in Finland, July 1995
ed. by Sadayoshi KOJIMA *et al.*
©1996 World Scientific Publishing Co.
pp. 221–236

PARABOLIZATION OF ELEMENTS OF KLEINIAN GROUPS

KEN'ICHI OHSHIKA

(Received December 25, 1995)

1. Introduction

In 1970 ([4]), Bers observed that on the boundary of the Bers embedding of the Teichmüller space, there exist Kleinian groups which are called cusps, i.e., those containing additional parabolic elements. By Maskit and Abikoff among others, it was proved that such a phenomenon of appearance of additional parabolic elements occurs for more general Kleinian groups, for example geometrically finite Kleinian groups. In [16], Maskit raised the following question: Is there any suitable criterion characterizing (conjugacy classes of) elements of a given Kleinian groups which can be simultaneously made parabolic elements in some Kleinian group on the boundary of the deformation space? He conjectured that the conjugacy classes of "primitive simple disjoint peripheral" elements are exactly such classes. More precisely, the conjecture can be formulated as follows.

Conjecture 1.1. *Let* Γ *be a Kleinian group with domain of discontinuity* Ω_Γ. *For a family of primitive loxodromic elements* $\{\gamma_1, \dots, \gamma_m\} \subset \Gamma$ *which satisfies the following condition. There would exist a Kleinian group* G *with an isomorphism* $\psi : \Gamma \to G$ *on the boundary of quasi-conformal deformation space of* Γ *such that the conjugacy classes of* $\{[\psi(\gamma_1)], \dots [\psi(\gamma_m)]\}$ *are exactly the additional parabolic classes, i.e., the conjugacy classes of elements which are parabolic but whose inverse image by* ψ *are not parabolic.*

- *The condition: In the Kleinian manifold* $(\mathbf{H}^3 \cup \Omega_\Gamma)/\Gamma$ *the free homotopy classes of* $\{\gamma_1, \dots, \gamma_m\}$ *are represented by a collection of disjoint simple closed curves on* Ω_Γ/Γ *no two of which are mutually homotopic in* $(\mathbf{H}^3 \cup \Omega_\Gamma)/\Gamma$.

Maskit proved that this conjecture is true for geometrically finite function groups ([16]). The author generalized Maskit's result to one for freely indecomposable Kleinian groups and geometrically finite Kleinian groups in general ([23], [25]). Naturally one can ask if the condition above is also necessary for loxodromic elements to be made parabolic in some Kleinian group on the boundary. In the present paper, we shall consider this problem in a special case where $m = 1$. In this case, the condition above is simplified to one demanding that $\gamma(= \gamma_1)$ is represented by a simple closed curve on the ideal boundary Ω_Γ/Γ. Our main theorem asserts that this condition is actually necessary.

This problem of parabolizing loxodromic elements is closely related to one to determine homeomorphism type of the quotient hyperbolic 3-manifold corresponding to a Kleinian group on the boundary of the deformation space. More precisely, the latter problem means the following. Let G be a Kleinian group which lies on the boundary of the deformation space of a geometrically finite Kleinian group Γ with isomorphism $\psi : \Gamma \to G$. Then the question is : " Is there a homeomorphism from $\mathbf{H}^3/\Gamma$ to $\mathbf{H}^3/G$ which induces the same isomorphism from Γ to G as ψ?" It is easy to see if the answer to this question is yes, then it follows that the condition in Conjecture 1.1 is also necessary. Although it used to be supposed that the answer would be yes, recently, Anderson and Canary ([3]) found remarkable examples for which there is no such a homeomorphism. Still even in their examples, the "new" parabolic elements are represented by disjoint simple closed curves on the ideal boundary. In other words, their examples show us that our problem for the necessity of the condition in Conjecture 1.1 is weaker than the homeomorphism problem above, and there still remains a possibility that the answer to our problem is yes in general.

2. Preliminaries

We call discrete subgroups of $PSL_2\mathbf{C}$ Kleinian groups. Note that $PSL_2\mathbf{C}$ is the group of linear fractional transformations of S^2, hence also the group of isometries of the hyperbolic space $\mathbf{H}^3$. Throughout the paper, we assume that Kleinian groups are finitely generated and torsion free. The limit set of a Kleinian group Γ, which will be denoted by Λ_Γ is the closure of the set of fixed points of elements of Γ on S^2. The complement of the limit set Λ_Γ is called the region of discontinuity and denoted by Ω_Γ. The Kleinian group Γ acts on Ω_γ properly discontinuously. Furthermore, Γ acts on $\mathbf{H}^3 \cup \Omega_\Gamma$ properly discontinuously if we regard S^2 as the ideal boundary of $\mathbf{H}^3$ by considering the Poincaré model of $\mathbf{H}^3$. The quotient $(\mathbf{H}^3 \cup \Omega_\Gamma)/\Gamma$ is called the Kleinian manifold associated with Γ. Ahlfors' finiteness theorem asserts that the Riemann surface Ω_Γ/Γ is of finite type if Γ is finitely generated. Punctures of Ω_Γ/Γ represent parabolic elements of Γ.

Let Γ be a Kleinian group. The space of conjugacy classes of faithful discrete representations from Γ to $PSL_2\mathbf{C}$, with the quotient topology induced from the representation space, is denoted by $AH(\Gamma)$. A homeomorphism $w : S^2 \to S^2$ is said to be quasi-conformal when w has L^2-distributional partial derivative with respect to z and $\bar{z}$ and $\|w_{\bar{z}}/w_z\|_\infty < 1$. A Kleinian group G is called a quasi-conformal deformation of Γ if there is a quasi-conformal homeomorphism w such that $w\Gamma w^{-1} = G$. The subspace of $AH(\Gamma)$ consisting of quasi-conformal deformations of Γ is called the quasi-conformal deformation space of Γ and denoted by $QH(\Gamma)$. By works of Ahlfors, Bers, Kra, Maskit, Marden, and Sullivan, among others, it is known that there is a covering projection from $\mathcal{T}(\Omega_\Gamma/\Gamma)$ to $QH(\Gamma)$.

The minimal closed convex set of $\mathbf{H}^3$ whose closure in the Poincaré disc $\mathbf{H}^3 \cup S^2_\infty$

contains the limit set Λ_Γ is called the Nielsen convex hull of Γ and denoted by H_Γ. Since the limit set Λ_Γ is Γ-invariant, so is H_Γ. The quotient H_Γ/Γ is called the convex core of $\mathbf{H}^3/\Gamma$ and denoted by C_Γ. We say that a Kleinian group Γ is geometrically finite when the convex core of $\mathbf{H}^3/\Gamma$ has a finite volume. If Γ is a geometrically finite Kleinian group without parabolics, then the convex core of $\mathbf{H}^3/\Gamma$ is compact. Furthermore a small regular neighbourhood of the convex core is homeomorphic to the Kleinian manifold $(\mathbf{H}^3 \cup \Omega_\Gamma)/\Gamma$ in this case.

For an open 3-manifold M, a compact codimension 0 submanifold which is a deformation retract is called a topological core of M. Scott proved in [27], that if the fundamental group of M is finitely generated, a core of M always exists. In particular, for a finitely generated (torsion-free) Kleinian group Γ, a core of $\mathbf{H}^3/\Gamma$ exists. In the case when Γ has parabolic elements, it is more convenient to consider a relative core as follows than an ordinary core. Let $(\mathbf{H}^3/\Gamma)_0$ denote the non-cuspidal part of $\mathbf{H}^3/\Gamma$, i.e., the complement of neighbourhoods of cusps corresponding to parabolic elements of Γ. Each boundary component of $(\mathbf{H}^3/\Gamma)_0$ is either an open annulus or a torus. A core C of $(\mathbf{H}^3/\Gamma)_0$ is called a relative core when $(C, C \cap \partial(\mathbf{H}^3/\Gamma)_0)$ is a relative deformation retract of $((\mathbf{H}^3/\Gamma)_0, \partial(\mathbf{H}^3/\Gamma)_0)$. By McCullough's result in [18], if Γ is finitely generated and torsion free, a relative core of $(\mathbf{H}^3/\Gamma)_0$ is proved to exist.

Let Γ be a Kleinian group and suppose that (G, ψ) is contained in $AH(\Gamma)$. A parabolic element in G is called an additional parabolic element when $\psi^{-1}(\gamma)$ is not parabolic in Γ. An element of Γ is said to be primitive when it is not a non-trivial power of another element in Γ.

3. Statement of the main theorem

In this section, we shall give statements of our main theorem and a proposition which is essential for its proof.

Theorem 3.1. *Let Γ be a geometrically finite Kleinian group without parabolic elements. Let (G, ψ) be a Kleinian group in $\partial\overline{QH(\Gamma)}$ which has only one primitive additional parabolic class. Let γ be an element of Γ such that $[\psi(\gamma)]$ is the primitive additional parabolic class. Then the free homotopy class of γ is represented by a primitive simple closed curve on Ω_Γ/Γ.*

As $\psi(\gamma)$ is primitive, it is obvious that γ is also primitive. Therefore the point which we have to prove is that the free homotopy class of γ is represented by a simple closed curve on Ω_Γ/Γ. We shall prove this by reductio ad absurdum. From now on until the end of the proof, we assume that the free homotopy class $[\gamma]$ cannot be represented by a simple closed curve on Ω_Γ/Γ to get a contradiction. This is equivalent to saying that γ is not freely homotopic to a simple closed curve on the boundary of the convex core of $\mathbf{H}^3/\Gamma$. Then the following proposition holds.

Proposition 3.2. *Let C be the convex core of $\mathbf{H}^3/\Gamma$. Let C' be a topological core of $\mathbf{H}^3/G$. Let $\psi|C : C \to C'$ be a homotopy equivalence induced from ψ. Suppose that in the situation of Theorem 3.1, there is no simple closed curve on the boundary of the convex core C which represents γ. Then $\psi|C$ is homotopic to a homeomorphism.*

A proposition analogous to Proposition 3.2 was proved in Ohshika [26]. Although it was assumed that G had no parabolic elements there, the proof is along the same line. Thus there is some part in the proof here, which overlaps with the argument developed in [26]. For the completeness of our exposition, we shall not cite the argument in [26] except for simple lemmata.

4. Proof of Proposition 3.2

Throughout this section we assume the following. The Kleinian group Γ is purely loxodromic and geometrically finite. The pair (G, ψ) is contained in $\partial \overline{QH(\Gamma)}$ and has only one primitive additional parabolic class. Let γ be an element of Γ such that $[\psi(\gamma)]$ is the primitive additional parabolic class. We assume that γ is *not* represented by a simple closed curve on the boundary of the convex core C of $\mathbf{H}^3/\Gamma$.

In the proof of Proposition 3.2, we shall use the notion of geometric limits of sequences of Kleinian groups. The definition of geometric limits is as follows.

Definition. A Kleinian group H is called the geometric limit of $\{\Gamma_i\}$ if every element of H is the limit of a convergent sequence $\{\gamma_i\}$ for $\gamma_i \in \Gamma_i$, and the limit of any convergent sequence $\{\gamma_{i_j} \in \Gamma_{i_j}\}$ for a subsequence $\{\Gamma_{i_j}\} \subset \{\Gamma_i\}$ is contained in H.

The geometric limit of Kleinian groups is also a (possibly infinitely generated) Kleinian group unless it is elementary (i.e., the limit set is finite). For any sequence of Kleinian groups, there exists a subsequence which converges geometrically. In particular, if $\{\Gamma_i\}$ converges algebraically, the algebraic limit is contained in the geometric limit. When the algebraic limit coincides with the geometric limit, we say that the sequence $\{(\Gamma_i, \phi_i)\}$ converges strongly to the algebraic limit, and the limit is called a strong limit. Since our Kleinian group (G, ψ) is contained in $\partial \overline{QH(\Gamma)}$, there is a sequence $\{(\Gamma_i, \phi_i)\}$ in $QH(\Gamma)$, which converges algebraically to (G, ψ). We can assume that ϕ_i converges to ψ as representations by taking conjugates. For this sequence $\{\Gamma_i\}$, we shall denote its geometric limit (after taking a subsequence) by Γ_∞.

Now we turn to the proof of Proposition 3.2. Consider a characteristic compression body H of the core C. (A characteristic compression body is an embedded compression body in C whose compressible boundary components are exactly those of C. See Bonahon [7] for the definition of the characteristic compression body.) Then the submanifold $W = \overline{C - H}$ is a possibly disconnected boundary-irreducible Haken manifold.

Lemma 4.1. *In this situation, the map $\psi|W$ is homotopic to an embedding into* $\mathbf{H}^3/G$.

Let $W_1, \ldots, W_k$ be the components of W. For each W_κ, let Γ^{W_κ} be a subgroup of Γ associated to $\pi_1(W_\kappa)$, let $\Gamma_i^{W_\kappa}$ be $\phi_i(\Gamma^{W_\kappa})$, and let G^{W_κ} be $\psi(\Gamma^{W_\kappa})$. Denote $\phi_i|\Gamma^{W_\kappa}$ by $\phi_i^{W_\kappa}$ and $\psi|\Gamma^{W_\kappa}$ by ψ^{W_κ}. Then the following lemmata hold.

Lemma 4.2. *Suppose that W_κ is not an I-bundle over a closed surface. Then the sequence $\{(\Gamma_i^{W_\kappa}, \phi_i^{W_\kappa})\}$ converges to $(G^{W_\kappa}, \psi^{W_\kappa})$ strongly.*

Proof. Note first that, by assumption, no simple closed curve on ∂C is mapped to a closed curve representing a parabolic element of G by ψ, and that because W_κ is a component of a characteristic compression body, every simple closed curve on ∂W_κ is freely homotopic to a simple closed curve on ∂C. Hence no simple closed curve on ∂W_κ is mapped to a closed curve representing a parabolic element by ψ. In Theorem 2.1 of Ohshika [24], it is proved that if $(\psi^{W_\kappa})^{-1}$ preserves the parabolicity, $\Gamma_i^{W_\kappa}$ converges to G^{W_κ} strongly. Although in our situation, there may be a loxodromic element of Γ^{W_κ} which is mapped to a parabolic element by ψ, we can see that the condition that such an element cannot be represented by a simple closed curve on the boundary of the convex core of $\mathbf{H}^3/\Gamma^{W_\kappa}$ is sufficient to prove Theorem 2.1 in [24]. Since W_κ is lifted homeomorphically to a topological core of $\mathbf{H}^3/\Gamma^{W_\kappa}$, which is isotopic to the convex core because Γ^{W_κ} is indecomposable, this condition is satisfied in our situation as remarked above. $\square$

Similarly, we can prove the following using the argument in [24].

Lemma 4.3. *The restriction of the homotopy equivalence ψ^{W_κ} to W_κ is homotopic to an embedding into $\mathbf{H}^3/G^{W_\kappa}$.*

Let S be a component of $\partial(\overline{C-H})$. Let Γ^S be a subgroup of Γ which corresponds to $\pi_1(S)$. We can assume that if W_κ is a component of W containing S as a boundary component, then Γ^S is a subgroup of Γ^{W_κ}.

Recall that we have a sequence of quasi-conformal deformations $\{(\Gamma_i, \phi_i)\}$ of Γ which converges algebraically to (G, ψ). As (Γ_i, ϕ_i) is a quasi-conformal deformation, the homotopy equivalence induced by ϕ_i, which we denote again by ϕ_i, is homotopic to a homeomorphism $\Phi_i : \mathbf{H}^3/\Gamma \to \mathbf{H}^3/\Gamma_i$. Let Γ_i^S be the subgroup $\phi_i(\Gamma^S)$ of Γ_i, and let ϕ_i^S be $\phi_i|\Gamma^S$. As (Γ_i, ϕ_i) converges algebraically to (G, ψ), the sequence $\{(\Gamma_i^S, \phi_i^S)\}$ converges algebraically to $(\psi(\Gamma^S), \psi|\Gamma^S)$, which we shall denote by (G^S, ψ^S).

Lemma 4.4. *If S is a component of ∂W which is not homotopic to a boundary component of C, then G^S is geometrically finite and $\{(\Gamma_i^S, \phi_i^S)\}$ converges to (G^S, ψ^S) strongly.*

Proof. Suppose that G^S is geometrically infinite. Then by Bonahon's theorem in [6], each end of $(\mathbf{H}^3/G^S)_0$ is geometrically tame. Since S is a boundary component of a characteristic compression body H, for any simple closed curve γ on S, the closed curve $\psi(\gamma)$ does not represent a parabolic element. This implies that $\mathbf{H}^3/G^S$ has no parabolic elements because any parabolic element must be homotopic to a simple closed curve on a topological core, and by homotoping ψ^S, we can assume that $\psi^S|S$ is an embedding and that a tubular neighbourhood of $\psi^S(S)$ is a core of $\mathbf{H}^3/G^S$.

Let W_κ and H_λ be components of W and H respectively which contain S as a boundary component. Denote the submanifold $W_\kappa \cup H_\lambda$ by V. Let Γ^V be a subgroup of Γ corresponding to $\pi_1(V)$ which contains Γ^{W_κ} as a subgroup. Let G^V be $\psi(\Gamma^V)$ and let Γ_i^V be $\phi_i(\Gamma^V)$.

Let $r : \mathbf{H}^3/G^S \to \mathbf{H}^3/G^V$ be the covering projection associated to the inclusion $G^S \subset G^V$. Let e be a geometrically infinite tame end of $\mathbf{H}^3/G^S$. As is proved in Thurston [30] and Lemma 2.2 of Ohshika [24], there is a neighbourhood E of e such that $r|E$ is a finite-sheeted covering map to a neighbourhood of an end of $\mathbf{H}^3/G^V$ which can be identified with $T \times (0, \infty)$ for a closed orientable surface T. We shall show that $r|E$ must be one-to-one. Suppose that $r|E$ is not one-to-one on the contrary. The surface S in $\mathbf{H}^3/\Gamma$ is lifted homeomorphically to $\tilde{S}$ in $\mathbf{H}^3/\Gamma^V$ which is also lifted homeomorphically to a core surface of $\mathbf{H}^3/\Gamma^S$. Since $r \circ \psi^S$ is homotopic to ψ^V, the map $\psi^V|\tilde{S}$ is homotopic to a non-trivial covering onto $T \times \{t\}$. This is impossible because $\tilde{S}$ is incompressible and embedded. (This can be seen for example by considering a covering associated to $\psi_*^{-1}(\pi_1(T))$ of $\mathbf{H}^3/\Gamma^V$ and lifting $\tilde{S}$ to the cover.) Thus $r|E$ is one-to-one and T is homeomorphic to S.

By the relative core theorem due to McCullough in [18], there exists a core C_V of $\mathbf{H}^3/G^V$ such that $C_V \cap E = T \times (0, t]$. Especially C_V has a boundary component $T \times \{t\}$.

Let $\tilde{V}$ is a homeomorphic lift of V to $\mathbf{H}^3/\Gamma^V$. Let Σ be an incompressible boundary component of $\tilde{V}$. By the same argument as before, we can see that there is no element of $\pi_1(\Sigma)$ which is mapped to a parabolic element by ψ. Then we can apply the same argument as in the proof of Theorem 2.1 in [24] to our case and prove that ψ^V is homotoped so that $\psi^V|\Sigma$ should be a homeomorphism to a boundary component of C_V. Let $\tilde{W}$ be the submanifold of V which is a homeomorphic lift of W_κ. Since we have already proved that $\psi^V|\tilde{S}$ is homotopic to a homeomorphism to a boundary component of C_V, we can see that ψ^V can be homotoped so that every boundary component of $\tilde{W}$ is mapped homeomorphically to a boundary component of C_V. By assumption, $\tilde{W}$ is not an I-bundle over a closed surface. Hence by Theorem 13.6 in Hempel [11], we can see that $\psi^V|\tilde{W}$ is homotopic to a homeomorphism to C_V. This contradicts the fact that $\pi_1(W_\kappa)$ is a proper subgroup of $\pi_1(V)$. $\square$

Let Σ be an incompressible component of ∂C, and let Γ^Σ, Γ_i^Σ, G^Σ be subgroups of Γ, Γ_i, G corresponding to $\pi_1(\Sigma)$ such that $\Gamma_i^\Sigma = \phi_i(\Gamma^\Sigma)$ and $G^\Sigma = \psi(\Gamma^\Sigma)$ as before. Recall that C' is a topological core of $\mathbf{H}^3/G$.

Lemma 4.5. *For each Σ, there exists a boundary component Σ' of $\partial C'$ such that $\psi|\Sigma$ is homotopic to a homeomorphism to Σ'. Moreover there exists a sequence of embedded surfaces Σ_i homotopic to $\phi_i(\Sigma)$ which converges geometrically to a surface Σ_∞ in $\mathbf{H}^3/\Gamma_\infty$ which can be lifted to an embedded surface in $\mathbf{H}^3/G$ isotopic to Σ'.*

Proof. First suppose that G^Σ is geometrically finite. By the assumption of Proposition 3.2, there is no element of Γ^Σ whose image by ψ represents a parabolic element. The group G_i^Σ corresponds to a point g_i in the quasi-conformal deformation space of Γ^Σ which can be identifed with $\mathcal{T}(\Sigma) \times \mathcal{T}(\Sigma)$. If $\{g_i\}$ does not converge inside the quasi-conformal deformation space, then either the limit G^Σ has a parabolic element or G^Σ is geometrically infinite. Thus in our case, $\{g_i\}$ must converge inside the quasi-conformal deformation space after taking a subsequence. Since Σ is a boundary component of C, there is a component Ω_i^Σ of the region of discontinuity Ω_{Γ_i} of Γ_i, whose stabilizer is Γ_i^Σ. Since $\{g_i\}$ converges inside $\mathcal{T}(\Sigma) \times \mathcal{T}(\Sigma)$, the marked conformal structure of $\Omega_i^\Sigma/\Gamma_i^\Sigma$, which is a factor of g_i, converges inside the Teichmüller space as $i \to \infty$(after taking a subsequence). This implies, by Lemma 3 of Abikoff [1], that there is a component Ω_∞^Σ of the region of discontinuity Ω_G, whose stabilizer is G^Σ.

Since there exists a boundary component of the convex core of $\mathbf{H}^3/G$ corresponding to $\Omega_\infty^\Sigma/G^\Sigma$ and we can assume that $\partial C'$ contains the boundary of the convex core by isotoping C', it follows that $\psi|\Sigma$ is homotopic to a homeomorphism to a boundary component of C' in this case.

Next suppose that G^Σ is geometrically infinite. By Bonahon's theorem in [6], the group G^Σ is geometrically tame. Hence there must exist a geometrically infinite tame end e^Σ of $\mathbf{H}^3/G^\Sigma$. Then by the same argument as the proof of Lemma 4.4 using Lemma 2.2 in [24], we can see that there exists a boundary component Σ' of the core C' and $\psi|\Sigma$ is homotopic to a homeomorphism to Σ'.

It remains to prove the latter sentence of our lemma. When G^S is geometrically finite, let Σ_i be the boundary component of the convex core of $\mathbf{H}^3/\Gamma_i$ homotopic to $\phi_i(S)$. Then as the marked hyperbolic structure on Σ_i converges inside the Teichmüller space, $\{\Sigma_i\}$ converges to a boundary component of the convex core of the geometric limit $\mathbf{H}^3/\Gamma_\infty$ which can be lifted to a boundary component of the convex core of $\mathbf{H}^3/G$ homotopic to $\psi(S)$ as is shown in Ohshika [24]. When G^S is geometrically infinite, we can choose a surface Σ'' isotopic to Σ so that its projection $\bar\Sigma$ to the geometric limit $\mathbf{H}^3/\Gamma_\infty$ is an embedding by Lemma 2.2 in [24]. Then by letting Σ_i be the pullback of $\bar\Sigma$ by an approximate isometry between $\mathbf{H}^3/\Gamma_i$ and $\mathbf{H}^3/\Gamma_\infty$ associated to the geometric convergence $\Gamma_i \to \Gamma_\infty$, we complete the proof. $\square$

To prove Lemma 4.1, we shall construct a submanifold homeomorphic to W in $\mathbf{H}^3/\Gamma_i$, which converges to a submanifold in the geometric limit $\mathbf{H}^3/\Gamma_\infty$ which can be lifted homeomorphically to $\mathbf{H}^3/G$. For that, we need to use a topological interpretation of Maskit's combination theorem using harmonic functions. An explanation of the basic idea of this interpretation can be found in Morgan [21].

Let $S_1, \dots, S_n$ be components of ∂W which are not homotopic to boundary components of ∂C. Then $\pi_1(C)$ is decomposed into amalgamated free products and amalgamated HNN-extensions of $\pi_1(W_1), \dots, \pi_1(W_k)$ and $\pi_1(H_1), \dots, \pi_1(H_l)$ amalgamated at $\pi_1(S_1), \dots, \pi_1(S_n)$, where $H_1, \dots, H_l$ are the components of H. Let Γ^{S_ν} be a subgroup of Γ corresponding to $\pi_1(S_\nu)$ and let $\Gamma_i^{S_\nu}$ be $\phi_i(\Gamma^{S_\nu})$. We can assume that if S_ν is a boundary component of W_κ, then Γ^{S_ν} is a subgroup of Γ^{W_κ}. Similarly we can assume that if S_ν is a boundary component of a component H_λ of H, then Γ^{S_ν} is a subgroup of Γ^{H_λ}, where Γ^{H_λ} is a subgroup of Γ corresponding to $\pi_1(H_\lambda)$.

To explain our idea, we shall first consider the case when $n = 1$ and W is not an I-bundle over a closed surface. Let S denote S_1. Then H and W are connected and $\pi_1(C) = \pi_1(W) *_{\pi_1(S)} \pi_1(H)$. The region of discontinuity of Γ_i^W has a component Ω_i^W whose stabilizer is Γ_i^S. Similarly, the region of discontinuity of Γ_i^H has a component Ω_i^H whose stabilizer is Γ_i^S. The region of discontinuity of Γ_i^S has two components Ω_i^S and $\Omega_i^{S'}$. By renaming if necessary, we can assume that Ω_i^S is equal to Ω_i^W and that $\Omega_i^{S'}$ is equal to Ω_i^H.

Let v_i^W be the visual measure function of Ω_i^W and let v_i^H be one for Ω_i^H, both of which are harmonic functions defined on $\mathbf{H}^3$. Let $D(r)_i^W$ be the subset of $\mathbf{H}^3$ consisting of the points where the function v_i^W takes a value greater than or equal to r. Similarly we define $D(r)_i^H$.

Lemma 4.6. *Suppose that W is not an I-bundle over a closed surface. For sufficiently small ϵ, the set $D(2\pi - \epsilon)_i^W$ (or $D(2\pi - \epsilon)_i^H$) is disjoint from its translates by Γ_i^W (or Γ_i^H) which are not equal to itself.*

Proof. Because the proof is completely the same for W and H, we shall only consider the case of W here. Since W is not an I-bundle over a closed surface, there are more than two cosets in $\Gamma^S \setminus \Gamma^W$. Take some coset $[\gamma]$ of $\Gamma^S \setminus \Gamma^W$ other than the one consisting of elements of Γ^S. Note that any translate of $\gamma\Omega_i^W$ by an element of Γ_i^S does not intersect Ω_i^W. Let $v(\gamma)$ be the visual measure of the union of the translates of $\gamma\Omega_i^W$ by elements of Γ_i^S.

Fix some small positive constant ϵ_0. (It is all right if $\epsilon_0 < \pi/2$.) Consider a set $T(\epsilon_0)$ consisting of the points where v_i^W takes values greater than or equal to $2\pi - \epsilon_0$ and smaller than or equal to $2\pi + \epsilon_0$. As v_i^W is Γ_i^S-invariant, so is $T(\epsilon_0)$. The quotient $T(\epsilon_0)/\Gamma_i^S$ is compact because the value of v_i^S approaches to 4π or 0 as a point tends to an end of $\mathbf{H}^3/\Gamma_i^S$, which is geometrically finite and corresponds to a component of the region of discontinuity of Γ_i^S. Since $v(\gamma)$ is also Γ_i^S-invariant, its value is well-defined (and continuous) in the compact set $T(\epsilon_0)/\Gamma_i^S$. Hence $v(\gamma)$ attains a minimum ϵ_1 on $T(\epsilon_0)$. As $v(\gamma)$ is a positive function, we have $\epsilon_1 > 0$.

Now take $\epsilon > 0$ so that $\epsilon < \epsilon_1/2$. Let γ' be an element of Γ_i^W whose coset is the same as neither that of γ nor that of 1. Such an element exists because $\Gamma_i^S \setminus \Gamma_i^W$ has more than two cosets. Suppose that there is a point $x \in D(2\pi - \epsilon)_i^W \cap \gamma'D(2\pi - \epsilon)_i^W$. Since Ω_i^W, $\Gamma_i^S\gamma\Omega_i^W$, and $\gamma'\Omega_i^W$ are mutually disjoint, the total visual measure, which must be equal to 4π, is at least $2(2\pi - \epsilon) + \epsilon_1$ which is greater than 4π. This is a contradiction. Therefore $D(2\pi - \epsilon)_i^W$ and $\gamma D(2\pi - \epsilon)_i^W$ are disjoint. To complete the proof of our lemma, we have only to show that, by changing ϵ smaller if necessary, $D(2\pi - \epsilon)_i^W$ and $\gamma'D(2\pi - \epsilon)_i^W$ are also disjoint. This can be done by retaking a coset, which we chose to be the one containing γ above, so that it should not contain γ, and applying the same argument as above. By taking the minimum of the old and the new ϵ's, we get a constant as we wanted. $\square$

Recall that $\Omega_i^S = \Omega_i^W$. By Lemma 4.6, we can see that $D(2\pi - \epsilon)_i^W/\Gamma_i^S$, which is a neighbourhood of a geometrically finite end of $\mathbf{H}^3/\Gamma_i^S$, is embedded in $\mathbf{H}^3/\Gamma_i^W$. Similarly, $D(2\pi - \epsilon)_i^H/\Gamma_i^S$, which is a neighbourhood of another geometrically finite end of $\mathbf{H}^3/\Gamma_i^S$, is embedded in $\mathbf{H}^3/\Gamma_i^H$. Let ι_W and ι_H be embeddings of $D(2\pi - \epsilon)_i^W/\Gamma_i^S$ into $\mathbf{H}^3/\Gamma_i^W$ and $D(2\pi - \epsilon)_i^H/\Gamma_i^S$ into $\mathbf{H}^3/\Gamma_i^H$. As the region of discontinuity of Γ_i^S is the union of Ω_i^W and Ω_i^H, the two sets $\partial D_i^W(2\pi + t)$ and $\partial D_i^H(2\pi - t)$ are equal. This gives rise to an identification h_i from $(D_i^W(2\pi - \epsilon) - \mathrm{Int}D_i^W(2\pi + \epsilon))/\Gamma_i^S$ to $(D_i^H(2\pi - \epsilon)$ $\mathrm{Int}D_i^H(2\pi + \epsilon))/\Gamma_i^S$. By Maskit's combination theorem, $\mathbf{H}^3/\Gamma_i$ is obtained by pasting $\mathbf{H}^3/\Gamma_i^W - D_i^W(2\pi + \epsilon)$ and $\mathbf{H}^3/\Gamma_i^H - D_i^H(2\pi + \epsilon)$ identifying $\iota_W(x)$ with $\iota_H(h_i(x))$ for all $x \in (D_i^W(2\pi - \epsilon) - \mathrm{Int}D_i^W(2\pi + \epsilon))$.

For a regular value r of the visual measure function, the set $\partial D_i^W(r)$ is an embedded surface. Hence we can choose $r \in (2\pi - \epsilon, 2\pi + \epsilon)$ such that $\tilde{F}_i(r) = \partial D_i^W(r)/\Gamma_i^S$ is an embedded surface, which may be compressible but is homologous to a homeomorphic lift $\tilde{S}$ of S in $\mathbf{H}^3/\Gamma_i^S$. Let F_i be the embedded image of $\tilde{F}_i(r)$ into $\mathbf{H}^3/\Gamma_i$ by Maskit's combination described above.

Lemma 4.7. *For any $\delta > 0$, by taking sufficiently small r, we can arrange so that the surface $\tilde{F}_i(r)$ should lie in the δ-neighbourhood of the convex core of $\mathbf{H}^3/\Gamma_i^W$.*

Proof. Let x be a point of $\mathbf{H}^3$ outside the δ-neighbourhood of the convex hull of the limit set $\Lambda_{\Gamma_i^W}$. Then there is a boundary component P of the convex hull separating x from the convex hull, and corresponding to P, there is a component Ω^P of the region of discontinuity $\Omega_{\Gamma_i^W}$. In this situation, the visual measure of Ω^P at x is greater than $2\pi + \epsilon(\delta)$, where $\epsilon(\delta) > 0$. Thus by taking sufficiently small ϵ and $r\epsilon(2\pi - \epsilon, 2\pi + \epsilon)$, we can see that the visual measure of Ω^P is larger than $2\pi + \epsilon$ and that the visual measure of Ω_i^W cannot be r whether or not Ω_i^W coincides with Ω^P. $\square$

Lemma 4.8. *We can isotope and compress F_i to get an incompressible surface F_i' which converges geometrically to an immersion S_∞ of S into $\mathbf{H}^3/\Gamma_\infty$ as $i \to \infty$. Furthermore we can assume that the geometric limit S_∞ is embedded after a small perturbation.*

Proof. We shall first show that we can compress $\tilde{F}_i(r)$ in $\mathbf{H}^3/\Gamma_i^W$ to obtain an incompressible surface converging to an immersion of S in the strong limit $\mathbf{H}^3/G^W$. Since $\tilde{F}_i(r)$ is the quotient of $\partial D_i^W(r)$ by Γ_i^S, it can be compressed to an embedded incompressible surface homotopic to $\phi_i(S)$. What we have to do is to control the diameters of compression discs so that the diameter of incompressible surface obtained by the compression should converge. As the convex core of $\mathbf{H}^3/\Gamma_i^W$ is a deformation retract, the compression can be performed in an arbitrary small regular neighbourhood of the convex core which is proved to contain $\tilde{F}_i(r)$ by Lemma 4.7.

Now by Lemma 4.2, the sequence $\{(\Gamma_i^W, \phi_i^W)\}$ converges strongly to (G^W, ψ^W), and by Bonahon's theorem [6], G^W is geometrically tame. Furthermore by Lemma 4.4, the end of $\mathbf{H}^3/G^W$ facing the boundary component of a core homotopic to $\psi(S)$ is geometrically finite. Let S_∞ be the boundary component of the convex core of $\mathbf{H}^3/G^W$ homotopic to $\psi(S)$. For each end e of $\mathbf{H}^3/G^W$ other than the one facing S_∞, choose an embedded surface Σ^e which is the boundary component of the convex core facing e if e is geometrically finite, and which is homotopic to pleated surfaces tending to e if e is geometrically infinite tame, so that S_∞ and the Σ^e's should be disjoint. Then $S_\infty \cup (\cup_e \Sigma^e)$ bounds a topological core C_*^W of $\mathbf{H}^3/G^W$. Since G^W is a geometric limit of $\{\Gamma_i^W\}$, there is an approximate isometry $\rho_i : B_{r_i}(\mathbf{H}^3/\Gamma_i^W, w_i) \to B_{r_i}(\mathbf{H}^3/G^W, w_\infty)$ where w_i and w_∞ are base frames which are the images of a fixed base frame on $\mathbf{H}^3$ by the universal covering projection. Then for sufficiently large i, the inverse image $C_i^W = \rho_i^{-1}(C_*^W)$ is a core of $\mathbf{H}^3/\Gamma_i^W$, and the boundary component of C_i^W homotopic to $\phi_i(S)$ is a boundary component of the convex core.

By Lemma 4.4, $\{(\Gamma_i^S, \phi_i^S)\}$ converges strongly to (G^S, ψ^S) which is geometrically finite. Therefore by Theorem 4.8 in Jørgensen-Marden [12], we can see that the domains Ω_i^W converge to a component $\Omega_{G^S}^0$ of the region of discontinuity of G^S in the sense of Carathéodory. Hence the surfaces $D_i^W(r)$ converge to a surface consisting of points at which the visual measure of $\Omega_{G^S}^0$ is r. It follows that the surface $\tilde{F}_i(r)$ converges geometrically to a (possibly singular) compact surface in $\mathbf{H}^3/G^W$. Therefore we can choose Σ^e as above so that the neighbourhood of the end e cut off by Σ^e does not intersect $\tilde{F}_i(r)$. Then by Lemma 4.7, we can see that $\tilde{F}_i(r)$ is contained in the δ-neighbourhood of the core C_i^W for small $\delta > 0$.

Let D be a compression disc for F_i. Then D can be lifted to a compression disc $\tilde{D}$ for $\tilde{F}_i(r)$. As was shown above $\tilde{D}$ can be homotoped to a disc $\tilde{D}'$ in the δ-neighbourhood of C_i^W. Now, let $q_i : \mathbf{H}^3/\Gamma_i^W \to \mathbf{H}^3/\Gamma_i$ be the covering projection associated to the inclusion. The disc $q_i(\tilde{D})$ may not be an embedded disc, but can be homotoped to an embedded disc D' in its arbitrarily small neighbourhood by analytic Dehn's lemma of Meeks-Yau [19] because it is homotopic to the embedding D. Although D' itself may intersect F_i, we can find a compression disc contained in D' by cut-and-paste techniques. This compression disc is renamed D'. On the other hand, by the observation above, D' is contained in the δ-neighbourhood of $q_i(C_i^W)$. Thus we can compress F_i to an incompressible surface F_i' which is contained in the δ-neighbourhood of $q_i(C_i^W)$. Then this surface F_i' converges geometrically to a compact surface in $\mathbf{H}^3/\Gamma_\infty$, which is an image of S, because $q_i(C_i^W)$ converges to $q_\infty(C_\cdot^W)$ where $q_\infty : \mathbf{H}^3/G^W \to \mathbf{H}^3/\Gamma_\infty$ is the covering projection associated to the inclusion. Since all of the F_i's are embedded, we can perturb the geometric limit to an embedding. $\square$

Continuation of Proof of Lemma 4.1 in the case when FrW *is connected.* Since F_i is the homeomorphic projection of the quotient of $D_i^W(r)$ by Γ_i^W, the incompressible surface F_i' which we obtained above is homotopic to $\phi_i(S)$ in the δ-neighbourhood of $q_i(C_i^W)$. Therefore the geometric limit S_∞ is homotopic to $q \circ \psi(S)$, where $q : \mathbf{H}^3/G \to \mathbf{H}^3/\Gamma_\infty$ is the covering projection associated to the inclusion, hence can be lifted to an embedded surface S' in $\mathbf{H}^3/G$ which is homotopic to $\psi(S)$. Recall that C' is a topological core of $\mathbf{H}^3/G$. We can choose C' so that C' should contain S'. By cutting C' along S', we obtain two compact submanifolds. By Lemma 4.5, we can homotope ψ so that for each component T_k of ∂W other than S, the restriction $\psi|T_k$ is a homeomorphism to a boundary component of C'. Further, as is shown in Lemma 4.5, we can assume that $\phi_i|T_k$ is an embedding which converges to a surface T_k^∞ in the geometric limit which can be lifted to a boundary component $\psi(T_k)$ of C'. Now, since S and the T_k's are lifted homeomorphically to $\mathbf{H}^3/\Gamma^W$, we can see that S' and $\psi(T_k)$ are lifted homeomorphically to embedded surfaces $\tilde{S}'$ and $\tilde{T}_k$ in $\mathbf{H}^3/G^W$. By the same argument as Lemma 4.5, we can see that $\tilde{S}' \cup (\cup_k \tilde{T}_k)$ are isotopic to the union of boundary components of a core $\tilde{C}_W$ of $\mathbf{H}^3/G^W$ since W is boundary-irreducible, i.e., G^W is freely indecomposable. Hence $\tilde{S}' \cup (\cup_k \tilde{T}_k)$ is null-homologous which implies that $S' \cup \cup T_k$ bounds a compact submanifold C_W' in C'. Then for sufficiently large i, the pull back $\rho_i^{-1} \circ q(C_W)$ is a compact 3-submanifold of $\mathbf{H}^3/\Gamma_i$ bounded by $F_i' \cup (\cup_k \phi_i(T_k))$, which must be homeomorphic to W by a homeomorphism homotopic to $\phi_i|W$. Hence C_W' is homeomorphic to W by a homeomorphism homotopic to $\psi|W$. Thus we have proved Lemma 4.1 in the case when FrW is connected. $\square$

Proof of Proposition 3.2 in the case when FrW *is connected.* We homotope ψ so that $\psi|W$ is a homeomorphism to C_W'. Let C_H' be the closure of $C' - C_W'$. What remains to prove is that $\psi|H$ is homotopic relatively to FrH to a homeomorphism to C_H'. As ψ is a homotopy equivalence, $\psi|W$ is a homeomorphism to C_W', and FrH is connected, we can see that we can homotope ψ so that $\psi|H$ should be a homotopy equivalence to C_H'

without changing $\psi|W$. As is shown in Lemma 3.1 of Ohshika [26], we can easily see that C'_H must be obtained from $S \times I$ by attaching 1-handles. If all the 1-handles are attached to the same component of $\partial S \times I$, we get a compression body homeomorphic to H. We can see that this is the case because the boundary component $\psi(S)$ of C'_W is incompressible. Thus C'_H is homeomorphic to H. As is shown in McCullough-Miller [20], the homotopy equivalence between homeomorphic compression bodies is homotopic to a homeomorphism, and we have completed the proof of Proposition 3.2 in the case when $\mathrm{Fr}W$ is connected. $\square$

Now we turn to prove Proposition 3.2 in general case. First we have to prove Lemma 4.1.

Proof of Lemma 4.1 *in general case.* First consider the case when $n = 1$ but W is an I-bundle over a closed surface. As before we denote S_1 by S. In this case, W is either a trivial I-bundle over S or a twisted I-bundle over a non-orientable closed surface T which is doubly covered by S.

First suppose that W is a trivial I-bundle over S. As no element of $\pi_1(S)$ is mapped to a parabolic element by ψ by assumption, either G^W is a quasi-Fuchsian group or $\mathbf{H}^3/G^W$ has a geometrically infinite tame end. Consider the case when G^W is a quasi-Fuchsian group. Let g_i be the element of $\mathcal{T}(S) \times \mathcal{T}(S)$ determined by the quasi-conformal deformation (Γ^W_i, ϕ^W_i) of Γ^W. If the sequence $\{g_i\}$ does not converge inside $\mathcal{T}(S) \times \mathcal{T}(S)$, then G^W either has parabolics or is geometrically infinite. Hence $\{g_i\}$ must converge inside $\mathcal{T}(S) \times \mathcal{T}(S)$. Let Ω^0_i be a component of the region of discontinuity Ω_{Γ_i}, whose stabilizer is $\Gamma^W_i = \Gamma^S_i$. Then since the marked conformal structure of Ω^0_i/Γ^W_i is a factor of g_i, it must converge in $\mathcal{T}(S)$ as $i \to \infty$. It follows that there is a component Ω^0_∞ of the region of discontinuity of Ω_G, whose stabilizer is $G^W = G^S$. Then Ω^0_∞/G^S corresponds to a boundary component of the convex core of $\mathbf{H}^3/G$ homotopic to $\psi(S)$. Taking its regular neighbourhood, we obtain an embedding homotopic to $\psi|W$.

Next consider the case when $\mathbf{H}^3/G^W$ has a geometrically infinite tame end. As before, by using Lemma 2.2 in [24], we can see that $\mathbf{H}^3/G$ has a geometrically infinite tame end with a neighbourhood homeomorphic to $S \times \mathbf{R}$ such that $\psi(S)$ is homotopic to $S \times \{t\}$. Then obviously, $\psi|W$ can be homotopic to an embedding corresponding to $S \times [0, 1]$.

Next suppose that W is a twisted I-bundle over a non-orientable surface T which is doubly covered by S. If some element of $\pi_1(T)$ is mapped to a parabolic element by ψ, there must be such an element also in $\pi_1(S)$. Hence G^W has no parabolic elements by our assumption in Proposition 3.2. Consider first the case when G^W is geometrically finite. let g_i be an element of $\mathcal{T}(S)$ which corresponds to the marked conformal structure of $\Omega_{\Gamma^W_i}/\Gamma^W_i$. Then $\{g_i\}$ converges inside $\mathcal{T}(S)$ by the same reason as before. (Otherwise either G^W would have parabolic elements or G^W would be geometrically infinite.) Lemma 4.6 can be applied to Γ^H_i, and we get an embedded surface $\tilde{F}_i(r)$ in $\mathbf{H}^3/\Gamma^H_i$ whose projection is embedded as F_i in $\mathbf{H}^3/\Gamma_i$. The surface F_i can be lifted homeomorphically to an embedded surface $\overline{F}_i$ in $\mathbf{H}^3/\Gamma^W_i$, and we can assume that $\overline{F}_i$ is contained in the δ-neighbourhood of the convex core of $\mathbf{H}^3/\Gamma^W_i$ by the same argument as Lemma 4.7. Since $\{(\Gamma_i^W, \phi_i^W)\}$ converges strongly to (G^W, ψ^W), as in the case when W is not an I-bundle, we can compress F_i so that we should get an incompressible surface F_i' converging geometrically to an embedded surface which can be lifted to an incompressible surface S' homotopic to $\psi(S)$. As $\psi(\pi_1(S))$ is contained in $\psi(\pi_1(T))$ as a subgroup of index 2, the surface S' must bound a twisted I-bundle in $\mathbf{H}^3/G$, which is homotopic to $\psi(W)$.

If G^W is geometrically infinite, as in the case when W is a trivial I-bundle, $\mathbf{H}^3/G$ has a geometrically tame end with a neighbourhood homeomorphic to $S \times \mathbf{R}$, where $S \times \{t\}$ is homotopic to $\psi|S$. Then $S \times \{t\}$ bounds a twisted I-bundle in $\mathbf{H}^3/G$, which is homotopic to $\psi(W)$. Thus we have completed the proof of Lemma 4.1 in the case when $n = 1$.

Next assume that $n > 1$. Recall that $W_1, \ldots, W_k$ are the components of W. Let W_κ be a component which is not an I-bundle over a closed surface. By renumbering the $S_1, \ldots, S_n$, we can assume that $S_1, \ldots S_{n_0}$ are the components of $\mathrm{Fr}W_\kappa$. For each $S_\nu(\nu = 1, \ldots, n_0)$, there is a component $\Omega_i^{W_\kappa}(S_\nu)$ of the region of discontinuity of $\Gamma_i^{W_\kappa}$, whose stabilizer is $\Gamma_i^{S_\nu}$, the subgroup of Γ_i associated to $\pi_1(S_\nu)$. Let $D(r)_i^{(W,\nu)}$ be the subset of $\mathbf{H}^3$ where the visual measure of $\Omega_i^{W_\kappa}(S_\nu)$ takes the value greater than or equal to r. We can generalize the argument in the proof of Lemma 4.6, and prove that for sufficiently small ϵ, the sets $D(2\pi - \epsilon)_i^{(W,1)}, \ldots, D(2\pi - \epsilon)_i^{(W,n_0)}$ and their translates by $\Gamma_i^{W_\kappa}$ are disjoint. Then by the same argument as the case when $n = 1$ using cut-and-paste techniques and by choosing compression discs of $\partial D_i^{(W,\nu)}(2\pi - \epsilon)/\Gamma_i^{S_\nu}$, we can prove that there are disjoint incompressible surfaces $F_i^1, \ldots F_i^{n_0}$ homotopic to $\phi_i(S_1), \ldots, \phi_i(S_{n_0})$ converging to surfaces $F_\infty^1, \ldots, F_\infty^{n_0}$ in $\mathbf{H}^3/\Gamma_\infty$ which can be lifted to incompressible surfaces $\tilde{F}_1, \ldots \tilde{F}_{n_0}$ in $\mathbf{H}^3/G$ homotopic to $\psi(S_1), \ldots, \psi(S_{n_0})$. We can assume that $\tilde{F}_1, \ldots, \tilde{F}_{n_0}$ are disjoint by slightly isotoping them because $F_\infty^1, \ldots F_\infty^{n_0}$ are approximated by disjoint surfaces in $\mathbf{H}^3/\Gamma_i$. As before, for each component T_k of $W_\kappa \cap \partial C$, the restriction $\psi|T_k$ can be homotoped to a homeomorphism onto a boundary component of the core C'. As before, $(\cup_\nu S_\nu) \cup (\cup_k T_k)$ is null-homologous and bounds a 3-submanifold W_κ' in $\mathbf{H}^3/G$. By the same argument as the case when $n = 1$, we can see that $\psi|W_\kappa$ is homotoped to a homeomorphism onto W_κ'.

When W_κ is an I-bundle over a surface, we can prove that $\psi|W_\kappa$ can be homotoped to an embedding by the same argument as the case when $n = 1$ and W is an I-bundle.

It remains to show that the W_κ's are taken to be disjoint. This can be proved as follows. Let W_κ and $W_{\kappa*}$ be distinct components of W. Let $S_1, \ldots, S_{n_0}$ be the components of $\mathrm{Fr}W_\kappa$ and $S_1^*, \ldots, S_{n_1}^*$ those of $\mathrm{Fr}W_{\kappa*}$. We can construct incompressible surfaces $F_i^1, \ldots, F_i^{n_0}$ and $F_i^{*1}, \ldots, F_i^{*n_1}$ homotopic to $\phi_i(S_1), \ldots \phi_i(S_{n_0})$ and $\phi_i(S_1^*), \ldots, \phi_i(S_{n_1}^*)$ as in the case when $n = 1$ so that they are all disjoint by choosing compression discs disjoint from the other F_i^ν's or $F_i^{*\nu}$'s in the process of compression. Note that $F_i^1, \ldots, F_i^{n_0}$ and $F_i^{*1}, \ldots, F_i^{*n_1}$ bound, together with boundary components of a core, disjoint submanifolds which are homotopic to $\phi_i(W_\kappa)$ and $\phi_i(W_{\kappa'})$. Then their geometric limits $F_\infty^1, \ldots, F_\infty^{n_0}, F_\infty^{*1}, \ldots, F_\infty^{*n_1}$ can be assumed to be disjoint after a perturbation if necessary, and they can be lifted homeomorphically to $\mathbf{H}^3/G$. Furthermore $F_\infty^1 \cup \ldots \cup F_\infty^{n_0}, F_\infty^{*1} \cup \ldots \cup F_\infty^{*n_1}$ bound, together with geometric limits of boundary components of a core of $\mathbf{H}^3/\Gamma_i$, compact submanifolds $\overline{W}_\kappa$, $\overline{W}_{\kappa*}$. By pulling back them to $\mathbf{H}^3/\Gamma_i$ by the approximate isometry ρ_i, we can see that $\overline{W}_\kappa$ and $\overline{W}_{\kappa*}$ can be assumed to be disjoint after a small perturbation because $F_i^1, \ldots, F_i^{n_0}$ and $F_i^{*1}, \ldots, F_i^{*n_0}$ bound disjoint submanifolds. The submanifolds $\overline{W}_\kappa$ and $\overline{W}_{\kappa*}$ are lifted homeomorphically to disjoint compact submanifolds of $\mathbf{H}^3/G$ homotopic to $\psi(W_\kappa)$ and $\psi(W_{\kappa'})$ respectively by the same argument as before. This completes the proof. $\square$

Proof of Proposition 3.2 *in general case.* By Lemma 4.1 we can assume that $\psi|W$ is an embedding. Since the core C is the union of W and compression bodies, C can be isotoped to the union of W and 1-handles $h_1, \ldots, h_p$ attached to W. Let $\alpha_1, \ldots \alpha_p$ be core arcs of $h_1, \ldots h_p$ whose endpoints lie on FrW. If we can prove that $\psi(\alpha_1), \ldots, \psi(\alpha_p)$ can be homotoped off Int$W$ preserving the endpoints, then by taking a regular neighbourhood of the union of $\psi(W)$ and the image of arcs, we can construct a core C' of $\mathbf{H}^3/G$ such that $\psi|C$ is homotoped to a homeomorphism onto C' and the proof will be completed.

Now, let us prove that $\psi(\alpha_1), \ldots, \psi(\alpha_p)$ can be homotoped off Int$\psi(W)$. Recall that we have the covering projection $q : \mathbf{H}^3/G \to \mathbf{H}^3/\Gamma_\infty$, and that the homotopy equivalence ϕ_i is homotopic to a homeomorphism Φ_i since it is induced from a quasi-conformal deformation. Furthermore, as was shown before, we can assume that $\Phi_i(W)$ converges geometrically to a compact submanifold homeomorphic to W in $\mathbf{H}^3/\Gamma_\infty$ which is lifted to $\psi(W)$. We can assume that $\rho_i \circ \Phi_i|W = q \circ \psi|W$ where ρ_i is an approximate isometry from $\mathbf{H}^3/\Gamma_i$ to $\mathbf{H}^3/\Gamma_\infty$ associated with the geometric convergence $\Gamma_i \to \Gamma_\infty$.

Lemma 4.9. *For sufficiently large i, the arc $\rho_i^{-1}(q \circ \psi(\alpha_\rho))$ is homotopic to the arc $\Phi_i(\alpha_\rho)$ for $\rho = 1, \ldots, p$.*

Proof. By isotoping Φ_i, we can assume that $\rho_i \circ \Phi_i(\partial\alpha_\rho) = q \circ \psi(\partial\alpha_\rho)$. Fix ρ and a point x in $\partial\alpha_\rho$, and let x_i be $\Phi_i(x)$. Furthermore, by conjugating the Γ_i's by elements in $PSL_2\mathbf{C}$ bounded independently of i, we can assume that x_i is the base point, with respect to a base frame on which $\mathbf{H}^3/\Gamma_i$ converges geometrically to $\mathbf{H}^3/\Gamma_\infty$ with a base frame on x_∞, which is equal to $q \circ \psi(x)$. Let y be an endpoint of α_ρ other than x. Let c be a loop based at y. Consider the loop $\alpha_\rho * c * \alpha_\rho^{-1}$ and denote by γ the element of Γ represented by this loop. Let c' be the geodesic loop based at $\psi(y)$ homotopic to $\psi(c)$ fixing the basepoint $\psi(y)$. Then the loop $\psi(\alpha_\rho) * c' * (\psi(\alpha_\rho))^{-1}$ represents $\psi(\gamma)$. Then the loop $\rho_i^{-1} \circ q(\psi(\alpha_\rho) * c' * (\psi(\alpha_\rho))^{-1})$ represents $\phi_i(\gamma)$ for sufficiently large i because ρ_i is a (K_i, r_i)-approximate isometry associated to the geometric convergence $G_i \to G_\infty$. On the other hand, we can see that for sufficiently large i, the loop $\rho_i^{-1} \circ q(c')$ is homotopic to $\Phi_i(c)$ preserving the basepoint $\Phi_i(y)$. This can be proved by changing the basepoints of the convergence from $\Phi_i(x)$ to $\Phi_i(y)$. Thus $\rho_i^{-1} \circ q(\psi(\alpha_\rho) * c' * (\psi(\alpha_\rho))^{-1})$ is homotopic to $\Phi_i(\alpha_\rho * c * \alpha_\rho^{-1})$ hence to $\Phi_i(\alpha_\rho) * \rho_i^{-1} \circ q(c') * \Phi_i(\alpha_\rho)^{-1}$. This implies that $\Phi_i(\alpha_\rho)^{-1} * (\rho_i^{-1} \circ q \circ \psi(\alpha_\rho))$ commutes with $\phi_i(\gamma)$ as elements of $\pi_1(\mathbf{H}^3/\Gamma_i, \Phi_i(y))$. Since the choice of γ was arbitrary and the centre of $\pi_1(\mathbf{H}^3/\Gamma_i, \Phi_i(y))$ is trivial, this shows that $\Phi_i(\alpha_\rho)$ is homotopic to $\rho_i^{-1}(q \circ \psi(\alpha_\rho))$ for sufficiently large i. $\square$

Continuation of Proof of Proposition 3.2. We suppose that $\psi(\alpha_\rho)$ cannot be homotoped off $\mathrm{Int}\psi(W)$ and we shall prove that this leads to a contradiction. Homotope $\psi(\alpha_\rho)$ to an arc transverse to $\partial\psi(W)$ so that the number of components of $\psi(\alpha_\rho) \cap \psi(W)$ is minimum among the arcs homotopic to $\psi(\alpha_\rho)$ fixing the endpoints. By Lemma 4.5, each component Σ of ∂W which is contained in ∂C is mapped homeomorphically to a boundary component of C' by ψ. Such a boundary component $\psi(\Sigma)$ of C' faces either a geometrically finite end or a geometrically infinite tame end, in particular the component of the complement of C' containing $\psi(\Sigma)$ is homeomorphic to $\psi(\Sigma) \times \mathbf{R}$. Hence if $\psi(\alpha_\rho)$ intersected $\psi(\Sigma)$, a part of $\psi(\alpha_\rho)$ going out of C' from $\psi(\Sigma)$ could be pushed into C', and this would reduce the number of components of $\psi(\alpha_\rho) \cap \psi(W)$. Thus $\psi(\alpha_\rho)$ intersects $\partial\psi(W)$ only at ψ (FrW).

Let W_0 be a component of W such that $\psi(\alpha_\rho)$ intersects $\mathrm{Int}\psi(W_0)$. For sufficiently large i, the arc $\rho_i^{-1}(q \circ \psi(\alpha_\rho))$ is homotopic to $\Phi_i(\alpha_\rho)$. Hence there is a homotopy $H_i : I \times I \to \mathbf{H}^3/\Gamma_i$ such that $H_i|(I \times \{0\})$ is $\rho_i^{-1} \circ q \circ \psi(\alpha_\rho)$ and $H_i|(I \times \{1\})$ is $\Phi_i(\alpha_\rho)$. Since $\psi(\alpha_\rho)$ intersects $\mathrm{Int}\psi(W_0)$, there is a component F of ∂W_0 such that $H_i^{-1}(\Phi_i(F)) \cap (I \times [0,1))$ contains an arc. Recall that $\rho_i^{-1} \circ q \circ \psi(W) = \Phi_i(W)$. Let β be an arc in $H_i^{-1}(\Phi_i(F)) \cap (I \times [0,1))$. Let α' be a subarc of $I \times \{0\}$ such that $\alpha' \cup \beta$ bounds a disc in $I \times I$. Then $H_i(\alpha')$ is an subarc of $\rho_i^{-1} \circ q \circ \psi(\alpha_\rho)$ which is homotopic to an arc on $\Phi_i(F)$.

As is shown in Lemma 4.4, the subgroup Γ_i^F associated to $\pi_1(F)$ converges strongly to a quasi-Fuchsian group G^F. Let $\tilde{\alpha}_i'$ and $\tilde{\beta}_i$ be lifts of $H_i(\alpha')$ and $H_i(\beta)$ to $\mathbf{H}^3/\Gamma_i^F$. Then $\tilde{\alpha}_i'$ and $\tilde{\beta}_i$ are homotopic in $\mathbf{H}^3/\Gamma_i^F$. Since the length of $\rho_i^{-1} \circ q \circ \psi(\alpha_\rho)$ is bounded above independently of i, the surface $\Phi_i(F)$ converges geometrically to a compact surface in $\mathbf{H}^3/\Gamma_\infty$ which is lifted to $\psi(F)$, and $\{\Gamma_i^F\}$ converges strongly to G^F, we can see that there is a homotopy between $\tilde{\alpha}'_i$ and $\tilde{\beta}_i$ whose diameter is bounded independently of i. Hence for sufficiently large i, the homotopy can be mapped by ρ_i and lifted to $\mathbf{H}^3/G$ which reduces the number of components of $\psi(W) \cap \psi(\alpha_\rho)$. This is a contradiction. Therefore $\psi(\alpha_\rho)$ can be homotoped off $\psi(W)$.

Now we have arcs $\psi(\alpha_1) \dots , \psi(\alpha_p)$ which intersect $\psi(W)$ only at endpoints. Obviously we can assume that $\psi(\alpha_1), \dots \psi(\alpha_p)$ are disjoint. By taking a rgular neighbourhood of $\psi(W) \cup (\cup_\rho \psi(\alpha_\rho))$, we have a core of $\mathbf{H}^3/G$ to which ψ can be homotoped to a homeomorphism. This completes the proof of Proposition 3.2. $\square$

Finally we are able to prove Theorem 3.1.

Proof of Theorem 3.1. Our assumption of the reductio ad absurdum was that the free homotopy class of γ is not represented by a simple closed curve on Ω_Γ/Γ. Under this assumption, we have proved that $\psi|C$ is homotopic to a homeomorphism to a core C' of $\mathbf{H}^3/G$. As $\psi(\gamma)$ is a parabolic element, it is represented by a simple closed curve on C' by for example applying McCullough's theorem in [18] to the non-cuspidal part of $\mathbf{H}^3/G$. This implies that γ is represented by a simple closed curve on C, hence a simjple closed curve on Ω_Γ/Γ. This contradicts our assumption of reductio ad absurdum. This completes the proof. $\square$

References

1. W. Abikoff : On boundaries of Teichmüller spaces and on Kleinian groups III. Acta Math. **134**, 211-237 (1975)
2. L. Ahlfors : Finitely generated Kleinian groups. Amer. J. Math. **86**, 413-423 (1964)
3. J. Anderson, R. Canary : Algebraic limits of Kleinian groups which rearrange the pages of a book. preprint
4. L. Bers, On boundaries of Teichmüller spaces and on kleinian groups I. Ann. of Math. **91** 570-600, (1970)
5. L. Bers : Spaces of Kleinian groups. Lecture Notes on Mathematics **155**,9-34 (1970)
6. F. Bonahon : Bouts de variétés hyperboliques de dimension 3. Ann. of Math. **124**, 71-158 (1986)
7. F. Bonahon : Cobordism of automorphisms of surfaces. Ann. Sci. Ecole Norm. Sup. (4) **16**, 237-270 (1983)
8. R. D. Canary, D. B. A. Epstein, P. Green : Notes on notes of Thurston. *Analytical and geometric aspects of hyperbolic spaces* London Math. Soc. Lecture Note Ser. **111**, Cambridge Univ. Press, 3-92 (1987)
9. M. Culler and P. Shalen : Variety of group representations and splittings of 3-manifolds. Ann. of Math. **117**, 109-146 (1983)
10. D. B. A. Epstein and A. Marden : Convex hulls in hyperbolic space, a theorem of Sullivan , and measured pleated surfaces. *Analytical and geometric aspects of hyperbolic spaces* London Math. Soc. Lecture Note Ser. **111**, Cambridge Univ. Press, 113-253 (1987)
11. J. Hempel : 3-Manifolds. Ann. of Math. Studies, No. 86. Princeton Univ. Press
12. T. Jørgensen and A. Marden : Algebraic and geometric convergence of Kleinian groups. Math.-Scand. **66**, 47-72 (1990)
13. W. Jaco and P. Shalen : Seifert fibered spaces in 3-manifolds. Mem. Amer. Math. Soc. **21**, (1979)
14. A. Marden : The geometry of finitely generated Kleinian groups. Ann. of Math. **99**, 465-496 (1974)
15. B. Maskit, On boundaries of Teichmüller spaces and on kleinian groups: II. Ann. of Math. **91** (1970), 607-639
16. B. Maskit : Parabolic elements on Kleinian groups. Ann. of Math. **117**, 659-668 (1983)
17. B. Maskit : Comparison of hyperbolic length and extremal length. Ann. Acad. Sci. Fenn. Ser. A I Math. **10**, 381-386 (1985)
18. D. McCullough : Compact submanifolds of 3-manifolds with boundary. Quart. J. Math. Oxford Ser. (2) **37**, 299-307 (1986)
19. W. Meeks III and S. T. Yau : The classical Plateau problem and the topology of three-dimensional manifolds. The embedding of the solution given by Douglas-Morrey and an analytic proof of Dehn's lemma. Topology **21**, 409-442 (1982).
20. D. McCullough, A. Miller and G.A. Swarup : Uniqueness of cores of non-compact 3-manifolds. J. London Math. Soc **32** , 548-556 (1985)
21. J. Morgan : On Thurston's uniformization theorem for three-dimensional manifolds. *The Smith Conjecture* . New York, London : Academic Press. 37-125 (1984)
22. J. Morgan and P. Shalen : Degeneration of hyperbolic structures III, action of 3-manifold groups on trees and Thurston's compactness theorem. Ann. Math. **120**, 457-519 (1988)
23. K. Ohshika : Ending laminations and boundaries for deformation spaces of Kleinian groups. J. London Math. Soc. (2) **42**, 111-121 (1990)
24. K. Ohshika : Geometric behaviour of Kleinian groups on boundaries for deformation spaces Quart. J. Math. Oxford (2) **43**, 97-111 (1992)
25. K. Ohshika : Geometrically finite Kleinian groups and parabolic elements. preprint
26. K. Ohshika : Kleinian groups which are limits of geometrically finite groups. preprint, Tokyo Institute of Technology
27. G. P. Scott : Compact submanifolds of 3-manifolds. J. London Math. Soc. (2) **7**, 246-250 (1973)
28. A. Shapiro and J.H.C. Whitehead : A proof and extension of Dehn's lemma. Bull. Amer. Math. Soc. **64**, 174-178 (1958)

29. D. Sullivan : A finiteness theorem for cusps. Acta Math. **147** , 289-299 (1980)

30. W. Thurston : The geometry and topology of 3-manifolds. lecture notes, Princeton Univ.

31. W. Thurston : Hyperbolic structures on 3-manifolds I : Deformation of acylindrical manifolds. Ann. Math. **124**, 203-246 (1986)

32. W. Thurston : Hyperbolic structures on 3-manifolds II : Surface groups and 3-manifolds which fiber over the circle. preprint, Princeton Univ.

33. W. Thurston : Hyperbolic structures on 3-manifolds III: Deformations of 3-manifolds with incompressible boundary. preprint, Princeton Univ.

34. F. Waldhausen : On irreducible 3-manifolds which are sufficiently large. Ann. of Math. **87**, 56-88 (1968)

DEPARTMENT OF MATHEMATICS, TOKYO INSTITUTE OF TECHNOLOGY, OH-OKAYAMA, MEGURO-KU, TOKYO 152, JAPAN

E-mail address: ohshika@math.titech.ac.jp

Proceedings of
THE 37TH TANIGUCHI SYMPOSIUM ON
TOPOLOGY AND TEICHMÜLLER SPACES
held in Finland, July 1995
ed. by Sadayoshi KOJIMA *et al.*
©1996 World Scientific Publishing Co.
pp. 237–252

THE SIMPLICIAL COMPACTIFICATION OF RIEMANN'S MODULI SPACE

R. C. PENNER

Dedicated to the memory of Claude Itzykson, an inspiration to so many

(Received March 1, 1996)

ABSTRACT. We describe a new compactification of Riemann's moduli space for punctured surfaces which is derived from a simplicial completion of Teichmüller space. The ideal points of this compactification admit a natural description intrinsically in terms of the geometry of the underlying surface. Furthermore, the compactification comes equipped with a canonical cell decomposition, where the cells in the decomposition are indexed by graphs plus extra structure, and this decomposition restricts to (a version of) the usual one on moduli space. We discuss the conjecture that our compactification is an orbifold and sketch the proof of this conjecture in the planar case.

1. Introduction and Statement of Main Results

For any integers $g \geq 0$ and $s \geq 1$ with $2g - 2 + s > 0$, let $F = F_g^s$ denote a fixed smooth oriented surface of genus g with s punctures (which we may sometimes regard as distinguished points). Let $MC = MC_g^s$ denote the *mapping class group* of homotopy classes of diffeomorphisms of F which setwise fix the collection of punctures, and let $PMC = PMC_g^s$ denote the *pure mapping class group*, whose underlying diffeomorphisms are required to fix each puncture. Define

$$\tilde{T} = \tilde{T}_g^s = \text{decorated Teichmüller space of } F_g^s,$$

$$\tilde{S} = \tilde{S}_g^s = \tilde{T}_g^s / \mathbf{R}_+ = \text{scaled decorated Teichmüller space of } F_g^s,$$

$$T = T_g^s = \text{Teichmüller space of } F_g^s,$$

$$\mathcal{M} = \mathcal{M}_g^s = T_g^s / MC_g^s = \text{Riemann moduli space of } F_g^s,$$

$$P\mathcal{M} = P\mathcal{M}_g^s = T_g^s / PMC_g^s = \text{pure Riemann moduli space of } F_g^s,$$

Partially supported by the National Science Foundation.

where $\mathcal{T}$ is the space of all complete finite-area Riemannian metrics on F of constant curvature -1 modulo pull-back by diffeomorphisms homotopic to the identity, $\mathcal{M} = \mathcal{T}/MC$ (and $P\mathcal{M} = \mathcal{T}/PMC$ respectively) is the same space of metrics modulo pull-back by arbitrary diffeomorphisms (and by diffeomorphisms fixing each puncture); $\tilde{\mathcal{T}}$ is the trivial $\mathbf{R}_+^s$-bundle consisting of all s-tuples of horocycles on F, one horocycle about each puncture (see [P1] for further information), and $\tilde{\mathcal{S}}$ is the quotient of $\tilde{\mathcal{T}}$ by the natural $\mathbf{R}_+$-action of homothety on the tuple of hyperbolic lengths of the horocycles. In particular in the once-punctured case, notice that $\tilde{\mathcal{S}}_g^1 = \mathcal{T}_g^1$. We also define the geometric realization

$$\tilde{\mathcal{A}} = \tilde{\mathcal{A}}_g^s = \text{arc complex of } F_g^s$$

of the partially ordered set ("poset") consisting of non-empty families Δ of arcs disjointly embedded in F and connecting punctures, where we demand that each arc in Δ is essential (i.e., no arc is homotopic into the punctures) and that no two arcs in Δ are parallel (i.e., no two arcs are homotopic rel endpoints in F).

As is well-known (see for instance [P1]), there are MC-invariant cell decompositions of $\tilde{\mathcal{T}}$ and $\tilde{\mathcal{S}}$ (which we shall recall presently), where the cells in the decomposition are indexed by arc families Δ as above and where each component of $F - \Delta$ is a polygon (i.e., simply connected); we say that such an arc family Δ *fills* the surface. Indeed, $\tilde{\mathcal{S}}$ is isomorphic to the subspace of $\tilde{\mathcal{A}}$ corresponding to arc families which fill F, and we henceforth simply identify these spaces. $\tilde{\mathcal{S}}$ is not a sub-complex of $\tilde{\mathcal{A}}$: Given an arc family which does fill, we can remove arcs from it (or in other words pass to faces of the corresponding cell) to produce an arc family which does not fill. It is thus natural to take the simplicial completion of $\tilde{\mathcal{S}}$ in $\tilde{\mathcal{A}}$ to define a completion of $\tilde{\mathcal{S}}$. Since the action of MC on $\tilde{\mathcal{S}}$ extends to its natural action on $\tilde{\mathcal{A}}$, this defines a completion, and indeed a compactification, of the quotient $\tilde{\mathcal{S}}/MC$ or $\tilde{\mathcal{S}}/PMC$.

All of this is quite obvious, but one would then like to control the topology, geometry, and combinatorics of this quotient (and the construction we just described works in this sense, as we shall see, only in the case of once-punctured surfaces); we shall find, however, that a variant of this naive and obvious construction leads to a suitable compactification, as follows.

We must choose, once and for all among the s punctures of F, a distinguished one and shall denote this distinguished puncture by π. Define the geometric realization

$$\tilde{\mathcal{A}}(\pi) = \tilde{\mathcal{A}}_g^s(\pi)$$

of the poset consisting of non-empty families Δ of arcs as before, where we now require the arcs to have their endpoints only at π (i.e., be "based" at π), and define

$$\mathcal{A}(\pi) = \mathcal{A}_g^s(\pi) = \tilde{\mathcal{A}}_g^s(\pi)/PMC_g^s.$$

We say that such an arc family Δ *quasi-fills* F if each component of $F - \Delta$ is either a polygon or a once-punctured polygon (so there are exactly $s - 1$ components of the latter type if Δ quasi-fills F_g^s).

Several of the main results in this note are summarized in

Theorem 1. *For any choice of puncture π, $\mathcal{T}$ is naturally isomorphic to the subspace of $\tilde{A}(\pi)$ corresponding to arc families based at π which quasi-fill F, so $A(\pi)$ is a compactification of $P\mathcal{M}$. In fact, the third barycentric subdivision of $\tilde{A}(\pi)$ descends to an honest simplicial complex on the compactification $A(\pi)$ of $P\mathcal{M}$. Furthermore, ideal points of this compactification (i.e., points of $A(\pi) - P\mathcal{M}$) are naturally identified with decorated hyperbolic structures (see §5 below) on proper subsurfaces of F, and cells in the compactification are naturally described by two-colored fatgraphs (see §6 below).*

One could also consider the mapping class group $MC(\pi)$ consisting of classes of homeomorphisms which are required to fix only π and may permute the other punctures. The cell decomposition of $\mathcal{T}$ in Theorem 1 is in fact invariant under $MC(\pi)$, and there is a compactification $\tilde{A}(\pi)/MC(\pi)$ of the corresponding moduli space, but we shall not further discuss this version of the theory in this paper.

Our main conjecture is

Conjecture A. $A(\pi)$ *is an orbifold.*

This admits a remarkable reformulation (Conjecture B) below as the sphericity of a related complex, but we shall only prove here

Theorem 2. *Conjecture A holds for multiply punctured spheres, i.e., for F_g^s with $g = 0$.*

This planar case of multiply punctured spheres is already of interest, for instance in the study of the absolute Galois group. More explicitly, we propose a treatment as in [LS] working with the tower of compactifications constructed here; this is our current motivation for this project and is what has prompted us to describe a proof of Theorem 2 here separately from considerations in the general case. At the same time, much of our proof of the conjecture in this setting actually applies to the general case as we shall see.

To put these results in some perspective, we observe that the (quotient of the) cell decomposition of $\mathcal{T}$ or $\tilde{\mathcal{T}}$ does not extend in any known way to the otherwise extremely useful Deligne-Mumford compactification [DM] of moduli space, which is of course itself an orbifold (and indeed a projective variety). Kontsevich [Ko] employs another compactification which is a different orbifold compactification than ours mapping continuously to the Deligne-Mumford compactification, and we shall prove in §7 below that our compactification here similarly maps continuously to the one of Deligne-Mumford. As alluded to above, Theorem 2 can be equivalently regarded as a sphericity result, and we mention that the spherical complex in Theorem 2 or Conjecture B is not a version of Teichmüller space, but rather a version of moduli space in contrast to various sphericity results of Harer [Ha]; indeed, this is what is remarkable about the reformulation of Conjecture A: there is a spherical compactification of a version of moduli space. Finally, our completion of $\mathcal{T}$ provides a sort of complement to Thurston's boundary of $\mathcal{T}$ (see [Th] or [PH]) in a sense we shall describe below.

Our starting point for Conjecture A was the request/challenge from Ed Witten in 1989 to find an orbifold compactification of moduli space together with a cell decomposition which restricts to the usual one on moduli space itself. This is certainly a reasonable mathematical goal, and we take a moment to recount further aspects (and our interpretation and elaboration) of Witten's ideas. Namely, correlation functions (i.e., integrals of top-dimensional forms over moduli spaces) should correspond via virtual Poincaré duality (here using that the compactifications are orbifolds) to intersection theory calculations. These intersection theory problems, in turn, should be expressed as corresponding graphical enumeration problems (here using that cells in the compactification are indexed by suitable graphs), which could presumably be calculated with matrix models. To do this effectively, one of course needs perturbation methods to put the cells in general position, and Witten proposed an explicit such perturbation for two-dimensional quantum gravity (as we shall further explain in §7 below). This is all just to say that in our opinion the compactification here is also a reasonable goal from the perspective of string theory; in fact the geometry here seems to suggest a constructive string field theory based on surfaces with boundary which are either topological or perhaps endowed with a hyperbolic structure.

Acknowledgements. The work described here has actually been a sporadically active project since 1985, and there have been helpful discussions at various points with Bill Harvey, Maxim Kontsevich, Jim Milgram, Walter Neumann, Dennis Sullivan, and Ed Witten. We are happy to acknowledge the kind support of Stanford University during the Spring of 1990 and of the Institut des Hautes Etudes Scientifique during the Spring of 1992.

2. The Convex Hull Construction

This section is dedicated to a discussion of the proof of the first two claims of Theorem 1, namely, that $\mathcal{T}$ is naturally isomorphic to $\tilde{\mathcal{A}}(\pi)$ and the third barycentric subdivision of $\tilde{\mathcal{A}}(\pi)$ descends to a simplicial complex on $P\mathcal{M}$. At the same time, we shall recall the construction of the usual cell decompositions of $\mathcal{T}$, $\tilde{\mathcal{T}}$, and $\tilde{\mathcal{S}}$ themselves.

Following [P1], let us uniformize hyperbolic surfaces in Minkowski space, so a conformal or hyperbolic structure on F is given as $\mathbf{H}/\Gamma$, where $\mathbf{H}$ denotes the upper sheet of the unit hyperboloid in Minkowski space, and Γ is a suitable discrete subgroup of $SO_+(1,2) \approx PSL(2, \mathbf{R})$. Also, identify the open positive light-cone with the collection of horocycles in $\mathbf{H}$ via affine duality.

In the usual construction, we begin with a point $\tilde{\Gamma} \in \tilde{\mathcal{T}}$ so that about each puncture of F_g^s, there is an associated horocycle. Via the affine duality above, each horocycle gives rise to a Γ-orbit of points in the light-cone, and we let $\mathcal{B}$ denote the union of these s disjoint Γ-orbits. The closed convex hull of $\mathcal{B}$ projects to a family of geodesics disjointly embedded in F which fills F, and we let $\Delta(\tilde{\Gamma})$ denote this arc family. Fixing the homotopy class Δ of such a family of arcs, we consider

$$C(\Delta) = \{\tilde{\Gamma} \in \tilde{\mathcal{T}} : \Delta(\tilde{\Gamma}) = \Delta\},$$

and then prove rather arduously in [P1] that each $\mathcal{C}(\Delta)$ is actually a cell in $\tilde{\mathcal{T}}$ (cf. the beginning of §5 below). In fact, since the construction is linear, the cell decomposition descends to $\tilde{\mathcal{S}}$ as well.

One method of restricting the cell decomposition to $\mathcal{T}$ is to simply fix horocycles at each puncture of F once and for all in order to define a section of the bundle $\tilde{\mathcal{T}} \to \mathcal{T}$ and restrict attention to PMC. Another approach is to take all the horocycles of the same length, say the length is unity (a canonical section), so that MC itself also sensibly acts. Each of these suffers from a certain non-naturality in its graphical interpretation (i.e., not all graphs actually arise) rendering enumerative problems somewhat awkward. Our next construction gives another more natural treatment of $\mathcal{T}$.

In the new construction given $\Gamma \in \mathcal{T}$, we have fixed once and for all the puncture π and now specify just a single horocycle about π. Again by affine duality, this gives rise to a Γ-orbit of points in the light-cone, and we may again take the convex hull but just of this single orbit. This gives rise to a collection Δ of arcs, as before, disjointly embedded in F always running from π back to π (that is, these arcs are based at π), and there is a corresponding simplex in $\tilde{A}(\pi)$. In fact, whereas in the usual construction, the support planes to the convex body are always elliptic, in our current case, the support planes are either elliptic or parabolic; it follows that complementary regions to Δ are either simply connected (elliptic case) or once-punctured (parabolic case), so Δ therefore quasi-fills F. Again since the construction is linear, it is independent of the choice of horocycle about π, so we have in fact associated a quasi-filling arc family $\Delta(\Gamma)$ to $\Gamma \in \mathcal{T}$. This was actually our first formulation of the convex hull construction in [P1, Addendum], and as mentioned there, the proof that each

$$\mathcal{C}(\Delta) = \{\Gamma \in \mathcal{T} : \Delta(\Gamma) = \Delta\}$$

is indeed a cell follows from an elaboration of the previous case.

This explains the intrinsic hyperbolic geometry underlying our new compactification (but see §5 below for the intrinsic description of the ideal points), and we close this section with various observations about the action of PMC on $\tilde{A}(\pi)$. First of all, let us remark explicitly that there are only finitely many PMC-orbits of cells in $\tilde{A}(\pi)$ since there are only finitely many combinatorially distinct ways (i.e., with distinct separation properties) to disjointly embed arcs as above in a given surface F; thus, $A(\pi)$ is indeed a compactification of $P\mathcal{M}$, completing the proof of the first part of Theorem 1, and we henceforth simply identify $\mathcal{T}$ with the corresponding subset of $\tilde{A}(\pi)$.

A cell of $\tilde{A}(\pi) - \mathcal{T}$ is said to lie "at infinity", and the points within it are said to be "ideal". Thus, a cell in $\tilde{A}(\pi)$ lies at infinity if and only if it does not quasi-fill F, which is in turn equivalent to the cell having infinite isotropy in PMC. In particular, $\tilde{A}(\pi)$ is not locally finite at infinity, but it is locally finite and has finite stabilizers except at infinity.

Furthermore, a general cardinality argument (about group actions on partially ordered sets) shows that the third barycentric subdivision of $\tilde{A}(\pi)$ descends to an honest simplicial complex on $\mathcal{A}(\pi)$, proving the second part of Theorem 1.

3. Reformulation of Conjecture A

To understand and reformulate Conjecture A and as a first step towards its proof, we introduce the following relative version of the arc complexes.

Consider a marked oriented smooth surface of genus $g \geq 0$ with $s \geq 0$ distinguished points and $r \geq 1$ boundary components, and let S denote the set of punctures (if any). In the ith boundary component, choose some number $\delta_i > 0$ of distinguished points, for each $i = 1, \ldots, r$, and let R denote the collection of all these distinguished points on the boundary. Define the r-tuple $\vec{\delta} = (\delta_1, \ldots, \delta_r)$ of natural numbers, and let $F = F_{g,\vec{\delta}}^s$ denote this surface together with its distinguished points. It will be convenient in the sequel to consider the surface $F' = F - R - S$ (and one imagines F' endowed with a complete hyperbolic structure whose boundary $\partial F - R$ is totally geodesic). The associated mapping class group $PMC = PMC(F)$ we consider (is the "purest" possible and) consists of the homotopy classes of orientation-preserving diffeomorphisms of F which preserve $R \cup S$ pointwise.

Suppose that α is an arc embedded in F with (not necessarily distinct) endpoints $\{p, q\} = \alpha \cap R$; thus, $\alpha \subset F' \cup \{p, q\}$, and we furthermore demand that α is not homotopic rel $\{p, q\}$ in $F' \cup \{p, q\}$ into $(\partial F - R) \cup \{p, q\}$. The homotopy class rel $\{p, q\}$ of such an arc α is an *ideal arc* in F, and an *ideal arc family* in F is a collection of distinct ideal arcs in F which admit representative arcs in F which are disjointly embedded; a representative of an ideal arc family in F will always be assumed to consist of disjointly embedded arcs (no pair of which are homotopic rel endpoints).

As before, we say an ideal arc family (or its representative a) *fills* F if each component of $F - a$ is a disk, whereas the arc family is said to *quasi-fill* F if each component of $F - a$ is either a disk or a once-punctured disk. In particular, the ideal arc family is called an *ideal triangulation* if each component of $F' - a$ is an ideal triangle, whereas it is a *quasi-triangulation* if each component of $F' - a$ is either an ideal triangle or a once-punctured monogon, i.e., a surface of type $F_{0,(1)}^1$.

Define another arc complex

$$\tilde{\mathcal{B}} = \tilde{\mathcal{B}}(F)$$

to be the geometric realization of the poset of such (not necessarily quasi-filling) arc families in the surface F with boundary, and take the quotient

$$\mathcal{B} = \mathcal{B}(F) = \tilde{\mathcal{B}}(F)/PMC(F).$$

Our reformulation of Conjecture A is

Conjecture B. *For any surface $F = F_{g,(\delta_1,\ldots,\delta_r)}^s$ with $r \geq 1$, $\mathcal{B}(F)$ is a triangulated topological sphere S^N of dimension $N = 6g - 7 + 3r + 2s + \delta_1 + \cdots + \delta_r$, provided $N \geq 0$.*

We shall prove that Conjecture B implies Conjecture A (and then concentrate on the former) in the next section, and for the remainder of this section, we consider several simple examples of Conjecture B. Notice first that the condition $N \geq 0$ just rules out the surface $F_{0,(1)}^1$ among surfaces with $r \geq 1$. Furthermore, if F is disconnected, say with components $F = F_1 \cup \cdots \cup F_n$, then $\mathcal{B}(F)$ is clearly isomorphic to the simplicial join $\mathcal{B}(F_1) * \cdots * \mathcal{B}(F_n)$.

Example -1. Conjecture B is true if $F = F^0_{0,(n)}$ is an n-gon, namely $\tilde{\mathcal{B}}(F) = \mathcal{B}(F) \approx S^{n-4}$. Indeed, $PMC(F)$ is the trivial group in this case, this example is well known, and there is a self-contained and elementary proof given in [PW].

Example 0. In case $F = F^1_{0,(2)}$, then again $PMC(F)$ is trivial, there are exactly two homotopy classes of arcs in F, and $\tilde{\mathcal{B}}(F) = \mathcal{B}(F) \approx S^0$.

Example 1a. In the annulus $F = F^0_{0,(1,1)}$, an ideal arc is determined by the (positive or negative) number of twists (right or left) about the core of the annulus, that is, the set of ideal arcs is in bijection with the set $\mathbf{Z}$ of integers. Furthermore, two such ideal arcs can be disjointly embedded if and only if their twisting numbers differ by unity, so in fact $\tilde{\mathcal{B}}(F) \approx \mathbf{R}$ in this case. $PMC(F)$ is generated by the Dehn twist about the core of the annulus, which acts on $\mathbf{R}$ as translation by unity, so $\mathcal{B}(F) = \tilde{\mathcal{B}}(F)/PMC(F) \approx S^1$, decomposed into one 0-cell and one 1-cell.

Example 1b. For the surface $F = F^2_{0,(1)}$, there are two types of homotopy classes of connected arcs; cutting on the arc decomposes F into a surface of type $F^1_{0,(1)}$ and a surface of type $F^1_{0,(2)}$, and the type of the arc is determined by which puncture lies in the former component. We can disjointly embed these two arc types in two distinct ways (depending on which type lies to the left near the distinguished point of F). It follows that in this case again $\tilde{\mathcal{B}}(F) \approx \mathbf{R}$ and $\mathcal{B}(F) \approx S^1$, this time decomposed into two 1-cells and two 0-cells.

In fact, we have checked by hand all of the complexes of dimension less than four (the only truly involved such example is the surface $F^0_{1,(1)}$ of genus one), and indeed, we have checked Conjecture B directly by hand for many further classes of surfaces including examples in genera two and three as well. We are pretty confident of the verity of Conjecture B at least for $g \leq 3$.

4. Conjectures A and B, and the Proof in the Planar Case

The connection between Conjectures A and B and a tool towards their proof is the following construction in which there are two basic cases A and B respectively. In the first case, suppose that $\sigma = \sigma(a)$ is a simplex in some complex $\tilde{\mathcal{A}}(\pi)$, where a is an arc family in a surface $F = F^s_g$ based at π; in the second case, suppose that $\sigma = \sigma(a)$ is a simplex in some complex $\tilde{\mathcal{B}}(F)$, where F is a surface each component of which has boundary, and a is an ideal arc family in F.

Let us cut F in either case along a to produce a surface we shall denote F_σ. A first basic point is that in either case, F_σ is a surface with boundary, where an arc α in a itself contributes (two copies of itself) to the boundary of F_σ, and there are four (possibly not distinct) distinguished points on the boundary, one such point for each side (left or right) of each end of α. Thus, in either case, we can sensibly consider the B-complex $\mathcal{B}(F_\sigma)$.

It is geometrically natural to just combine a point of σ (i.e., a positive-real projective weight on the arcs in a) with a point of $\mathcal{B}(F_\sigma)$ (i.e., a positive-real projective weight on a $PMC(F_\sigma)$-orbit of arc families in $\mathcal{B}(F_\sigma)$) by "joining" the two points.

Precisely, let us denote by X' the first barycentric subdivision of a simplicial complex X; thus an n-simplex in X' is a "flag" $\sigma_0 < \cdots < \sigma_n$ of simplices of X, where $<$ denotes the proper face relation of X. Pass to first barycentric subdivisions $\tilde{\mathcal{A}}'(\pi)$ in case A and $\tilde{\mathcal{B}}'(F)$ in case B, so in either case, there is a corresponding "flag" $a_0 \subset \cdots \subset a_n$ of suitable arc families in F. The assignment

$$\left(a_0 \subset \cdots \subset a_n\right) \mapsto \left((a_0 \cup a) \subset \cdots \subset (a_n \cup a)\right)$$

gives rise to a simplex in $\tilde{\mathcal{A}}'(\pi)$ or $\tilde{\mathcal{B}}'(F)$ from a simplex in $\tilde{\mathcal{B}}'(F_\sigma)$, and one defines embeddings

$$\sigma * \tilde{\mathcal{B}}'(F_\sigma) \to \tilde{\mathcal{A}}'(\pi),$$
$$\sigma * \tilde{\mathcal{B}}'(F_\sigma) \to \tilde{\mathcal{B}}'(F),$$

where the binary operation $*$ again denotes the join.

Composing finally with the natural projections $\tilde{\mathcal{A}}'(\pi) \to \mathcal{A}'(\pi)$ or $\tilde{\mathcal{B}}'(F) \to \mathcal{B}'(F)$ and restricting to the interior $Int[\sigma * \mathcal{B}'(F_\sigma)]$ in either case, we find mappings

$$E_A : Int[\sigma * \mathcal{B}'(F_\sigma)] \to \mathcal{A}'(\pi),$$
$$E_B : Int[\sigma * \mathcal{B}'(F_\sigma)] \to \mathcal{B}'(F).$$

Further cardinality arguments about group actions on the barycentric subdivision of a poset give

Lemma 3. *In case A, there is a finite group $\mathcal{G}_\sigma$ acting on $\sigma * \mathcal{B}'(F_\sigma)$ as well as on its interior, and E_A induces a homeomorphism between the quotient $Int[\sigma * \mathcal{B}'(F_\sigma)]/\mathcal{G}_\sigma$ and the image of E_A.*
In case B, E_B is an embedding.

The local groups $\mathcal{G}_\sigma$ in the orbifold structure of $\mathcal{A}(\pi)$ are easily described and recognized as (a generalization of) a standard generalization of Dehn twists as follows. Consider the stabilizer $F_\sigma^{PMC(F)}$ of the subsurface $F_\sigma \subset F$ in $PMC(F)$. We evidently have a normal subgroup $PMC(F_\sigma)$ of $F_\sigma^{PMC(F)}$, and the quotient

$$\mathcal{G}_\sigma = F_\sigma^{PMC(F)}/PMC(F_\sigma)$$

is a finite group (indeed, it is a subgroup of the permutation group of the sides of the arcs in σ). For instance, a component of F_σ might be an annulus of type $F_{0,(1,n)}^0$, $\phi \in PMC(F)$ might cyclically permute the n distinguished points on a boundary component (if there is such a homeomorphism of the supersurface), and the resulting coset corresponds to a "fractional Dehn twist".

The following absolutely simple remark explains the difference between the A-complexes and the B-complexes and illuminates the previous and subsequent discussions. The distinction between the two cases is that in the case of a B-complex, the arcs in an ideal arc family come in a well-defined linear ordering as in Example 1b above; thus, if $\phi \in PMC(F)$ fixes an arc family setwise (i.e., given isotropy of the corresponding cell), it must do so pointwise (so any isotropy of the cell fixes the cell pointwise). This is just not so in the case of an A-complex (the arcs come in only a cyclic order), and cells do indeed sometimes have finite isotropy.

By the first part of Lemma 3, sphericity of the B-complexes proves that the A-complexes are orbifolds, that is, Conjecture B indeed implies Conjecture A.

Turning to the proof of Conjecture B in general, we might proceed by topological induction using the second part of Lemma 3 to guarantee that the "next" B-complex in the induction is a manifold. One can prove directly (with some involved but elementary arguments about arcs embedded in surfaces) simple connectivity of every A-complex and of each B-complex other than those of dimension zero and one (which were described in the examples above).

Insofar as we have checked Conjecture B by hand for all complexes of dimension at most three, we might then apply the Poincaré Conjecture [Fr] and [Sm] to conclude that Conjecture B follows if we can prove that the B-complexes are homology spheres (i.e., show that the homology vanishes through the middle dimension).

Thus, the proof of Conjecture B has been reduced here to a homology calculation, on which we have worked extensively at several points to no complete satisfaction. We also hope for a simpler proof (perhaps adapting arguments of [Iv] or [Ha] to the B-complexes) of simple connectivity.

As a practical matter, we now turn from the general case to discuss Conjecture A in the planar case (that is, the proof of Theorem 2 above), where we sketch the remaining homology calculation. In fact, simple connectivity also follows from our discussion below in this case, so we actually sketch a complete proof here in the planar case. The proof depends upon two simplifications as well as a normal form for arc families in surfaces of type $F_{0,(1)}^s$ for $s \geq 2$.

The first simplification is just that if a is an arc family in $F = F_0^s$ based at π in case A (or an ideal arc family in F in case B), then a separates F, and indeed each component of $F - a$ is a surface with only one boundary component, that is, of type $F_{0,(n)}^t$ for some $t < s$; the proof is an elementary combinatorial topology argument involving separating families of arcs on punctured spheres and disks. Thus, Theorem 2 follows from Conjecture B in the special case of surfaces of type $F_{0,(n)}^s$.

The second simplification is the (completely general) observation of Harer [Ha] that if $F_1 = F_{g,(\delta_1,\ldots,\delta_r)}^s$ is a surface with $\delta_i > 1$ and we choose a distinguished point of F_1 on the ith boundary component and remove it from the set of distinguished points to produce another surface F_2, then $\mathcal{B}(F_1) \approx \Sigma \mathcal{B}(F_2)$, where ΣX denotes the suspension of X (i.e., the join of X with S^0). Thus, the proof of Theorem 2 is reduced to Conjecture B for surfaces of type $F_{0,(1)}^s$.

As to the case $F = F_{0,(1)}^s$, we introduce a "normal form" for ideal arc families. The basic idea (as in Example 1b above) is that each arc decomposes F into surfaces of type $F_{0,(1)}^{s_1}$ and $F_{0,(2)}^{s_2}$ (where $s_1 + s_2 = s$), and we might take the subset S_1 of $S = \{\text{distinguished points of} F\}$ corresponding to the punctures contained in the former component as the "name" for the original arc family and choose standard models of each arc with the property that the models are disjointly embedded if and only if the corresponding subsets of S are disjoint.

More generally, we define a *tableaux* to be a rooted tree embedded in the plane, where: the root is a formal vertex; the other vertices are labeled by subsets of S, and for any n, the vertices at distance n from the root are pairwise disjoint subsets of S and occur in a counter-clockwise order in the plane corresponding to the counter-clockwise order near the distinguished point of F (again as in Example 1b above); if a simple path from the root passes consecutively through vertices labeled S' and S'', then $S'' \subset S'$. It follows from the topological classification of surfaces that

Proposition 4. *$PMC(F)$-orbits of cells in $\tilde{B}(F)$ are in one-to-one correspondence with isomorphism classes (as labeled graphs in the plane) of tableaux.*

Using this description of $\mathcal{B}(F)$, it is quite easy to prove that that the complement of a vertex in $\mathcal{B}(F)$ is contractible. It follows from this that $\mathcal{B}(F)$ is simply connected and a homology sphere, completing our discussion of the proof of Theorem 2.

It is worth emphasizing that the only missing ingredient for the proof of the general conjecture is the homology calculation.

Setting $F = F^s_{g,(1,\dots,1)}$ where F has $r \geq 1$ boundary components (i.e., there are r ones in the vector), let $\tilde{B}_f = \tilde{B}_f(F)$ denote the subspace of $\tilde{B} = \tilde{B}(F)$ corresponding to arc families which quasi-fill F. One can show rather easily using train tracks (see analogously [Ha]) that $\tilde{B}_f$ is contractible. This is not surprising as it can also be identified with a "Fricke" space, where we keep track of not only a marked hyperbolic structure on F as for Teichmüller space, but also a twisting parameter for each boundary component, i.e., the isotopies fix the boundary of F pointwise; see also Proposition 6 below.

In fact, the action of $PMC = PMC(F)$ on $\tilde{B}_f$ furthermore has finite isotropy, so the quotient

$$\mathcal{B}_f = \mathcal{B}_f(F) = \tilde{B}_f(F)/PMC(F)$$

is actually an Eilenberg-Maclane space $K(PMC, 1)$ in the rational sense. Taking the quotient by PMC, we thus have the commutative diagram

$$\begin{array}{ccc} \tilde{B}_f & \subset & \tilde{B} \\ \downarrow & & \downarrow \\ K(PMC, 1) \approx \mathcal{B}_f & \subset & \mathcal{B} \approx S^N, \end{array}$$

where the bottom-right homeomorphism follows from our Conjecture B with $N = 6g - 7 + 4r + 2s$. Thus, assuming the conjecture we might apply Alexander duality to (rationally) calculate the group cohomology of PMC in general in terms of the homology of the complement of $\mathcal{B}_f$ in $\mathcal{B}$, that is, the subcomplex corresponding to arc families which do not quasi-fill. In the planar case, these remarks compare favorably with Arnold's calculations [Ar] for the pure braid groups.

Finally, let us mention that there is a more elaborate stratified version of our Conjecture B, where in the first stratum with no punctures allowed at the endpoints of arcs, we have already conjectured a manifold. If we were to allow one puncture, then we conjecture a manifold with boundary, and allowing two, then we conjecture a manifold with corners (whose boundary has boundary), and so on. This may be a useful formalism for an elementary proof of (this stratified generalization of) Conjecture B.

5. Asymptotics of the Arithmetic Problem and the Geometry of Ideal Points

Just for emphasis, we point out here that our ensuing discussion of the geometry (in this section) and the combinatorics (in the next section) of the compactification is in the general context of not necessarily planar surfaces.

In order to understand the intrinsic degeneration of structure associated with our compactification, we must recall a technical but basic aspect of the decorated Teichmüller theory, namely, the "arithmetic problems" discussed briefly in [P2] (as a necessary part of our calculations there) and systematically in [P3] (where we give geometric and physical interpretations of these problems). We shall just survey aspects of [P2] or [P3] (to be reviewed only briefly here) and refer the reader to these references for further formulas and more information.

Fix some surface $F = F_g^s$ with corresponding decorated bundle $\tilde{\mathcal{T}} \to \mathcal{T}$ once and for all for this section, and suppose that Δ is an ideal triangulation of F, say with edges ϵ_i for $i = 1, \ldots, N = 6g - 6 + 3s$. There are actually several different geometrically natural coordinatizations of $\tilde{\mathcal{T}}$, two of which are as follows:

- There are global coordinates, called "lambda-lengths", one such positive real lambda-length e_i for each edge $\epsilon_i \in \Delta$. These are essentially inner products in Minkowski space, and the specification of Δ should be thought of as analogous to the specification of a basis for a vector space.
- There are local coordinates on the top-dimensional cell $\mathcal{C}(\Delta)$ corresponding to Δ, called "simplicial coordinates", one such non-negative real simplicial coordinate E_j for each edge $\epsilon_j \in \Delta$. These arise as certain scaled volumes of simplices in Minkowski space in [P2], and an intrinsic geometric interpretation is given in [P3].

It is much more natural at the moment to consider Poincaré duals of ideal triangulations (cf. [P2] or [P4] for a discussion of these dual "fatgraphs"), and we let $G = G(\Delta)$ denote the trivalent fatgraph dual to Δ. We may regard lambda lengths and simplicial coordinates as assigned to the edges of the dual fatgraph in the natural way.

There are furthermore explicit rational formulas (which are "local" in G) for the simplicial coordinates E_j in terms of the lambda lengths e_i, as follows. Suppose first that the edge ϵ of G has distinct endpoints, say edges of lambda lengths a, b and c, d are incident on the respective endpoints of ϵ, and let e denote the lambda length of ϵ itself. The simplicial coordinate E corresponding to ϵ is given by the formula

$$E = \frac{a^2 + b^2 - e^2}{abe} + \frac{c^2 + d^2 - e^2}{cde}.$$

(In case ϵ is a loop in G, then we actually just apply the identical formula in the universal cover of G.) The arduous argument mentioned above proves

Theorem ([P1] Theorem 5.1). *For any strictly positive tuple E_j of simplicial coordinates on G, there is a unique tuple e_i of lambda lengths inverting the formulas. In fact, we may also allow the vanishing of certain of the E_j subject to the constraint that there are no "vanishing cycles" of edges of G.*

The "arithmetic problem" is the explicit calculation of the functions $e_i = e_i(E_1, \cdots, E_N)$, that is, the explicit inversion of the coupled non-linear equations above. Many examples as well as a general discussion can be found in [P3]. Part of our progress to be reported here in this realm is a complete calculation (requiring explicit estimates on "h-lengths", which we suppress) on the asymptotics of these arithmetic problems.

Define a *degeneration* to be a continuous one-parameter family $E_j(t) \geq 0$ for $t \in [0,1]$, where for $t \neq 1$ we demand that $E_j(t)$ satisfy the no vanishing cycle condition above, i.e., there is no cycle in G with vanishing simplicial coordinates at any time $t < 1$. There is by Theorem A a corresponding continuous one-parameter family $e_i(t)$ for $t \neq 1$ whose asymptotics as $t \to 1$ we next describe. We shall study degenerations on the canonical section (discussed above) $\mathcal{S}$ of the bundle $\tilde{\mathcal{T}} \to \mathcal{T}$, where by definition each horocycle in the decoration of a point in $\mathcal{S}$ is taken to have hyperbolic length unity. It is easy to see from the formulas that on $\mathcal{S}$, the simplicial coordinates are bounded above by one half and the lambda lengths are bounded below by unity, so perhaps $E_j \to 0$ or $e_i \to \infty$ during a degeneration.

Theorem 5. *Consider a degeneration where $E_i(t) \to 0$ for the indices $i \in I$, and suppose $e_j \to \infty$ for $j \in J$. Then $J \subset I$. Furthermore, denoting by G_I the subfatgraph of G comprised of the edges of G whose indices lie in I, each component C of G_I satisfies one of the following conditions:*

- *C is a tree, and $e_i(t)$ has finite limits as $t \to 1$ for each edge $\epsilon_i \in C$.*
- *C has no terminal nodes, and $e_i(t) \to \infty$ as $t \to 1$ for each edge $\epsilon_i \in C$.*

Finally, a formal such specification of edges on which we might demand vanishing simplicial coordinates and/or diverging lambda lengths actually arises for some degeneration on $\mathcal{S}$.

In order to apply these results to our current situation, given a quasi-triangulation Δ of $F = F_g^s$ based at π, we may formally add (in an essentially unique way) an edge to Δ in F in each punctured component of $F - \Delta$ to produce an ideal triangulation Δ' of F. Thus, on the level of dual fatgraphs, there is one loop of $G = G(\Delta')$ for each puncture of F other than π. These loops are treated "formally" in this section for convenience (and are disposed of entirely in the next section); by this we mean that we shall formally associate a lambda length to each loop of G, where each such lambda length is taken to be arbitrarily large in our calculations and the relative sizes of these formal lambda lengths are otherwise immaterial. (One can think of this locally as a different section of the bundle $\tilde{\mathcal{T}} \to \mathcal{T}$ where we demand that all the horocycles except the one about π have the same very small hyperbolic length.)

Now, given a surface F' of type $F_{g,\vec{\delta}}^s$, define a *decorated hyperbolic structure* on F' to be a complete finite-area Riemannian metric on F' of constant curvature -1 and totally geodesic boundary plus a specification of a horocycle at each cusp on the boundary of F', i.e., double F' along its boundary (so that cusps on the boundary give rise to punctures of the double) and specify a horocycle as usual at each resulting puncture. We shall also consider a *decorated topological structure* on a surface F' to be simply the specification of a single positive real parameter, the hyperbolic length of the corresponding horocycle, one such specification for each cusp on the boundary of F'.

Proposition 6. *Let ∂ denote the arcs comprising the boundary of F'. A decorated topological structure on F' is uniquely determined by the assignment of an arbitrary positive real tuple of lambda lengths, one such lambda length for each arc in ∂. Fixing a quasi-triangulation Δ of F', a decorated hyperbolic structure on F' is uniquely determined by the assignment of an arbitrary positive real tuple of lambda lengths, one lambda length for each arc in $\Delta \cup \partial$.*

Pulling together the results of this section, we find the promised intrinsic description of the ideal geometry.

Theorem 7. *Fix a quasi-triangulation Δ of F_g^s with associated ideal triangulation Δ', let $E_j(t)$ be a degeneration of simplicial coordinates on Δ', and let Δ_f denote the collection of arcs in Δ' so that the corresponding lambda lengths e_i do not diverge. Fixing a horocycle about π, each component of $F - \Delta_f$ comes equipped with a canonical decoration at the cusps on the boundary, so there is a well-defined decorated topological structure on each non simply connected component of $F - \Delta_f$. The (possibly once-punctured) simply connected components of $F - \Delta_f$ combine into connected subsurfaces of F_g^s, and there is a well-defined limiting decorated hyperbolic structure on each such component.*

The point is that we shall simply remove from consideration any arc whose lambda length diverges and recover a well-defined decorated hyperbolic structure on a (possibly empty) subsurface and a well-defined decorated topological structure on the (possibly empty) complement of this geometric part. It is interesting that the data forgotten in Thurston's boundary for $\mathcal{T}$ is exactly the data we record here and conversely; specifically, the limiting lamination (see for instance [Th] or [PH]) in the Thurston theory keeps track of the divergent hyperbolic lengths, and we keep track here of the non-divergent ones.

Corollary 8. *Given a point of $\tilde{A}(\pi)$ and a choice of horocycle at π, there is a naturally corresponding decomposition of $F = F_G \cup F_T$ into (possibly disconnected and possibly empty) subsurfaces with disjoint interiors, where F_G is a geometric surface with decorated hyperbolic structure together with a decomposition into (perhaps once-punctured) polygons, and F_T is a topological surface with decorated topological structure (but no specification of polygonal decomposition). A point is ideal if and only if the topological part is non empty.*

6. Combinatorics and Graph Theory of the Compactification

The combinatorics of our compactification is clear on the level of families of arcs based at a puncture π in a surface $F = F_g^s$. Cells simply correspond to arc families, which may or may not quasi-fill, and the topological part of the decomposition of F, if any, arises from the complementary components which are different from (perhaps once-punctured) polygons.

The graphical formulation is somewhat complicated but interesting nonetheless, so we continue and describe it. Given the quasi-triangulation Δ of F, complete it to an ideal triangulation Δ' as before with corresponding dual marked fatgraph $G' = G(\Delta')$. In contrast to the last section, we here simply remove all the loops to produce a marked fatgraph $G = G(\Delta)$, so G has $s - 1$ external nodes, and we label these external nodes with the respective indices $2, \ldots, s$ of the corresponding punctures (taking the distinguished puncture π to be the first one).

Of course, G contains a characteristic subgraph

$$G' = \{\text{edges } \epsilon \text{ of } G : \exists \text{ simple cycle in } G \text{ through } \epsilon\},$$

which is identified with a fatgraph corresponding to a surface F_g^1, and each component of $G - G'$ is a tree.

One considers Whitehead moves (i.e., contraction/expansion of edges of the fatgraphs) as usual, and now also another elementary move supported on a surface of type $F_{0,(2)}^2$. This new move interchanges the two possible quasi-triangulations of this surface, so on the dual level of fatgraphs, given a puncture i, it just flips the edge containing the terminal node labeled "i" by altering the fattening at its other vertex. In fact, this new elementary move is just the composition of two Whitehead moves on ideal triangulations of $F_{0,(2)}^2$.

We now two-color the edges of G by specifying "regular" and "ghost" edges, where we think of ghost edges as "missing", and there must be at least one regular edge.

Take equivalence classes of these two-colored fatgraphs with labeled terminal nodes under the equivalence relation generated by Whitehead moves and the new elementary move, where we are only allowed to perform moves along the ghost edges. Furthermore, once an edge is a ghost edge, it remains a ghost, and otherwise the color of an edge of G is unaffected. For instance, imagine collapsing a single edge with distinct vertices. Rather than produce a four-valent vertex as usual, we now just put a ghost icon on the edge to be collapsed; collapse and expand it to produce another trivalent fatgraph with a ghost icon on the new edge, and identify these two ghostly fatgraphs.

Theorem 9. *The dual of the simplicial complex $\tilde{A}(\pi)$ is naturally isomorphic to the geometric realization of these (marked) two-colored labeled fatgraph equivalence classes.*

It is implicit in the statement of the theorem that the partial ordering of faces is compatible with the equivalence relation.

It would be interesting to describe a suitable two- or many-matrix model for the combinatorics of this complex, say to calculate its virtual Euler characteristic.

Observe also how naturally this graphical description stabilizes in s, namely, just erase the edge containing the last terminal node.

7. Closing Remarks

First, let us describe Witten's 1989 idea for perturbing (which fit well with our understanding of the compactification at the time) and was intended for the proof of his (then) conjectures about two-dimensional quantum gravity (cf. [Ko]). It is simply that generically in the convex hull construction for the puncture π (which, as we said above, was our original 1985 formulation), every complementary region is either a triangle or a once-punctured monogon. In a component of the latter type containing the puncture p, take the canonical geodesic decomposition into an ideal triangle. This determines a direction near p, and hence a fiberwise tangent or cotangent vector in the universal curve. One might try to use these canonical fiberwise trivializations to perturb cells in a controllable way to allow the intersection theory problems of two-dimensional quantum gravity explicit descriptions as enumerative problems as in the introduction.

Finally, we turn to the continuous mapping of our compactification $\mathcal{A}(\pi)$ of $P\mathcal{M}$ to the (pure) Deligne-Mumford compactification $\overline{P\mathcal{M}}$ of $P\mathcal{M}$, and unfortunately, we must make choices to define this as follows. Choose in each surface $F^s_{g,\vec{\delta}}$ a maximal family of essential non puncture parallel disjointly embedded curves no two of which are homotopic; in fact, only the combinatorial type of the decomposition is relevant, but you might as well make these choices compatibly under some fixed family of inclusions of these surfaces for definiteness. Notice that each such curve family has exactly one component curve for each boundary component of $F^s_{g,\vec{\delta}}$.

Given the decomposition $F = F^s_g = F_T \cup F_G$, the curves in F arising from the boundary of F_T (the only canonical curves in our construction) need not all be distinct, and we formally combine them whenever they do happen to be homotopic. Now, pinch in the sense of Deligne-Mumford along the chosen curves in each component of F_T as well as along the curves arising from the boundary of F_T. This defines a mapping $\mathcal{A}(\pi) \to P\overline{\mathcal{M}}$, and it is evidently continuous.

Notice that the ideal locus is of (real) codimension two in $\mathcal{A}(\pi)$ just as for $\overline{P\mathcal{M}}$. Furthermore, the mapping $\mathcal{A}(\pi) \to P\overline{\mathcal{M}}$ restricts to an isomorphism of the codimension-two cells at infinity in $\mathcal{A}(\pi)$ onto the top-dimensional ideal locus in $\overline{P\mathcal{M}}$ (those points in $P\mathcal{M}$ corresponding to pinching a single curve); that is, $\mathcal{A}(\pi)$ agrees with $\overline{P\mathcal{M}}$ in codimension two.

References

[Ar] V. I. Arnold, *The cohomology ring of the colored braid group*, Math. Notes of Acad. Sci. USSR **5** (1969), 138–140.

[DM] P. Deligne and D. Mumford, *Irreducibility of the space of curves of a given genus*, Publ. IHES **36** (1979), 75–110.

[Fr] M. H. Freedman, *The topology of four-dimensional manifolds*, Jour. Diff. Geom. **17** (1982), 357–453.

[Ha] J. L. Harer, *Stability of the homology of the mapping class group*, Ann. Math. **121** (1985), 215–249.

252   R. C. PENNER

[Iv] N. Ivanov, *Complexes of curves and the Teichmüller modular group*, Russian Math. Surveys **42** (1987), 55–107.

[Ko] M. Kontsevich, *Intersection theory on the moduli space of curves and the matrix Airy function*, Comm. Math. Phys. **147** (1992), 1–23.

[LS] P. Lochak and L. Schneps, *The universal Ptolemy-Teichmüller groupoid*, Proceedings of Luminy (1995), to appear, (*eds.* Pierre Lochak and Leila Schneps).

[P1] R. C. Penner, *The decorated Teichmüller space of punctured surfaces*, Comm. Math. Phys. **113** (1987), 299–339.

[P2] ______ , *Weil-Petersson volumes*, Jour. Diff. Geom. **35** (1992), 559–608.

[P3] ______ , *An arithmetic problem in surface geometry*, The Moduli Space of Curves (1995), Birkhäuser, 427–466, (*eds.* R. Dijgraaf, C. Faber, G. van der Geer).

[P4] ______ , *Perturbative series and the moduli space of Riemann surfaces*, Jour. Diff. Geom. **27** (1988), 35–53.

[PH] ______ with J. L. Harer, *Combinatorics of Train Tracks*, Annals of Math. Studies **125**, Princeton Univ. Press (1992).

[PW] ______ and M. S. Waterman, *Spaces of RNA secondary structures*, Adv. Math. **101** (1993), 31–49.

[Sm] S. Smale, *The generalized Poincaré conjecture in higher dimensions*, Bull. Amer. Math. Soc. **66** (1960), 373–375.

[Th] W. P. Thurston, *Geometry and Topology of Three-Manifolds*, Princeton Univ. Press (to appear).

DEPARTMENTS OF MATHEMATICS AND PHYSICS/ASTRONOMY,, UNIVERSITY OF SOUTHERN CALIFORNIA, LOS ANGELES, CA 90089

E-mail address: rpenner@mtha.usc.edu

Proceedings of
THE 37TH TANIGUCHI SYMPOSIUM ON
TOPOLOGY AND TEICHMÜLLER SPACES
held in Finland, July 1995
ed. by Sadayoshi KOJIMA *et al.*
©1996 World Scientific Publishing Co.
pp. 253–264

CHARACTER VARIETY OF REPRESENTATIONS OF A FINITELY GENERATED GROUP IN SL_2

KYOJI SAITO

(Received January 10, 1996)

This is a partial exposition of [S4, S5, S6], which study the space of representations of a (finitely generated) group Γ in SL_2 and GL_2 in an attempt of its application in geometry: Teichmüller spaces, knot theory, hyperbolic manifolds, moduli spaces, $\cdots$, etc. (see for instance, [A], [Be], [Br], [C-C-G-L-S], [C-S], [F-K], [G], [H], [H-L-M], [H-K], [J-W], [K], [Kj], [Ko], [Kr], [Ma], [Mu], [N-Z], [O], [S], [S-S], [T], [W], [We], [Wo], [Y], $\cdots$, etc). For simplicity, we omit the case for GL_2 in the present exposition. Let us explain the main result of the present paper.

Let Γ be a group. The purpose of the present paper is to introduce the character variety $Ch(\Gamma, SL_2)$ in order to parameterize conjugacy classes of representations of Γ in SL_2 in a functorial way. At first, the character variety is introduced as a scheme over $\mathbb{Z}$, independent of the coefficient ring of representations in question. Then, the scaler field is specialized to $\mathbb{R}$ to obtain results on representations in $SL_2(\mathbb{R})$ and in $SU(2)$ with respect to the classical topology. Let us explain this briefly.

Let $Hom(\Gamma, SL_n)$ be the functor $R \in \{\text{commutative rings with } 1\} \mapsto Hom(\Gamma, SL_n(R)) \in \{\text{sets}\}$. The functor is representable (see §1.3 Lemma) and so, for an abuse of notation, we denote by the same $Hom(\Gamma, SL_n)$ the scheme over $\mathbb{Z}$ representing the functor. The group scheme PGL_n acts on $Hom(\Gamma, SL_n)$. Whether the universal categorical quotient $Hom(\Gamma, SL_n)//PGL_n$ (Mumford [Mu1]) defined over $\mathbb{Z}$ exists or not seems to be a hard and unsolved question. Instead of asking directly for the quotient space, we introduce *i*) the *character variety* $Ch(\Gamma, SL_2)$ together with its *discriminant subvariety* D_Γ as schemes over $\mathbb{Z}$ abstractly, and *ii*) the PGL_2-*invariant morphism* π_Γ: $Hom(\Gamma, SL_2) \to Ch(\Gamma, SL_2)$, for which we prove that *i*) *the restriction of π_Γ on the complement of $\pi_\Gamma^{-1}(D_\Gamma)$ is a principal PGL_2-bundle with respect to the etal topology, and ii*) *the inverse image $\pi_\Gamma^{-1}(D_\Gamma)$ is a subfunctor of* $Hom(\Gamma, SL_2)$ *consisting of abelian or reducible representations.* This implies that the complement $Hom^*(\Gamma, SL_2) := Hom(\Gamma, SL_2) \backslash \pi^{-1}(D_\Gamma)$ consists of absolutely irreducible representations, and that $Hom^*(\Gamma, SL_2)$ has the universal categorical quotient space $Ch^*(\Gamma, SL_2) := Ch(\Gamma, SL_2) \backslash D_\Gamma$ defined over $\mathbb{Z}$. Then the result is specialized to $\mathbb{R}$-coefficient. Namely, *the complement* $\text{Ch}^*(\Gamma, \text{SL}_2)(\mathbb{R}) := \text{Ch}(\Gamma, \text{SL}_2)(\mathbb{R}) \backslash D_\Gamma(\mathbb{R})$ *of the real discriminant decomposes into a disjoint union of two semialgebraic open components* $H_\Gamma(\mathbb{R})$ *and* $T_\Gamma(\mathbb{R})$ *such that i)* $\text{Hom}^*(\Gamma, \text{SL}_2)(\mathbb{R})$ *is a principal* $\text{PGL}_2(\mathbb{R})$ *bundle over* $H_\Gamma(\mathbb{R})$*, and ii)* $\text{Hom}^*(\Gamma, \text{SU}_2(\mathbb{C}))$ *is a principal* $\text{PU}(2)$ *bundle over* $T_\Gamma(\mathbb{R})$*, respectively.* This fact has an application: *the Teichmüller space* $\mathcal{T}_{g,n}$ *carries a natural semi-algebraic structure defined over* $\mathbb{Z}$ [S5].

The §1 prepares notations of a representation variety $\text{Hom}(\Gamma, \text{SL}_2)$ defined over $\mathbb{Z}$. The §2 studies the PGL_2-invariants of $M_2 \times M_2$ as the building block. The universal character ring $R(\Gamma, \text{SL}_2)$ is introduced in §3, and we put $\text{Ch}(\Gamma, \text{SL}_2) := \text{Spec}(R(\Gamma, \text{SL}_2))$. The principal PGL_2-bundle structure on $\text{Hom}^*(\Gamma, \text{SL}_2)$ over $\text{Ch}^*(\Gamma, SL_2)$ with respect to the etal topology is formulated in §4 Theorem A, and that with respect to the classical topology is formulated in §5 Theorem C.

§1. Universal representation of a group Γ in SL_n

This section is devoted for a preparation of notion and terminology for the representation varieties. One is referred to [Pr1, Pr2] [L-M] etc.

1.1. Let Γ be a group. As in the introduction, for a fixed $n \in \mathbb{Z}_{>0}$, $\text{Hom}(\Gamma, \text{SL}_n)$ is the functor: $R \in \{\text{commutative rings with } 1\} \mapsto \text{Hom}(\Gamma, \text{SL}_n(R)) \in \{\text{sets}\}$. In order to fix the notation for the present paper, we state precisely the representability of the functor in the next lemma.

Lemma (the representability of $\text{Hom}(\Gamma, \text{SL}_n)$).

(1) *For the given* Γ *and* $n \geq 1$*, there exists a pair* $(A(\Gamma, \text{SL}_n), \sigma)$ *of a commutative ring* $A(\Gamma, \text{SL}_n)$ *with 1 and representation* $\sigma \colon \Gamma \to \text{SL}_n(A(\Gamma, \text{SL}_n))$ *such that for any commutative ring* R *with 1, the correspondence:*

$$(1.1.1) \qquad \varphi \in \text{Hom}^{ring}(A(\Gamma, \text{SL}_n), R) \mapsto \varphi \circ \sigma \in \text{Hom}^{gr}(\Gamma, \text{SL}_n(R))$$

is a bijection.

(2) *The pair* $(A(\Gamma, \text{SL}_n), \sigma)$ *is unique up to an isomorphism of the ring* $A(\Gamma, \text{SL}_n)$ *which commutes with the universal representation* σ*.*

(3) *If* Γ *is a finitely generated group, then* $A(\Gamma, \text{SL}_n)$ *is a finitely generated ring over* $\mathbb{Z}$ *and hence is noetherian.*

The lemma is proven by routine arguments. Here we give an explicit description of $(A(\Gamma, \text{SL}_n), \sigma)$ without a proof. For each $\gamma \in \Gamma$, consider a $n \times n$ matrix:

$$(1.1.2) \qquad\qquad \sigma(\gamma) := (a_{ij}(\gamma))_{i,j=1,\cdots,n}$$

Then $A(\Gamma, \text{SL}_n)$, called the *universal representation algebra*, is generated by the entries $a_{ij}(\Gamma)$ $i, j = 1, \cdots, n$ $\gamma \in \Gamma$, and divided by the ideal generated by all entries of the matrices $\sigma(e) - I_n$ (I_n = the $n \times n$ unit matrix) and $\sigma(\gamma\delta) - \sigma(\gamma)\sigma(\delta)$ and $\det(\sigma(\gamma)) - 1$.

$$(1.1.3) \qquad\qquad A(\Gamma, \text{SL}_n) := \mathbb{Z}[a_{ij}(\gamma) \text{ for } \gamma \in \Gamma \text{ and } 1 \leq i, j \leq n]/I,$$

where

$$I := \Big(a_{ij}(e) - \delta_{ij},\ a_{ij}(\gamma\delta) - \sum_k a_{ik}(\gamma)a_{kj}(\delta),\ \det(\sigma(\gamma)) - 1$$

$$\text{for } 1 \leq i,j \leq n \text{ and } \gamma, \delta \in \Gamma \Big).$$

By definition, the map $\sigma \colon \gamma \in \Gamma \to \sigma(\gamma) \in \mathrm{SL}_n(A(\Gamma, \mathrm{SL}_n))$ is a representation of Γ, which is called the *universal representation* of Γ in SL_n.

1.2. Let PGL_n be the group scheme ([SGAIII]), whose coordinate ring $A(\mathrm{PGL}_n)$ is given by the subring $A_0(\mathrm{GL}_n)$ of the coordinate ring $A(\mathrm{GL}_n) := \mathbb{Z}[x_{ij}\ 1 \leq i, j \leq n]_{\det(X)}$ (here $X := (x_{ij})_{i,j=1}^n$) of GL_n consisting of homogeneous elements of degree 0. The adjoint action of PGL_n on $\mathrm{Hom}(\Gamma, \mathrm{SL}_n)$ is written in terms of its dual action:

$$(1.2.1) \qquad\qquad \mathrm{Ad} \colon A(\Gamma, \mathrm{SL}_n) \to A(\Gamma, \mathrm{SL}_n) \otimes_{\mathbb{Z}} A(\mathrm{PGL}_n),$$

sending an entry of $\sigma(\gamma)$ to the same entry of $X^{-1}\sigma(\gamma)X$. Obviously, the coefficients of characteristic polynomial of $\sigma(\gamma)$ for $\gamma \in \Gamma$ are invariants under the adjoint action.

1.3. From now on, we switch to the case $n = 2$. Let us list up some relations among the PGL_2-invariants $\mathrm{tr}(\sigma(\gamma))$ for $\gamma \in \Gamma$. The first one is trivial:

$$(1.3.1) \qquad\qquad\qquad \mathrm{tr}(\sigma(e)) = 2.$$

The next one follows from the Cayley-Hamilton relation: $\sigma(\gamma)^2 + \det(\sigma(\gamma)) \cdot I_2 = \mathrm{tr}(\sigma(\gamma)) \cdot \sigma(\gamma)$. Multiply $\sigma(\gamma^{-1}\delta)$ from right and take traces. So we obtain:

$$(1.3.2) \qquad \mathrm{tr}(\sigma(\gamma\delta)) + \mathrm{tr}(\sigma(\gamma^{-1}\delta)) = \mathrm{tr}(\sigma(\gamma)) \cdot \mathrm{tr}(\sigma(\delta))$$

for γ and $\delta \in \Gamma$ (cf. [F-K, formulas (2), pp.338]).

§2. PGL_2-invariants for pairs of 2×2 matrices

We study the invariants of the diagonal adjoint action of PGL_2 on the space $M_2 \times M_2$ of pair (A, B) of 2×2 matrices. The morphism $\tilde{\pi} \colon M_2 \times M_2 \to \mathbf{A}^5$ given by $(A, B) \mapsto (\mathrm{tr}(A), \mathrm{tr}(B), \mathrm{tr}(AB), \det(A), \det(B))$ is shown to be the universal quotient map (§2.3 Lemma A,B). The discriminant Δ for $\tilde{\pi}$ is introduced (§2.4 Lemma C) so that the $\tilde{\pi}$ is a principal PGL_2-bundle on the complement of the discriminant loci $\Delta = 0$ (§2.5 Lemma D). These are building blocks of character variety defined over $\mathbb{Z}$.

2.1. Let $M_2 \times M_2$ be the space of pairs of 2×2 matrices. The $X \in \mathrm{PGL}_2$ acts on $(A, B) \in M_2 \times M_2$ from the right diagonally by letting $(A, B) \cdot \mathrm{Ad}(X) := (X^{-1}AX, X^{-1}BX)$. So we have the dual action on the coordinate ring:

$$(2.1.1) \qquad\qquad \mathrm{Ad} \colon \mathbb{Z}[M_2 \times M_2] \to \mathbb{Z}[M_2 \times M_2] \otimes_{\mathbb{Z}} A(\mathrm{PGL}_2)$$

sending an entry of (A, B) to the corresponding entry of $(A, B) \cdot \mathrm{Ad}(X)$ where $\mathbb{Z}[M_2 \times M_2]$ is the polynomial ring generated by 8 entries of (A, B).

2.2. Let us consider the morphism

$$(2.2.1) \qquad\qquad \tilde{\pi}\colon M_2 \times M_2 \to \mathbf{A}^5 := \mathrm{Spec}(\mathbb{Z}[\underline{T}, \underline{D}]),$$

where $\mathbb{Z}[\underline{T}, \underline{D}]$ denotes the polynomial ring $\mathbb{Z}[T_1, T_2, T_3, D_1, D_2]$ of the 5 indeterminates and $\tilde{\pi}$ is associated to the ring homomorphism:

$$(2.2.2) \qquad\qquad \iota\colon \mathbb{Z}[\underline{T}, \underline{D}] \to \mathbb{Z}[M_2 \times M_2],$$

given by $\iota(T_1) := \mathrm{tr}(A)$, $\iota(T_2) := \mathrm{tr}(B)$, $\iota(T_3) := \mathrm{tr}(AB)$, $\iota(D_1) := \det(A)$ and $\iota(D_2) := \det(B)$. The morphism $\tilde{\pi}$ is $\mathbf{G}_m$-equivariant and PGL_2-invariant, since ι is homogeneous w.r.t the weights $\deg(T_1) = \deg(T_2) = 1$ and $\deg(T_3) = \deg(D_1) = \deg(D_2) = 2$, and the $\mathrm{Image}(\iota)$ is fixed by the PGL_2-action pointwisely.

2.3. The next lemma is easily proven by a use of Euclid division algorithms.

Lemma A. $\mathbb{Z}[M_2 \times M_2]$ *is a free module over* $\mathbb{Z}[\underline{T}, \underline{D}]$.

As a consequence of the Lemma A, $\mathbb{Z}[M_2 \times M_2]$ *is faithfully flat over* $\mathbb{Z}[\underline{T}, \underline{D}]$. In particular, the map ι (2.2.2) is injective, and we regard $\mathbb{Z}[\underline{T}, \underline{D}]$ as a subring of $\mathbb{Z}[M_2 \times M_2]$. More strongly, the next lemma says that it is the universal invariant subring with respect to the PGL-action.

Lemma B. *Let R be any $\mathbb{Z}[\underline{T}, \underline{D}]$-algebra with 1. Then*

$$(2.3.1) \qquad\qquad R \simeq \left(R \otimes_{\mathbb{Z}[\underline{T},\underline{D}]} \mathbb{Z}[M_2 \times M_2] \right)^{\mathrm{PGL}_2}.$$

Here, the PGL_2 action on $\mathbb{Z}[M_2 \times M_2]$ is extended to the tensor product by letting PGL_2 act trivially on R.

Remark. (1) It is classically well known that $\mathbb{Q}[M_2 \times M_2]^{\mathrm{PGL}_2}$ is generated by traces $\mathrm{tr}(W)$ for $W \in \{$ the monoid generated by the A and $B\}$ ([G-Y], [Pr1, Pr2]). This is not true for the $\mathbb{Z}$-coefficient. For instance, the relation $2\det(A) = \mathrm{tr}(A)^2 - \mathrm{tr}(A^2)$ implies the algebraic dependence of $\mathrm{tr}(A^2)$ and $\mathrm{tr}(A)$ if char $= 2$, whereas $\det(A)$ and $\mathrm{tr}(A)$ are universally algebraically independent as shown in Lemma B.

(2) Donkin [D] has shown that $\mathbb{Z}[M_n \times \cdots \times M_n]^{\mathrm{PGL}_n}$ is generated by $\mathrm{tr}(\overset{i}{\wedge} W)$ for $i = 1, \cdots, n$ and $W \in \{$ the monoid generated by $A_1, \cdots, A_m\}$, where we denote by $M_n \times \cdots \times M_n$ the space of m-tuple $n \times n$ matrices $(A_1, \cdots, A_m)$.

2.4. For a matrix $A = \begin{bmatrix} a & b \\ c & d \end{bmatrix}$, the adjoint matrix A^* is the matrix $\begin{bmatrix} d & -b \\ -c & a \end{bmatrix}$ so that $A^*A = AA^* = \det(A)I_2$ and $A + A^* = \mathrm{tr}(A)I_2$.

Definition. The *discriminant* for $\tilde{\pi}$ is the polynomial in $\mathbb{Z}[M_2 \times M_2]$ given by:

$$(2.4.1) \qquad \Delta(A, B) := \mathrm{tr}(ABA^*B^*) - \mathrm{tr}(AA^*BB^*)$$
$$= T_1^2 D_2 + T_2^2 D_1 + T_3^2 - T_1 T_2 T_3 - 4 D_1 D_2.$$

A justification for this name is given in the next Lemma C. Let J be the ideal of $\mathbb{Z}[M_2 \times M_2]$ generated by all 5×5 minors of the Jacobian matrix of $\tilde{\pi}$:

$$\frac{\partial(T_1, T_2, T_3, D_1, D_2)}{\partial(a, b, c, d, e, f, g, h)} = \begin{bmatrix} 1 & 0 & 0 & 1 & 0 & 0 & 0 & 0 \\ 0 & 0 & 0 & 0 & 1 & 0 & 0 & 1 \\ e & g & f & h & a & c & b & d \\ d & -c & -b & a & 0 & 0 & 0 & 0 \\ 0 & 0 & 0 & 0 & h & -g & -f & e \end{bmatrix}.$$

Lemma C. *The intersection of the Jacobian ideal J with the invariant subring $\mathbb{Z}[\underline{T}, \underline{D}]$ is a principal ideal generated by the discriminant.*

$$(2.4.2) \qquad\qquad J \cap \mathbb{Z}[\underline{T}, \underline{D}] = (\Delta).$$

2.5. Let us denote by D_Δ the divisor in $\mathbf{A}^5 = \mathrm{Spec}(\mathbb{Z}[\underline{T}, \underline{D}])$ defined by the principal ideal (Δ). Owing to the (2.4.2), the Jacobian criterion implies that the restriction of the morphism $\tilde{\pi}$ (2.2.1) on the complement of $\tilde{\pi}^{-1}(D_\Delta)$

$$(2.5.1) \qquad\qquad \tilde{\pi}_\Delta \colon M_2 \times M_2 \setminus \pi^{-1}(D_\Delta) \to \mathbf{A}^5 \setminus D_\Delta$$

is smooth. More strongly, we show in the next lemma, to which the proof of main theorem A in §4 is reduced.

Lemma D. *The morphism (2.5.1) is a principal PGL_2-bundle with respect to the etal topology.*

§3.   The universal character ring $R(\Gamma, \mathrm{SL}_2)$

The universal character ring $R(\Gamma, \mathrm{SL}_2)$ is introduced in this § by generators and relations in terms of Γ. The goal of this § is the formula (3.4.3), which is used in the proof of the main theorem B in §4 essentially.

3.1.

Definition. The *universal character ring $R(\Gamma, \mathrm{SL}_2)$ of representations of Γ in SL_2 is* generated by the indeterminates $s(\gamma)$ for $\gamma \in \Gamma$ and divided by the ideal generated by $s(e) - 2$ ($e = $ the unit of Γ) and by $s(\gamma)s(\delta) - s(\gamma\delta) - s(\gamma^{-1}\delta)$ for all $\gamma, \delta \in \Gamma$.

$$(3.1.1) \quad R(\Gamma, \mathrm{SL}_2) := \mathbb{Z}[s(\gamma), \gamma \in \Gamma] \big/ \big(s(e) - 2, s(\gamma)s(\delta) - s(\gamma\delta) - s(\gamma^{-1}\delta) \big).$$

3.2.   The ring $R(\Gamma, \mathrm{SL}_2)$ is finitely generated over $\mathbb{Z}$, if Γ is finitely generated as shown in the next lemma. Such finiteness was asserted for the ring of traces of SL_2 by Fricke [F-K] and proven in [H1, H2, (2.f)], [Ho1, Ho2, Theorem 3.1] and [C-S]. We formulate the finiteness in terms of the universal character ring.

Proposition. *Let A be a linearly ordered subset of Γ, which generates Γ. Then $R(\Gamma, \mathrm{SL}_2)$ is generated by $G := \bigcup_{m \in \mathbb{N}} \{ s(\alpha_1 \cdots \alpha_m) | \alpha_i \in A, \alpha_1 < \cdots < \alpha_m \}$ over $\mathbb{Z}$.*

Corollary 1. *If the group Γ is finitely generated, then the $R(\Gamma, \mathrm{SL}_2)$ is a finitely generated ring over $\mathbb{Z}$. Hence it is noetherian.*

3.3. For α and $\beta \in \Gamma$, define the *discriminant* $\Delta(\alpha, \beta) \in R(\Gamma, \mathrm{SL}_2)$ by

$$(3.3.1) \qquad \Delta(\alpha, \beta) := s(\alpha\beta\alpha^{-1}\beta^{-1}) - 2$$
$$= s(\alpha)^2 + s(\beta)^2 + s(\alpha\beta)^2 - s(\alpha)s(\beta)s(\alpha\beta) - 4.$$

This definition of Δ is parallel to that in (2.4.1). In fact, we shall confuse them in a proof of main theorem B in §4. The polynomial has been studied extensively by R. Fricke [F] and others.

3.4. An importance of the discriminant is explained in the next lemma, which plays an essential role in a proof of the main theorem B in §4.

Definition. Let M be a $R(\Gamma, \mathrm{SL}_2)$-module. A map $h \colon \Gamma \to M$ is called a *form* with values in M, if for any γ and $\delta \in \Gamma$ one has a relation:

$$(3.4.1) \qquad h(\gamma\delta) + h(\gamma^{-1}\delta) = s(\gamma)h(\delta).$$

Lemma. *Let h be a form with values in M. Suppose $h(e) = h(\alpha) = h(\beta) = h(\alpha\beta) = 0$ for some α and $\beta \in \Gamma$. Then for any $\gamma \in \Gamma$ one has*

$$(3.4.2) \qquad \Delta(\alpha, \beta)h(\gamma) = 0.$$

Therefore, the images of the values of h by the localization $M \to M_{\Delta(\alpha,\beta)}$ are zero.

Corollary. *For any α, β, γ and $\delta \in \Gamma$, one has the bilinear expression of $s(\gamma\delta)$*

$$(3.4.3)$$
$$s(\gamma\delta) = -\frac{1}{\Delta(\alpha\beta)}(s(\gamma), s(\gamma\alpha), s(\gamma\beta), s(\gamma\alpha\beta)) \cdot T \cdot {}^t(s(\delta), s(\alpha\delta), s(\beta\delta), s(\alpha\beta\delta))$$

in the localization $R(\Gamma, \mathrm{SL}_2)_{\Delta(\alpha,\beta)}$, where $T \in M_4(R(\Gamma, \mathrm{SL}_2))$ is a 4×4 matrix such that $T \cdot (s(\xi_i\xi_j))_{i,j=1}^4 = -\Delta(\alpha, \beta)I_4$ for $\xi_1 = e, \xi_2 = \alpha, \xi_3 = \beta$ and $\xi_4 = \alpha\beta$.

This corollary is the key step to obtain the universality of the character ring, since it implies that the system $(s(\gamma), \gamma \in \Gamma)$ satisfies any algebraic relations which is satisfied by the system $(\mathrm{tr}(\sigma(\gamma)), \gamma \in \Gamma)$ of characters, so far as $\Delta(\alpha, \beta)$ is invertible for some $\alpha, \beta \in \Gamma$.

3.5.

Remark. The study of the algebra of traces (=characters) of representations of a group Γ into SL_2 started by Voigt and Fricke and is developed by many authors Helling, Horowitz, Magnus, Bass, Lubotzky, Procesi, Platonov and others. See references of the quoted papers.

· Let F_n be a free group generated by n elements. Magnus called the homomorphic image of the ring $R(F_n, \mathrm{SL}_2)$ in the ring of functions on the representation space $\mathrm{Hom}(F_n, \mathrm{SL}_2(\mathbb{C}))$ the ring of Fricke characters [Ma]. We do not know whether the homomorphism has non trivial kernel or not.

§4. The invariant morphism π_Γ

In the present §4.1, we introduce the character variety $\mathrm{Ch}(\Gamma, \mathrm{SL}_2)$, the PGL_2-invariant morphism $\pi_\Gamma\colon \mathrm{Hom}(\Gamma, \mathrm{SL}_2) \to \mathrm{Ch}(\Gamma, \mathrm{SL}_2)$ and the discriminant D_Γ. The inverse image $\pi_\Gamma^{-1}(D_\Gamma)$ consist of the representations ρ such that $\rho(\Gamma)$ is either abelian or reducible (§4.2 Assertion). Then the first main results of the present paper are formulated in 4.3 theorem A and 4.4 theorem B, which say that the complement $\mathrm{Ch}^\bullet(\Gamma, \mathrm{SL}_2) := \mathrm{Ch}(\Gamma, \mathrm{SL}_2) \backslash D_\Gamma$ of the discriminant is the regular orbit space of PGL_2-action on absolutely irreducible representations $\mathrm{Hom}^\bullet(\Gamma, \mathrm{SL}_2) := \mathrm{Hom}(\Gamma, \mathrm{SL}_2) \backslash \pi^{-1}(D_\Gamma)$.

4.1. Let Γ be a group. Put

$$(4.1.1) \qquad\qquad \mathrm{Ch}(\Gamma, \mathrm{SL}_2) := \mathrm{Spec}(R(\Gamma, \mathrm{SL}_2)),$$

where $R(\Gamma, \mathrm{SL}_2)$ is defined in §3.1. Recall the fact that $\mathrm{Hom}(\Gamma, \mathrm{SL}_2)$ is identified with the affine scheme for the universal representation ring $A(\Gamma, \mathrm{SL}_2)$ (cf. §1.3 Lemma). Then the invariant morphism

$$(4.1.2) \qquad\qquad \pi_\Gamma\colon \mathrm{Hom}(\Gamma, \mathrm{SL}_2) \to \mathrm{Ch}(\Gamma, \mathrm{SL}_2)$$

is defined through the ring homomorphism:

$$(4.1.3) \qquad\qquad \begin{array}{rccc} \Phi\colon & R(\Gamma, \mathrm{SL}_2) & \to & A(\Gamma, \mathrm{SL}_2) \\ & s(\gamma) & \mapsto & \mathrm{tr}(\sigma(\gamma)) \end{array} .$$

Comparing relations (3.1.1) with (1.3.1) and (1.3.2), Φ is well defined. The *discriminant subvariety* in $\mathrm{Ch}(\Gamma, \mathrm{SL}_2)$ of the morphism π_Γ is defined as

$$(4.1.4) \qquad\qquad D_\Gamma := \bigcap_{\alpha,\beta\in\Gamma} V(\Delta(\alpha,\beta))$$

where the discriminant $\Delta(\alpha, \beta) \in R(\Gamma, \mathrm{SL}_2)$ is defined in (3.3.1).

4.2. First, in the next lemma we characterize the inverse image $\pi_\Gamma^{-1}(D_\Gamma)$ of the discriminant loci in terms of representation of Γ.

Assertion. *Let $p \in \mathrm{Spec}(A(\Gamma, \mathrm{SL}_2))$. Then p belongs to $\pi_\Gamma^{-1}(D_\Gamma)$, if and only if the image $\sigma_p(\Gamma)$ in $\mathrm{SL}_2(k_p)$ of the group Γ is either abelian or reducible, where k_p is the fraction field of the integral domain $A(\Gamma, \mathrm{SL}_2)/p$ and $\sigma_p\colon \Gamma \to \mathrm{SL}_2(k_p)$ is a representation obtained by a specialization of the universal σ at p.*

4.3. Let us state the first main result of the present paper.

Theorem A. *The restriction of the invariant morphism π_Γ to the complement of the inverse image $\pi_\Gamma^{-1}(D_\Gamma)$ of the discriminant is a principal PGL_2-bundle with respect to the etal topology defined over $\mathbb{Z}$.*

Proof. By definition of D_Γ, one has an affine open covering $\mathrm{Ch}(\Gamma, \mathrm{SL}_2) \backslash D_\Gamma = \bigcup_{\alpha,\beta\in\Gamma} \mathrm{Spec}(R(\Gamma, \mathrm{SL}_2)_{\Delta(\alpha,\beta)})$. Then the proof is reduced to each affine open piece, stated as a consequence of the next Theorem B. $\square$

4.4. For any fixed pair α and β of Γ, consider the PGL_2-equivariant morphism $\mathrm{Hom}(\Gamma, \mathrm{SL}_2) \to M_2 \times M_2$ and a morphism $h_{\alpha\beta} \colon \mathrm{Ch}(\Gamma, \mathrm{SL}_2) \to \mathbf{A}^5 = \mathrm{Spec}(\mathbb{Z}[\underline{T}, \underline{D}])$ defined by the coordinate ring homomorphisms given by

$$(4.4.1) \qquad A \mapsto \sigma(\alpha) \text{ and } B \mapsto \sigma(\beta),$$

$$(4.4.2) \qquad T_1 \mapsto s(\alpha), T_2 \mapsto s(\beta), T_3 \mapsto s(\alpha\beta), D_1 \mapsto 1 \text{ and } D_2 \mapsto 1.$$

Then the next diagram becomes commutative

$$(4.4.3) \qquad
\begin{array}{ccc}
\mathrm{Hom}(\Gamma, \mathrm{SL}_2) & \longrightarrow & M_2 \times M_2 \\
\downarrow{\scriptstyle \pi_\Gamma} & & \downarrow{\scriptstyle \tilde{\pi}} \\
\mathrm{Ch}(\Gamma, \mathrm{SL}_2) & \longrightarrow & \mathbf{A}^5 = \mathrm{Spec}(\mathbb{Z}[\underline{T}, \underline{D}])
\end{array}$$

Here we remark that the discriminant $\Delta(\alpha, \beta)$ (3.3.1) is the pull back of $\Delta(A, B)$ (2.4.1). For an abuse of notation, we shall denote both of them by Δ.

Theorem B. *The diagram (4.4.3) is Cartesian on the complement of the loci $\Delta = 0$. That is: the localization by Δ of the homomorphism*

$$(4.4.4) \qquad \Psi_{\alpha,\beta} \colon R(\Gamma, \mathrm{SL}_2) \otimes_{\mathbb{Z}[\underline{T},\underline{D}]} \mathbb{Z}[M_2 \times M_2] \to A(\Gamma, \mathrm{SL}_2)$$

(obtained from (4.1.3), (4.4.1) and (4.4.2)) is an isomorphism.

Theorem A follows from the theorem B applied with §2.5 Lemma D. The theorem B is proved by a construction of the inverse homomorphism for the localization $(\Psi_{\alpha,\beta})_\Delta$ of (4.4.4) by Δ. This is equivalent to the *construction of a representation σ^* for a prescribed system of "characters" $s(\gamma)$ $\gamma \in \Gamma$ and a pair of matrix $(A, B) \in M_2 \times M_2$ (over the same point of $\mathbf{A}^5 \setminus \{\Delta = 0\}$) such that $\mathrm{tr}(\sigma^*(\gamma)) = s(\gamma)$ and $(\sigma^*(\alpha), \sigma^*(\beta)) = (A, B)$.* Actually, this is achieved by the formula:

$$\sigma^*(\gamma) := -\frac{1}{\Delta(\alpha\beta)}(I_2, A, B, AB) \cdot T \cdot {}^t(s(\gamma), s(\alpha\gamma), s(\beta\gamma), s(\alpha\beta\gamma)),$$

where T is the same matrix in $M_4(R(\Gamma, \mathrm{SL}_2))$ as in (3.4.3). The multiplicativity $\sigma^*(\gamma\delta) = \sigma^*(\gamma)\sigma^*(\delta)$ is shown by an essential use of the formula (3.4.3).

§5. Representation variety with real coefficients

We specialize the result in the previous § to $\mathbb{R}$-coefficient case. The second main result of the present paper is formulated in 5.5 theorem C, which says that the complement $\mathrm{Ch}(\Gamma, \mathrm{SL}_2)(\mathbb{R})$ of real discriminant $D_\Gamma(\mathbb{R})$ decomposes into two semialgebraic sets, which are regular orbit spaces of $\mathrm{PGL}_2(\mathbb{R})$ action on $\mathrm{Hom}^*(\Gamma, \mathrm{SL}_2(\mathbb{R}))$ and $\mathrm{PU}(2)$ action on $\mathrm{Hom}^*(\Gamma, \mathrm{SU}_2(\mathbb{C}))$, respectively. For the purpose we analyze the discriminant Δ over $\mathbb{R}$ (§5.3 Lemma E). See also [K1, K2, K3, K4], [G1], [H1] and [Ko] for the geometry of the discriminant.

5.1. Consider the invariant map $\tilde{\pi}$ (2.2.1) over the real number field $\mathbb{R}$.

$$(5.1.1) \qquad \tilde{\pi}(\mathbb{R}) \colon M_2(\mathbb{R}) \times M_2(\mathbb{R}) \to \mathbf{A}^5(\mathbb{R}) := \mathrm{Hom}(\mathbb{Z}[\underline{T}, \underline{D}], \mathbb{R})$$

given by $\tilde{\pi}(A, B) := (\mathrm{tr}(A), \mathrm{tr}(B), \mathrm{tr}(AB), \det(A), \det(B))$. Consider also the real loci of the discriminant Δ (§2.4).

5.2. Let us introduce an open semialgebraic subset of $\mathbf{A}^5(\mathbb{R})$:

$$(5.2.1) \qquad \widetilde{D_\Delta}(\mathbb{R}) := \left\{ \varphi \in \mathbf{A}^5(\mathbb{R}) : \Delta(\varphi) := \varphi(\Delta) = 0 \right\}.$$

where

$$(5.2.2) \qquad \delta_1 = T_1^2 - 4D_1,\ \delta_2 = T_2^2 - 4D_2 \text{ and } \delta_3 = T_3^2 - 4D_1D_2.$$

are the discriminants for the characteristic polynomials of A, B and AB respectively. The following facts are shown by direct elementary calculations.

Lemma E. *i)* $\widetilde{T_\Delta}$ *is a convex connected component of* $\mathbf{A}^5(\mathbb{R}) \setminus \widetilde{D_\Delta}$.
 ii) $\mathbf{A}^5(\mathbb{R}) \setminus \widetilde{T_\Delta} = \tilde{\pi}(\mathbb{R})\,(M_2(\mathbb{R}) \times M_2(\mathbb{R}))$.

Let us decompose the space $\mathbf{A}^5(\mathbb{R})$ into semialgebraic sets:

$$(5.2.3) \qquad \mathbf{A}^5(\mathbb{R}) = \widetilde{D_\Delta} \amalg \widetilde{H_\Delta} \amalg \widetilde{T_\Delta},$$

where $\widetilde{H_\Delta} := \mathbf{A}(\mathbb{R}) \setminus (\widetilde{D_\Delta} \amalg \widetilde{T_\Delta})$ is a union of connected components of $\mathbf{A}(\mathbb{R}) \setminus \widetilde{D_\Delta}$. Then the lemma E E can be paraphrased as: $\mathrm{Image}(\tilde{\pi}(\mathbb{R})) = \widetilde{D_\Delta} \amalg \widetilde{H_\Delta}$.

5.3. For any given $\alpha, \beta \in \Gamma$, consider the homomorphism $F_2 \to \Gamma$ associating the two generators of F_2 to α and β. This induces the diagram (4.4.3) over $\mathbb{R}$:

$$(5.3.1)$$

$$\begin{array}{ccc}
\mathrm{Hom}(\Gamma, \mathrm{SL}_2(\mathbb{R})) & \longrightarrow & \mathrm{Hom}(F_2, \mathrm{SL}_2(\mathbb{R})) \subset M_2(\mathbb{R}) \times M_2(\mathbb{R}) \\
\big\downarrow{\scriptstyle \pi_\Gamma(\mathbb{R})} & & \big\downarrow{\scriptstyle \tilde{\pi}(\mathbb{R})} \\
\mathrm{Ch}(\Gamma, \mathrm{SL}_2)(\mathbb{R}) & \xrightarrow[h_{\alpha,\beta}]{} & \mathbf{A}^5(\mathbb{R})
\end{array}$$

where the morphism $h_{\alpha,\beta}$ is defined by (4.4.2) and $\mathrm{Ch}(\Gamma, \mathrm{SL}_2)(\mathbb{R}) = \mathrm{Hom}(R(\Gamma, \mathrm{SL}_2), \mathbb{R})$ is the real character variety. The fact that the diagram (5.3.1) is Cartesian outside of the loci $\Delta = 0$ (§4.4 Theorem B) together with Lemma E implies the following disjoint decomposition:

$$(5.3.2) \qquad \mathrm{Ch}(\Gamma, \mathrm{SL}_2)(\mathbb{R}) = D_\Gamma(\mathbb{R}) \amalg H_\Gamma(\mathbb{R}) \amalg T_\Gamma(\mathbb{R}),$$

where

$$(5.3.3)$$

$$D_\Gamma(\mathbb{R}) := \left\{ t \in \mathrm{Hom}(R(\Gamma, \mathrm{SL}_2), \mathbb{R}) \mid h_{\alpha,\beta}(t) \in D_\Delta(\mathbb{R}) \text{ for } {}^\vee\alpha, \beta \in \Gamma \right\}$$

$$H_\Gamma(\mathbb{R}) := \left\{ t \in \mathrm{Hom}(R(\Gamma, \mathrm{SL}_2), \mathbb{R}) \mid {}^\vee\alpha, \beta \in \Gamma \text{ such that } h_{\alpha,\beta}(t) \in \widetilde{H_\Delta} \right\}$$

$$T_\Gamma(\mathbb{R}) := \left\{ t \in \mathrm{Hom}(R(\Gamma, \mathrm{SL}_2), \mathbb{R}) \mid {}^\vee\alpha, \beta \in \Gamma \text{ such that } h_{\alpha,\beta}(t) \in \widetilde{T_\Delta} \right\}.$$

By definition, $D_\Gamma(\mathbb{R})$ is Zariski closed. $H_\Gamma(\mathbb{R})$ and $T_\Gamma(\mathbb{R})$ are open semialgebraic (Due to the basis theorem of Hilbert, one can find a finite system $\{(\alpha_i, \beta_i)\}_{i \in I}$ such that $D_\Gamma(\mathbb{R}) = \bigcap_{i \in I} \{\Delta(\alpha_i, \beta_i) = 0\}$. Then for the same index set I, one can show the equalities: $H_\Gamma(\mathbb{R}) = \bigcup_{i \in I} h^{-1}_{\alpha_i,\beta_i}(\widetilde{H_\Delta})$, $T_\Gamma(\mathbb{R}) = \bigcup_{i \in I} h^{-1}_{\alpha_i,\beta_i}(\widetilde{T_\Delta}))$.

5.4. Let us determine the image set of $\pi_\Gamma(\mathbb{R})$ as a semialgebraic set (up to the discriminant) in the next lemma. It is proven by a use of the fact that (5.3.1) is Cartesian (§4.4 Theorem B) together with the lemma E.

Lemma F. *Let $\pi_\Gamma(\mathbb{R})$ be the invariant morphism (4.1.2) defined over $\mathbb{R}$. Then*

$$(5.4.1) \qquad \text{Image}(\pi_\Gamma(\mathbb{R})) \setminus D_\Gamma(\mathbb{R}) = H_\Gamma(\mathbb{R}).$$

A meaning of the set $T_\Gamma(\mathbb{R})$ is given by the next lemma. Since $\text{SU}(2) \subset \text{SL}_2(\mathbb{C})$, the restriction of $\pi_\Gamma(\mathbb{C})$ induces the following map:

$$(5.4.2) \qquad \begin{array}{cccc} u_\Gamma & \text{Hom}(\Gamma, \text{SU}(2)) & \to & \text{Ch}(\Gamma, \text{SL}_2)(\mathbb{R}). \\ & \rho & \mapsto & s(\gamma) \mapsto \text{tr}(\rho(\gamma)) \end{array}$$

Lemma G. *The image set of the morphism u_Γ is given by*

$$(5.4.3) \qquad \text{Image}(u_\Gamma) \setminus D_\Gamma(\mathbb{R}) = T_\Gamma(\mathbb{R}).$$

5.5. This is the goal of the present paper. Combining the above (5.4.1), (5.4.2) with the §4 Theorem A, we obtain principal bundles by the natural adjoint action with respect to the classical topology.

Theorem C. *The restrictions of the maps $\pi_\Gamma(\mathbb{R})$ (4.1.2) and u_Γ (5.4.2) to the subset of absolutely irreducible representations $\text{Hom}^*(\Gamma, \text{SL}_2(\mathbb{R})) := \text{Hom}(\Gamma, \text{SL}_2(\mathbb{R})) \setminus \pi_\Gamma^{-1}(D_\Gamma(\mathbb{R}))$ and $\text{Hom}^*(\Gamma, \text{SU}(2)) := \text{Hom}(\Gamma, \text{SU}(2)) \setminus u_\Gamma^{-1}(D_\Gamma(\mathbb{R}))$ are a principal $\text{PGL}_2(\mathbb{R})$-bundle and a principal $\text{U}(2)/\text{U}(1)$-bundle over the open semialgebraic sets $H_\Gamma(\mathbb{R})$ and $T_\Gamma(\mathbb{R})$ (5.3.2) of the real character variety $\text{Ch}(\Gamma, \text{SL}_2)(\mathbb{R})$ respectively.*

$$\pi_\Gamma(\mathbb{R}) \colon \text{Hom}^*(\Gamma, \text{SL}_2(\mathbb{R})) \xrightarrow{\text{PGL}_2(\mathbb{R})} H_\Gamma(\mathbb{R}),$$

$$u_\Gamma \colon \text{Hom}^*(\Gamma, \text{SU}(2)) \xrightarrow{\text{U}(2)/\text{U}(1)} T_\Gamma(\mathbb{R}).$$

The value of a coordinate $s(\gamma) \in R(\Gamma, \text{SL}_2)$ at a point of the base space is equal to the character $\text{tr}(\rho(\gamma))$ for any representation ρ in its fiber.

References

[A]   W. Abikoff, *The real analytic theory of Teichmüller space*, Lect. Notes in Math. **820**, Springer, (1980).

[B-L]   Hyman Bass and Alexander Lubotzky, *Automorphism of groups and of schemes of finite type*, Israel J. of Math. **44** No. 1 (1983), 1–22.

[Be]   Alan F. Beardon, *The geometry of discrete groups*, GTM, Springer Verlag, 197.

[Br]   G. W. Brumfiel, *The real spectrum compactification of Teichmüller space*, Contem. Math. **74** (1988), 51–75.

[C-C-G-L-S] D. Cooper, M. Culler, H. Gillet, D. D. Long and P. Shalen, *Plane curves associated to character varieties of 3-manifolds*, Invent. math. **118** (1994), 47–84.

[C-W]   Paula Cohen and Jürgen Wolfart, *Modular Embedding for some nonarithmetic Fuchsian groups*, preprint, IHES/M/88/57, Nov. 1988.

[C-S]   Marc Culler and Peter B. Shalen, *Varieties of group representations and splittings of 3-manifolds*, Annals of Math. **117** (1983), 109–146.

[D]   Stephen Donkin, *Invariants of several matrices*, Invent. Math. **110** (1992), 389–401.

[F]   G. Faltings, *Real projective structures on Riemann surfaces*, Compos. Math. **48** (1983), 223–369.

[F-K]   Robert Fricke and Felix Klein, *Vorlesungen uber die Theorie der Automorphen Funktionen*, Vol. 1, pp. 365–370, Leipzig: B.G. Teubner 1987, Reprint: New York Johnson Reprint Corporation (Academic Press) 1965.

[G1]   William Goldman, *Topological components of spaces of representations*, Invent. Math. **93** (1988), 557–607.

[G2]   William Goldman, *Geometric structures on manifolds and varieties of representations*, Contemporary Math. **74** (1988), 169–198.

[G-M]   F. González-Acuña and José María Montesinos-Amilibia, *On the character variety of group representations in* SL(2, C) *and* PSL(2, C), Math. Z. **214** (1993), 627–652.

[G-Y]   J. H. Grace and A. Young, *The ALGEBRA of INVARIANTS*, Cambridge Univ. Press., New York, 1903.

[H1]   Heinz Helling, *Diskrete Untergruppen von* $SL_2(\mathbb{R})$, Inventiones math. **17** (1972), 217–229.

[H2]   Heinz Helling, *Ueber den Raum der kompakten Riemannschen flächen vom geschlecht 2*, J. reine angew. Math. **268/269**:286–293, (1974).

[H-L-M] M. Hilden, M. Lozano and J. Montesinos-Amilibia, *Volumes and Chern-Simons invariant of cyclic coverings over rational knots*, These Proceedings, 31–55.

[H-K]   C. Hodgson and S. Kerckhoff, *Rigidity of hyperbolic cone-manifolds and and hyperbolic Dehn surgery*, preprint.

[Ho1]   Robert Horowitz, *Characters of free groups represented in the two dimensional linear group*, Comm. Pure Appl. Math. **25** (1972), 635–649.

[Ho2]   Robert Horowitz, *Induced automorphisms of Fricke characters of free groups*, Trans. Am. Math. Soc. **208** (1975), 41–50.

[J-W]   Lisa Jeffrey and Jonathan Weitsman, *Bohr-Sommerfeld orbits in the moduli space of flat connections and the Verlinde dimension formula*, Comm. Math. Phys. **150** (1992), 593–630.

[J]   T. Jørgensen, *On discrete groups of Möbius transformations*, Amer. J. Math. **98** (1976), 739–749.

[K1]   Linda Keen, *Intrinsic Moduli*, Ann. of Math. **84** (1966), 404–420.

[K2]   Linda Keen, *On Fricke Moduli, Advances in the theory of Riemann surfaces*, Ann. of Math. Studies **66** (1971), 205–2024.

[K3]   Linda Keen, *A correction to "On Fricke Moduli"*, Proc. Amer. Math. Soc. **40** (1973), 60–62.

[K4]   Linda Keen, *A rough fundamental domain for Teichmüller spaces*, Bull. Amer. Math. Soc. **83** (1977), 1199–1226.

[Kj]   S. Kojima *Deformations of hyperbolic 3-cone-manifolds*, preprint.

[Ko]   Yohei Komori, *Semialgebraic description of Teichmüller space*, Thesis, Rims, Jan. 1994.

[Kr1]   Irvin Kra, *Automorphic Forms and Kleinian Groups*, W. A. Benjamin Inc., (1972).

[Kr2]   Irvin Kra, *On ligring of Kleinian groups to* SL(2, C) *in Differential Geometry and Complex Analysis*, (H. E. Rouch, Memorial Volume)., Springer, Berlin, Heidelberg, New York, 1985, 181–193.

[Kr3]   Irvin Kra, *Maskit coordinate*, Holom. Funct. & Moduli 2, Math. Sci. Research Inst. Publication **11**, Springer, 1992.

[L-M]   Alexander Lubotzky and Andy R. Magid, *Varieties of representations of finitely generated groups*, Memoirs of the AMS, Vol. 58 No. 336, (1985).

[Ma]   William Magnus, *Rings of Fricke characters and automorphism groups of free groups*, Math. Z. **170** (1980), 91–103.

[Mu1]   David Mumford, *Geometric Invariant Theory*, Springer Verlag, Berlin, Heidelberg, 1965, Library of Congress Catalog Card Number 65-16690.

[Mu2]   David Mumford, *Curves and their Jacobians*, ©by University of Michigan, 1975, All right reserved, ISBN 0-472-66000-4, Library of Congress Catalog Card No. 75-14899.

[N-Z]   W. Neumann and D. Zagier, *Volumes of hyperbolic 3-manifolds*, Topology **24** (1985), 307–332.

[O1]   Yoshihide Okumura, *Fricke moduli and Keen moduli for Fuchsian groups and a certain class of quasi-Fuchsian groups*, Master Thesis, Shizuoka Univ., 1986.

[O2]   Yoshihide Okumura, *Global real analytic coordinates for Teichmüller spaces*, Thesis, Kanazawa Univ., 1989.

[O3]   Yoshihide Okumura, *On lifting problem of Kleinian group into* $SL(2, \mathbb{C})$, Summer Seminar on Function Theory, July 1992.

[O4]   Yoshihide Okumura, *On the global real analytic coordinates for Teichmüller spaces*, J. Math. Soc. Japan **42** (1990), 91–101.

[P-B-K] V. P . Platonov and V. V. Benyash-Krivets, *Characters of representations of finitely generated groups*, Proc. of Steklov Inst. of Math. (1991), 203–213.

[Pr1]   Claudio Procesi, *Finite dimensional representations of algebras*, Israel J. of Math. **19** (1974), 169–182.

[Pr2]   Claudio Procesi, *Invariant theory of* $n \times n$ *matrices*, Adv. Math. **19** (1976), 306–381.

[S1]   Kyoji Saito, *Moduli Space for Fuchsian Groups, Algebraic Analysis*, Vol. II, Academic Press, (1988), 735–787

[S2]   Kyoji Saito, *The Limit Element in the Configuration Algebra for a Discrete Group*, A précis: Proc. Int. Congr. Math., Kyoto 1990, 931–942, Preprint: RIMS–726, Nov. 1990.

[S3]   Kyoji Saito, *The Teichmüller space and a certain modular function from a view point of group representations*, Alg. Geom. and related Topics, Proc. Int. Symp., Inchoen, Republic of Korea, 1992.

[S4]   Kyoji Saito, *Representation variety of a finitely generated group into* SL_2 *or* GL_2, Preprint RIMS–958, Dec. 1993.

[S5]   Kyoji Saito, *Algebraic Representation of the Teichmüller spaces*, LMS. Lecture Note Series **200** (1994), 255–288.

[S6]   Kyoji Saito, *Algebraic Representation of the Teichmüller spaces*, Kodai Math. J. **17** (1994), 609–626.

[S-S1]   Mika Seppälä and Tuomas Sorvali, *Parametrization of Möbius groups acting in a disk*, Comment. Math. Helvetici **61** (1986), 149–160.

[S-S2]   Mika Seppälä and Tuomas Sorvali, *Affine coordinates for Teichmüller spaces*, Math. Ann. **284** (1989), 169–176.

[S-S3]   Mika Seppälä and Tuomas Sorvali, *Trace commutators of Möbius transformations*, Math. Scand. **68** (1991), 53–58.

[S-S4]   Mika Seppälä and Tuomas Sorvali, *Geometry of Riemann Surfaces and Teichmüller Spaces*, North-Holland Mathematics Studies **169** (1992).

[T]   W. Thurston, *The geometry and topology of 3-manifolds*, Lecture Notes, Princeton Univ., 1977/78.

[V]   H. Vogt, *Sur les invariants, fondamentaux des équations différentielles linéaires du second ordre*, Ann. Sci. Ecole Nrom. sup. **6(3)**, Suppl.3–72 (1889)(Thèse, Paris).

[W1]   Andre Weil, *On discrete subgroups of Lie Groups I*, Ann. Math. **72** (1960), 369–384,

[W2]   Andre Weil, *On discrete subgroups of Lie Groups II*, Ann. Math. **75** (1962), 578–602,

[We]   Jonathan Weitsman, *Geometry of the intersection ring of the moduli space of flat connections and the conjectures of Newstead and Witten*, preprint 1993.

[Wf]   Jürgen Wolfart, *Eine arithmetische Eigenschaft automorpher Formen zu gewiisen nicht-arithmetischen Gruppen*, Math. Ann. **62** (1983), 1–21.

[Wo1]   Scott Wolpert, *The Fenchel-Nielsen deformation*, Ann. of Math. **115** (1982), 501–528.

[Wo2]   Scott Wolpert, *On the Symplectic Geometry of deformation of a hyperbolic surface*, Ann. of Math. **117** (1983), 207–234.

[Wo3]   Scott Wolpert, *Geodesic length functions and the Nielsen problem*, J. Diff. Geom. **25** (1987), 275–296.

[Y]   T. Yoshida, *On ideal points of deformation curves of hyperbolic 3-manifold with one cusp*, Topology **30** (1991), 155–170.

[Z-V-C] Ziechang, Vogt and Colderway, Lect. Notes in Math. **122, 835** and **875**.

RIMS, KYOTO UNIVERSITY
E-mail address: saito@kurims.kyoto-u.ac.jp

Proceedings of
THE 37TH TANIGUCHI SYMPOSIUM ON
TOPOLOGY AND TEICHMÜLLER SPACES
held in Finland, July 1995
ed. by Sadayoshi KOJIMA *et al.*
©1996 World Scientific Publishing Co.
pp. 265–277

THE THIRD BOUNDED COHOMOLOGY AND KLEINIAN GROUPS

TERUHIKO SOMA

(Received December 21, 1995)

Introduction

The notion of bounded cohomology was introduced for discrete groups by P. Trauber and for topological spaces by M. Gromov [7]. Besides, Gromov proved that the bounded cohomology $H_b^*(X; \mathbf{R})$ of any topological space X coincides with the bounded cohomology $H_b^*(\pi_1(X), \mathbf{R})$ of its fundamental group (see also Ivanov [8]). Up to the present, various results on the second bounded cohomology have been obtained by many authors, for example see the references of Grigorchuk [6]. However, the author does not know any published papers on the third bounded cohomology except Yoshida [19], where he proved that the $\mathbf{R}$-dimension of the third bounded cohomology of closed surface of genus $g > 1$ is infinite.

In this paper, we will survey some results on the third bounded cohomology which will be appeared in the series of papers [15], [16], [17], and give sketches (or outlines) of their proofs. This may be helpful for the reader to understand that arguments on the 3-dimensional hyperbolic geometry are very useful for the study of the third bounded cohomology.

This paper is organized as follows. In §1, some definitions, notations and fundamental theorems needed later sections will be given. In §2, we will present examples of bounded cohomology whose pseudonorms are not norms. In fact, Gromov [7], Ivanov [8] and others asked whether there exists a discrete group G such that the pseudonorm $\| \cdot \|$ on $H_b^n(G; \mathbf{R})$ is a norm. It is well known that $\| \cdot \|$ is a norm if and only if $(H_b^n(G; \mathbf{R}), \| \cdot \|)$ is a Banach space. In the case of $n < 3$, Matsumoto-Morita [10] and Ivanov [9] proved independently that, for any discrete group G, the n-th bounded cohomology $(H_b^n(G; \mathbf{R}), \| \cdot \|)$ is a Banach space. Contrary, our Theorems 2.1 and 2.2 imply that, for any G in a wide class of non-amenable discrete groups, the third bounded cohomology $(H_b^3(G; \mathbf{R}), \| \cdot \|)$ is not a Banach space. In §3, we will study the third bounded cohomology of closed, connected, orientable surface Σ_g of genus $g > 1$, and give the rigidity theorem for pairs (M, f) such that M are hyperbolic 3-manifolds with $\mathrm{inj}(M) > 0$ and $f : \Sigma_g \longrightarrow M$ are homotopy equivalent maps. For such pairs (M_0, f_0) and (M_1, f_1) with M_0 doubly-degenerate, it will be seen sketchily that, if their fundamental classes $[f_0^*(\omega_{M_0})]$, $[f_1^*(\omega_{M_1})]$ in $H_b^3(\Sigma_g; \mathbf{R})$ are sufficiently close to each other with respect to the pseudonorm $\| \cdot \|$, then M_0 and M_1 have the same ending invariants, and hence M_0 is isometric to M_1 by Minsky's Ending Lamination Theorem [11], [12]. As an application of this rigidity theorem, it is proved that the dimension of the $\mathbf{R}$-vector space $H_b^3(\Sigma_g; \mathbf{R})$ is the cardinarity of continuum. In §4, we will study interactions between bounded cohomology and topologically tame Kleinian groups. Canary's results [2] on such Kleinian groups and his Covering Theorem [3], [4] will play there important roles.

§1. Definitions and some fundamental results

First of all, we define the bounded cohomology of topological spaces. Let X be a topological space and $(C_*(X), \partial_*)$ the singular chain complex of real coefficient. The *Gromov norm* of a k-chain $c = \sum_{i=1}^n a_i \sigma_i^k \in C_k(X)$ is defined by

$$\|c\| = \sum_{i=1}^n |a_i|.$$

Consider the norm completion $C_*^{l_1}(X)$ of $(C_*(X), \|\cdot\|)$, that is, an element c of $C_k^{l_1}(X)$ is a series $c = \sum_{i=1}^\infty a_i \sigma_i^k$ with $\|c\| = \sum_{i=1}^\infty |a_i| < \infty$. Note that the boundary operator $\partial_k : C_k(X) \longrightarrow C_{k-1}(X)$ is naturally extended to

$$\partial_k^{l_1} : C_k^{l_1}(X) \longrightarrow C_{k-1}^{l_1}(X).$$

We often set $\partial_k^{l_1} = \partial_k$ for simplicity. The normed space $(C_*^{l_1}(X), \| \cdot \|)$ is a Banach space. The dual Banach space $(C_b^*(X), \| \cdot \|)$ together with the coboundary operator $\delta_b^k : C_b^k(X) \longrightarrow C_b^{k+1}(X)$ defines the *bounded cohomology* of X by

$$H_b^*(X; \mathbf{R}) = H^*(C_b^*(X)).$$

The pseudonorm $\|\alpha\|$ of $\alpha \in H_b^k(X; \mathbf{R})$ is given by

$$\|\alpha\| = \inf\{\|c\|; c \in Z_b^k(X) = \mathrm{Ker}(\delta_b^k) \text{ with } [c] = \alpha\}.$$

As will be seen in §2, in general, the pseudo-norm on $H_b^k(X; \mathbf{R})$ is not a norm. So, it is covenient to consider the quotient space $HB^k(X; \mathbf{R}) = H_b^k(X; \mathbf{R})/\{[c] \in H_b^k(X; \mathbf{R}); \||[c]\|| = 0\}$. Note that $HB^k(X; \mathbf{R})$ is a Banach space with the norm $\| \cdot \|$. The element of $HB^k(X; \mathbf{R})$ corresponding to $[c] \in H_b^k(X; \mathbf{R})$ is denoted by $[c]_B$.

For any (discrete) group G, the bounded cohomology $H_b^k(G; \mathbf{R})$ of G is defined with the normed space $(C_b^k(G), \| \cdot \|)$ such that $C_b^k(G) = \{0\}$ if $k < 0$, $C_b^0(G) = \mathbf{R}$ with the euclidean norm, and $C_b^k(G)$ $(k > 0)$ is the $\mathbf{R}$-vector space of bounded functions

$$c : \underbrace{G \times \cdots \times G}_{k} \longrightarrow \mathbf{R}$$

with the norm $\|c\| = \sup\{|c(g_1, \cdots, g_k)|; g_i \in G \ (i = 1, \cdots, k)\}$. The coboundary coperator $\delta_b^k : C_b^k(G) \longrightarrow C_b^{k+1}(G)$ is given by

$$\delta_b^k(c)(g_1, \cdots, g_{k+1}) = c(g_2, \cdots, g_{k+1}) + \sum_{i=1}^{k}(-1)^i c(g_1, \cdots, g_{i-1}, g_i g_{i+1}, \cdots, g_{k+1})$$
$$+(-1)^{k+1} c(g_1, \cdots, g_k)$$

Then, the bounded cohomology $H_b^*(G; \mathbf{R})$ of G is defined with the cochain complex $(C_b^*(G), \delta_b^*)$. The pseudonorm $\| \cdot \|$ on $H_b^*(G; \mathbf{R})$ is given as in the previous case.

Theorem 1.1 is one of fundamental results on bounded cohomology.

Theorem 1.1 (Gormov [7], Ivanov [8]). *For any arcwise connected space X, $(H_b^*(X, \mathbf{R}), \| \cdot \|)$ is isometrically isomorphic to $(H_b^*(\pi_1(X); \mathbf{R}), \| \cdot \|)$.*

For an $\mathbf{R}$-vector space V, $\dim V = \#\mathbf{R}$ (resp. $\dim V \leq \#\mathbf{R}$) denotes that the dimension of V is (resp. is not greater than) the cardinariy of continuum. If a group G is countable as a set, then the cardinality of the set $C_b^k(G)$ is at most the continuum. In particular, $\dim H_b^k(G; \mathbf{R}) \leq \dim C_b^k(G) \leq \#\mathbf{R}$. Thus, Theorem 1.1 implies:

Corollary 1.2. *For any arcwise conneced space X with the finitely generated fundamental group, $\dim H_b^k(X; \mathbf{R}) \leq \#\mathbf{R}$.*

The chain complex $(C_*^{l_1}(X), \partial_*^{l_1})$ is said to satisfy *k-uniform boundary condition* (for short *k*-UBC^{l_1}) if there exists a constant $L > 0$ such that, for any $z \in B_k^{l_1} = \mathrm{Im}(\partial_{k+1}^{l_1})$, there is a $(k+1)$-chain $c \in C_{k+1}^{l_1}(X)$ with $\partial_{k+1}(c) = z$ and $\|c\| \leq L\|z\|$. In §2, we will use the following theorem.

Theorem 1.3 (Matsumoto-Morita [10]). *The pseudonorm $\| \cdot \|$ on $H_b^k(X; \mathbf{R})$ is a norm (or equivalently $(H_b^k(X; \mathbf{R}), \| \cdot \|)$ is a Banach space) if and only if $(C_*^{l_1}(X), \partial_*^{l_1})$ satisfies $(k-1)$-UBC^{l_1}.*

For a topological space X, we denote the direct product space $X \times \cdots \times X$ of m-copies of X by $\Pi^m X$. For any $Y \subset X$, the subspace $\nabla^m Y$ of $\Pi^m Y \subset \Pi^m X$ is defined by

$$\{(y_1, \cdots, y_m) \in \Pi^m Y; y_i = y_j \text{ for some } i, j \text{ with } 1 \leq i < j \leq m\}.$$

We denote by $\mathbf{B}^n$ the unit n-ball model for the union $\mathbf{H}^n \cup S_\infty^{n-1}$ of the hyperbolic n-space and the $(n-1)$-sphere at infinity. Let Γ be a torsion-free, discrete subgroup of $\mathrm{Isom}^+(\mathbf{H}^n)$, so that the quotient space $M^n = \mathbf{H}^n/\Gamma$ is a hyperbolic n-manifold and the quotient map $q : \mathbf{H}^n \longrightarrow M^n$ is the universal covering. Since $\nabla^4 S_\infty^{n-1}$ is invariant under the diagonal action of Γ on $\Pi^4 \mathbf{B}^n$, the quotient map

$$P_{M^n} : \Pi^4 \mathbf{B}^n - \nabla^4 S_\infty^{n-1} \longrightarrow S(M^n) = (\Pi^4 \mathbf{B}^n - \nabla^4 S_\infty^{n-1})/\Gamma$$

is well defined.

Consider the standard 3-simplex (or tetrahedron) Δ^3 with vertices v_0, v_1, v_2, v_3. For any $(x_0, x_1, x_2, x_3) \in \Pi^4 \mathbf{B}^n - \nabla^4 S_\infty^{n-1}$, there exists a unique straight (ideal) 3-simplex $\bar{\sigma} : \Delta^3 \longrightarrow \mathbf{B}^n$ with $\bar{\sigma}(v_i) = x_i$ for $i = 0, 1, 2, 3$, and such that $\bar{\sigma}$ is an affine map with respect to the quadratic form model for $\mathbf{H}^n$, see [18, §6.1]. Hence, the set of all straight 3-simplices is identified with $\Pi^4 \mathbf{B}^n - \nabla^4 S_\infty^{n-1}$. For a singular simplex $\sigma : \Delta^3 \longrightarrow \mathbf{B}^n$, the straight simplex $\bar\sigma : \Delta^3 \longrightarrow \mathbf{B}^n$ with $\bar\sigma(v_i) = \sigma(v_i)$ ($i = 0, 1, 2, 3$) is said to be obtained by *straightening* σ and denoted by $\mathrm{straight}(\sigma)$.

For any straight simplex $\tilde\sigma : \Delta^3 \longrightarrow \mathbf{B}^n$, the composition $\sigma = q \circ \tilde\sigma|_{\Delta^3 - \mathcal{V}_\sigma} : \Delta^3 - \mathcal{V}_\sigma \longrightarrow M^n$ is a *straight* simplex in M^n, where $\mathcal{V}_\sigma = \tilde\sigma^{-1}(S_\infty^{n-1})$ is the set of ideal vertices of $\tilde\sigma$. Through the one-to-one correspondence between σ and $P_{M^n}(\tilde\sigma(v_0), \tilde\sigma(v_1), \tilde\sigma(v_2), \bar\sigma(v_3))$, one can identify the set of all straight 3-simplices in M^n with $\mathcal{S}(M^n)$. For any singular 3-simplex $\sigma : \Delta^3 \longrightarrow M^n$, the *straightened* simplex $\mathrm{straight}(\sigma)$ for σ is defined by $\mathrm{straight}(\sigma) = q \circ \mathrm{straight}(\tilde\sigma)$, where $\tilde\sigma : \Delta^3 \longrightarrow \mathbf{H}^n$ is a lift of σ.

Let $\mathcal{B}(M^n)$ be the set of all Borel measures μ on $\mathcal{S}(M^n)$ with compact support and with bounded total variation $\|\mu\|$, that is,

$$\|\mu\| = \sup\left\{ \int_{\sigma \in \mathcal{S}(M^n)} f(\sigma)d\mu(\sigma) \,;\, f \text{ is a measurable function with } |f| \le 1 \right\} < \infty.$$

Here, we consider the case of $n = 3$ and set $M^3 = M$. The *inner center* of a non-degenerate straight 3-simplex $\sigma : \Delta^3 \longrightarrow \mathbf{B}^3$ is the center of the greatest hyperbolic 3-ball in $\sigma(\Delta_0^3)$, where $\Delta_0^3 = \Delta^3 - \{v_0, v_1, v_2, v_3\}$. The *inner center* of a straight 3-simplex $\sigma : \Delta^3 - \mathcal{V}_\sigma \longrightarrow M$ is $x_\sigma = q(\tilde x_{\tilde\sigma})$, where $\tilde x_{\tilde\sigma} \in \mathbf{H}^3$ is the inner center of a lift $\tilde\sigma : \Delta^3 - \mathcal{V}_\sigma \longrightarrow \mathbf{H}^3$ of σ.

For a non-degenerate, straight simplex $\sigma : \Delta^3 \longrightarrow \mathbf{B}^3$, $\mathrm{smear}_M(\sigma)$ is the Borel measure on $\mathcal{S}(M)$ defined in [18, Chapter 6] such that $\mathrm{supp}(\mathrm{smear}_M(\sigma)) = P_M(\{\gamma \circ \sigma : \Delta^3 \longrightarrow \mathbf{B}^3 ; \gamma \in \mathrm{Isom}^+(\mathbf{H}^3)\})$, and, for any Borel subspace L of M, $\mathrm{smear}_M(\mathcal{L}) = \mathrm{vol}(L)$, where $\mathcal{L} = \{\tau \in \mathrm{supp}(\mathrm{smear}_M(\sigma)) ; \text{the inner center of } \tau \in L\}$. A *reflected* simplex $\sigma_- : \Delta^3 \longrightarrow \mathbf{B}^3$ of σ is defined by $r \circ \sigma$, where $r : \mathbf{B}^3 \longrightarrow \mathbf{B}^3$ is an orientation-reversing involution with $r|_{\mathbf{H}^3} : \mathbf{H}^3 \longrightarrow \mathbf{H}^3$ isometric. For a regular ideal simplex $\sigma_{\mathrm{reg}} : \Delta^3 \longrightarrow \mathbf{B}^3$, we consider the Borel measure

$$z(M) = \frac{1}{2}(\mathrm{smear}_M(\sigma_{\mathrm{reg}}) - \mathrm{smear}_M(\sigma_{\mathrm{reg},-}))$$

on $\mathcal{S}(M)$. We set $z(L) = z(M)|_{\mathcal{L}}$. If L is compact, then $z(L)$ has a compact support $\mathcal{L}$ and $\|z(L)\| = \mathrm{vol}(L) < \infty$. In particular, $z(L)$ is an element of $\mathcal{B}(M)$.

A *Kleinian group* Γ is a discrete subgroup of $\mathrm{PSL}_2(\mathbf{C})$, the group $\mathrm{Isom}^+(\mathbf{H}^3)$ of all orientation preserving isometries on the hyperbolic 3-space $\mathbf{H}^3$. We usually consider the case where Γ is finitely generated and torsion free. Then, $M_\Gamma = \mathbf{H}^3/\Gamma$ is a hyperbolic 3-manifold and the natural quotient map $p : \mathbf{H}^3 \longrightarrow M_\Gamma$ is the universal covering. Let ω_Γ (or ω_{M_Γ}) be the 3-cocycle on M_Γ defined by

$$\omega_\Gamma(\sigma) = \Omega_\Gamma(\mathrm{straight}(\sigma)) = \int_{\Delta^3} \mathrm{straight}(\sigma)^*(\Omega_\Gamma),$$

where $\sigma : \Delta^3 \longrightarrow M_\Gamma$ is any singular 3-simplex and Ω_Γ is the volume form on M_Γ. Since the volume of any straight simplex in M_Γ is less than the volume $\mathbf{v}_3$ of a regular ideal simplex in $\mathbf{H}^3$. Thus, $|\omega_\Gamma(\sigma)| = |\Omega_\Gamma(\mathrm{straight}(\sigma))| < \mathbf{v}_3$. This implies that $[\omega_\Gamma]$ is an element of $H_b^3(M_\Gamma; \mathbf{R})$, called the *funtamental class* of M_Γ, with

$$\|[\omega_\Gamma]\| \le \|\omega_\Gamma\| \le \mathbf{v}_3.$$

If Γ is non-elementary, then the convex hull H_Γ of the limit set L_Γ of Γ in $\mathbf{H}^3$ is non-empty, and the image $C_\Gamma = p(H_\Gamma)$ is the smallest, closed, convex core of M_Γ. Let $\mathcal{P}_\Gamma$ be a maximal union of mutually disjoint, parabolic cusps in M_Γ. Here $\mathcal{P}_\Gamma$ *maximal* means that any cusp of M_Γ meets $\mathcal{P}_\Gamma$ non-trivially. Let X_Γ be a 3-dimensional, closed submanifold of C_Γ with $\partial X_\Gamma \supset \partial C_\Gamma$ and such that the inclusion $X_\Gamma \longrightarrow C_\Gamma$ is homotopy equivalent. We say that X_Γ is a *finite core* of C_Γ (or of M_Γ) if X_Γ satisfies the following (1.1) and (1.2). Here, for a subspace B of a topological space A, we denote by $\mathrm{cl}(A - B)$ the closure of $A - B$ in A.

(1.1) The closure $\mathrm{cl}(X_\Gamma - \mathcal{P}_\Gamma \cap X_\Gamma)$ is compact, and for each component P of $\mathcal{P}_\Gamma$, $P \cap X_\Gamma$ is non-empty. If $P \not\subset X_\Gamma$, then $P \cap X_\Gamma$ is a pared $\mathbf{Z}$-cusp, and if $P \subset X_\Gamma$, then P is a $\mathbf{Z} \times \mathbf{Z}$-cusp.

(1.2) For each component E of $\mathrm{cl}(C_\Gamma - X_\Gamma)$, if a non-contractible loop l in $F = \partial E$ is homotopic in E to a loop in $E \cap \mathcal{P}_\Gamma$, then l is homotopic in F to a loop in $F \cap \mathcal{P}_\Gamma$. In other words, E contains no "accidental" parabolic loops.

A component E of $\mathrm{cl}(M_\Gamma - X_\Gamma)$ is *topologically tame* if E is homeomorphic to $\partial E \times [0, \infty)$. A component E_0 of $\mathrm{cl}(M_\Gamma - C_\Gamma)$ is called a *geometrically finite end*. Note that each geometrically finite end is topologically tame, for example see the proof of [18, Proposition 8.3.4]. The group Γ is called *geometrically finite* if $X_\Gamma = C_\Gamma$, or equivalently, each end of M_Γ is geometrically finite, and otherwise *geometrically infinite*. We say that Γ is a *topologically tame* Kleinian group if each end of $\mathrm{cl}(M_\Gamma - X_\Gamma)$ is topologically tame. According to Canary [2], Γ is topologically tame if and only if each component E of $\mathrm{cl}(C_\Gamma - X_\Gamma)$ is *simply degenerate*. That is, there exists a sequence $\{f_n : (\Sigma_g, \tau_n) \longrightarrow E\}$ of simplicial hyperbolic surfaces properly homotopic to the inclusion $\partial E \longrightarrow E$ and tending toward the end of E. Here, $f_n : (\Sigma_g, \tau_n) \longrightarrow E$ *simplicial hyperbolic* means that Σ_g admits a hyperbolic triangulation τ_n such that the cone-angle of Σ_g at each vertex of τ_n is at least 2π and such that, for each 2-simplex Δ of τ_n, $f_n|_\Delta$ is a locally isometric, totally geodesic immersion.

Now, we consider the case where Γ is a topologically tame Kleinian group without parabolic elements. For any subset A of $M_\Gamma = \mathbf{H}^3/\Gamma$ and $r > 0$, let $\mathcal{N}_r(A)$ denote the r-neighborhood of A in M_Γ. For any simply degenerate end E of M_Γ, one can have a sequence $\{f_n : (\Sigma_g, \tau_n) \longrightarrow M_\Gamma; n \in \mathbf{N}\}$ of simplicial hyperbolic surfaces satisfying the following (1.3) and (1.4).

(1.3) For all $n, n' \in \mathbf{N}$ with $n \neq n'$, $\mathcal{N}_2(f_n(\Sigma_g)) \cap \mathcal{N}_2(f_{n'}(\Sigma_g)) = \emptyset$.

(1.4) For any two integers n, n' with $0 < n < n'$, $f_n(\Sigma_g)$ is contained in the component of $E - f_{n'}(\Sigma_g)$ containing ∂E.

Since M_Γ has no parabolic cusps, one can have a sequence $\{B_n\}_{n=1}^\infty$ of compact, connected subsets of E, called *blocks*, satisfying that (i) $\mathrm{cl}(E - \cup_{n=1}^\infty B_n)$ is a compact, connected set containing ∂E, (ii) $\partial B_n \subset f_n(\Sigma_g) \cup f_{n+1}(\Sigma_g)$, (iii) $B_n \cap B_{n+1} \subset f_{n+1}(\Sigma_g)$, and (iv) $B_n \cap B'_n = \emptyset$ for all $n, n' \in \mathbf{N}$ with $n' - n \geq 2$. For any $n, n' \in \mathbf{N}$ with $n < n'$, we say that the union $L(n, n') = \cup_{i=n}^{n'-1} B_i$ is the (n, n')-*bank* for $\{B_n\}$ of *block-length* $n' - n$. Set $z(n, n') = z(L(n, n')) \in \mathcal{B}(M_\Gamma)$. Then, we have

$$\|\partial z(n, n')\| \leq 4\mathrm{vol}(\mathcal{N}_2(f_n(\Sigma_g) \cup f_{n'}(\Sigma_g))).$$

This is easily derived from the following two facts. The first is that each 3-simplex in

$\operatorname{supp}(z(n, n'))$ has four 2-faces. The second is that, for any 2-face $\tau : \Delta^2 \longrightarrow M_\Gamma$ of $\sigma \in$ $\operatorname{supp}(z(L(n, n') - \mathcal{N}_2(f_n(\Sigma_g) \cup f_{n'}(\Sigma_g))))$, there exists a 3-simplex $\sigma' \in \operatorname{supp}(z(n, n'))$ such that $-\tau$ is one of 2-faces of σ'. In [2, Lemmas 7.1 and 8.2], Canary showed that there exists a constant $K > 0$ such that, for any $n \in \mathbf{N}$, $\operatorname{vol}(\mathcal{N}_2(f_n(\Sigma_g))) \le K$. This implies that $\|\partial z(n, n')\| \le 8K = K_1$. According to the proof of [15, Lemma 3.2], there exists a sequence $\{z_m(n, n')\}_{m=1}^\infty$ of elements in $C_3(M_\Gamma)$ converging weakly to $z(n, n')$ in the sense of [15, §3] such that each $z_m(n, n')$ is represented by a finite linear combination $\sum_j \alpha_j \sigma_j$ with $\alpha_j > 0$ and satisfying that (i) $\|z_m(n, n')\| = \sum_j \alpha_j = \operatorname{vol}(L(n, n'))$, (ii) $\sigma_j : \Delta^3 \longrightarrow M_\Gamma$ are straight simplices with $\operatorname{vol}(\sigma_j) > \mathbf{v}_3 - 1/m$ and (iii) $\|\partial z_m(n, n')\| \le \|\partial z(n, n')\| \le K_1$.

Theorem 1.4 ([16]). *For any geometrically infinite and topologically tame Kleinian group Γ, $[\omega_\Gamma]$ is a non-trivial element of $H_b^3(M_\Gamma, \mathbf{R})$ with $\|[\omega_\Gamma]\| = \mathbf{v}_3$.*

Sketch of Proof. Here, we only consider the special case where Γ contains no parabolic elements. Then, one can use the notation in the paragraph preceding this theorem. We refer to [16, Theorem 1] (see also [15, Proposition 3.3]) for the proof of the general case.

If $\|[\omega_\Gamma]\|$ were strictly less than $\mathbf{v}_3$, then there would exist $\varepsilon > 0$ and a bounded 2-cochain $d \in C_b^2(M_\Gamma)$ such that $\|\omega_\Gamma + \delta(d)\| < \mathbf{v}_3 - \varepsilon$. If $m \in \mathbf{N}$ is sufficiently large, then for any straight simplex $\sigma_j : \Delta^3 \longrightarrow M_\Gamma$ in $z_m(1, n) = \sum_j \alpha_j \sigma_j$,

$$\omega_\Gamma(\sigma_j) = \Omega_\Gamma(\sigma_j) > \mathbf{v}_3 - \frac{\varepsilon}{2}.$$

This shows that $\omega_\Gamma(z_m(1, n)) > (\mathbf{v}_3 - \varepsilon/2) \sum_j \alpha_j = (\mathbf{v}_3 - \varepsilon/2)\|z_m(1, n)\|$. Since

$$\|\omega_\Gamma + \delta(d)\| \, \|z_m(1, n)\| \ge |(\omega_\Gamma + \delta(d))(z_m(1, n))|,$$

and since $\|z_m(1, n)\| = \operatorname{vol}(L(n, n')) \to \infty$ as $n \to \infty$, we have

$$
\begin{aligned}
\|\omega_\Gamma + \delta(d)\| \;&\ge\; \limsup_{n \to \infty} \frac{|(\omega_\Gamma + \delta(d))(z_m(1, n))|}{\|z_m(1, n)\|} \\
&=\; \limsup_{n \to \infty} \frac{|\omega_\Gamma(z_m(1, n)) + d(\partial z_m(1, n))|}{\|z_m(1, n)\|} \\
&\ge\; \limsup_{n \to \infty} \frac{|\omega_\Gamma(z_m(1, n))| - K_1\|d\|}{\|z_m(1, n)\|} \\
&\ge\; \mathbf{v}_3 - \frac{\varepsilon}{2}.
\end{aligned}
$$

This contradiction implies $\|[\omega_\Gamma]\| = \mathbf{v}_3$. $\quad\square$

§2.   Existence of non-Banach bounded cohomology

The following theorem shows that, for any G in a wide class of non-amenable discrete groups including all free groups of rank > 1 and the fundamental groups of all closed, orientable surfaces of genus > 1, the third bounded cohomology $(H_b^3(G; \mathbf{R}), \|\cdot\|)$ is not a Banach space.

Theorem 2.1 ([17]). *For any discrete group G admitting a surjective homomorphism $f : G \longrightarrow \mathbf{Z} * \mathbf{Z}$, the pseudonorm $\| \cdot \|$ on $H_b^3(G; \mathbf{R})$ is not a norm. That is, $(H_b^3(G; \mathbf{R}), \| \cdot \|)$ is not a Banach space.*

For any surjective homomorphism $f : G \longrightarrow \mathbf{Z} * \mathbf{Z}$, there exists a homomorphism $g : \mathbf{Z} * \mathbf{Z} \longrightarrow G$ with $f \circ g = \mathrm{id}_{\mathbf{Z}*\mathbf{Z}}$. Since $(f \circ g)^* : H_b^3(\mathbf{Z} * \mathbf{Z}; \mathbf{R}) \longrightarrow H_b^3(\mathbf{Z} * \mathbf{Z}; \mathbf{R})$ is the identity, $f^* : H_b^3(\mathbf{Z} * \mathbf{Z}; \mathbf{R}) \longrightarrow H_b^3(G; \mathbf{R})$ is injective. So, it suffices to consider the case of $G = \mathbf{Z} * \mathbf{Z}$.

Here, we will give a sketch of the proof of the following theorem, which is a special case of Theorem 2.1. Its proof is simple comparing with that of Theorem 2.1 because the model manifold discussed there is a closed hyperbolic 3-manifold.

Theorem 2.2. *For any closed, connected, orientable surface Σ_g of genus $g > 1$, $(H_b^3(\Sigma_g; \mathbf{R}), \| \cdot \|)$ is not a Banach space.*

Sketch of Proof. Let M be a closed, orientable, hyperbolic 3-manifold admitting a surface bundle structure over S^1 with fiber Σ_g. We may assume that Σ_g is piecewise totally geodesic, that is, Σ_g admits a triangulation $\tilde{\tau}$ consisting of totally geodesic 2-simplices in M. Let $p : \widetilde{M} \longrightarrow M$ be the infinite cyclic covering associated to $\pi_1(\Sigma_g) \subset \pi_1(M)$. Let $\{\widetilde{\Sigma}_k\}_{k=-\infty}^{\infty}$ be the set of connected components of $p^{-1}(\Sigma_g)$ such that each $\widetilde{\Sigma}_k$ is adjacent to both $\widetilde{\Sigma}_{k-1}$ and $\widetilde{\Sigma}_{k+1}$ in $\widetilde{M}$. We denote by τ_k the triangulations of $\widetilde{\Sigma}_k$ with $p_*(\tau_k) = \tau$. Let s_0, s_n be the elements of $C_2(\widetilde{M})$ represented by the triagulations τ_0, τ_n respectively such that $z_n = s_n - s_0 \in B_2(\widetilde{M}) \subset B_2^{l_1}(\widetilde{M})$.

If $(C_*^{l_1}(\widetilde{M}), \partial_*)$ satisfied 2-UBC^{l_1}, then there would exist a constant $L > 0$ such that, for any $n \in \mathbf{N}$, there is $c_n \in C_2^{l_1}(\widetilde{M})$ with $\partial_3 c_n = z_n$ and $\|c_n\| \leq L\|z_n\| \leq 2L\|s_0\|$. We set $c_n = \sum_{i=1}^{\infty} \alpha_i \sigma_i$ and $\bar{c}_n = \sum_{i=1}^{\infty} \alpha_i \bar{\sigma}_i$, where $\bar{\sigma}_i = \mathrm{straight}(\sigma_i)$. Then,

$$(2.1) \qquad \|\bar{c}_n\| \leq \|c_n\| = \sum_{i=1}^{\infty} |\alpha_i| \leq 2L\|s_0\|.$$

Since $\mathrm{straight}(s_0) = s_0$ and $\mathrm{straight}(s_n) = s_n$, we have $\partial_3 \bar{c}_n = \partial_3 c_n = s_n - s_0$. For any $\varepsilon > 0$, there exists $m = m(n) \in \mathbf{N}$ with $\sum_{i=m+1}^{\infty} |\alpha_i| < \varepsilon$. We set $\bar{c}_n^m = \sum_{i=1}^{m} \alpha_i \bar{\sigma}_i \in C_3(\widetilde{M})$. Then,

$$\partial_3 \bar{c}_n^m = \partial_3 (\bar{c}_n - \sum_{i=m+1}^{\infty} \alpha_i \bar{\sigma}_i) = s_n - s_0 - \sum_{i=m+1}^{\infty} \alpha_i \partial_3 \bar{\sigma}_i.$$

The boundary cycle $d_n \in B_2(\widetilde{M})$ is defined by $d_n = \sum_{i=m+1}^{\infty} \alpha_i \partial_3 \bar{\sigma}_i = s_n - s_0 - \partial_3 \bar{c}_n^m$. Since the Gromov norm $\|\partial \bar{\sigma}_i\| = 4$,

$$(2.2) \qquad \|d_n\| \leq 4 \sum_{i=m+1}^{\infty} |\alpha_i| < 4\varepsilon.$$

Some arguments in the proof of [17, Theorem 1] together with the inequality (2.1) implies the existence of a constant $K > 1$ independent of n and satisfying

$$\sum_{k=1}^{n-1} \sum_{i=1}^{m} |\alpha_i| \mathrm{Area}(\bar{\sigma}_i | \widetilde{\Sigma}_k) < KL\|s_0\|.$$

Thus, if we take n sufficiently large, then there exists an integer k, $0 < k < n$, satisfying

$$\text{Area}(\bar{c}_n^m|\tilde{\Sigma}_k) = \sum_{i=1}^m |\alpha_i|\text{Area}(\bar{\sigma}_i|\tilde{\Sigma}_k) < \varepsilon.$$

We denote by $\widetilde{M}_+$ and $\widetilde{M}_-$ the closures in $\widetilde{M}$ of the components of $\widetilde{M}-\tilde{\Sigma}_k$ containing $\tilde{\Sigma}_0$ and $\tilde{\Sigma}_n$ respectively. Consider a subdivision $\bar{c}_n^{m'}$ of $\bar{c}_n^m$ such that the image of each singular 3-simplex in $\bar{c}_n^{m'}$ is contained in either $\widetilde{M}_+$ or $\widetilde{M}_-$. Let

$$\bar{c}_n^{m'}|_{\widetilde{M}_+} = \sum_{i=1}^m \sum_{u=1}^{u(i)} \alpha_i \rho_{iu} \in C_3(\widetilde{M})$$

be the subchain of $\bar{c}_n^{m'}$ consisting of 3-simplices ρ_{iu} with $\rho_{iu}(\Delta^3) \subset \widetilde{M}_+$. Then, we have $\partial_3(\bar{c}_n^{m'}|_{\widetilde{M}_+}) = s_k' - s_0' - d_n'$, where s_0', $d_n'|_{\widetilde{M}_+}$ are subdivisions of s_0, $d_n|_{\widetilde{M}_+}$ respectively, and s_k' is a 3-chain with $\text{supp}(s_k') \subset \tilde{\Sigma}_k$ and such that $\text{Area}(s_k') \le \text{Area}(\bar{c}_n^m|\tilde{\Sigma}_k) < \varepsilon$. By (2.2), $\text{Area}(d_n') \le \text{Area}(d_n) < 4\pi\varepsilon$. Since s_0' (and hence s_0) are homologous to $s_k' - d_n'$ in $\widetilde{M}$, $p_*(s_0)$ is homologous to $p_*(s_k' - d_n')$ in M. Since $[p_*(s_0)] \ne 0$ in $H_2(M;\mathbf{R})$, there exists a constant $Q > 0$ depending only on $[p_*(s_0)]$ with $\text{Area}(p_*(s_k' - d_n')) > Q$. On the other hand,

$$\text{Area}(p_*(s_k' - d_n')) \le \text{Area}(s_k') + \text{Area}(d_n') < (4\pi + 1)\varepsilon.$$

Since one can choose $\varepsilon > 0$ arbitrarily small, this gives a contradiction. Thus, $(C_2^{l_1}(\widetilde{M}), \partial_*)$ cannot satisfy 2-UBC^{l_1}. Then, Theorem 1.3 implies that $H_b^3(\widetilde{M};\mathbf{R}) = H_b^3(\Sigma_g;\mathbf{R})$ is not a Banach space. $\square$

Some arguments used in the proof of Theorem 2.3 are applicable to prove the following:

Theorem 2.3 ([17]). *For any integer $q \ge 5$, there exists a finitely generated, discrete group G such that $(H_b^q(G;\mathbf{R}), \|\cdot\|)$ is not a Banach space.*

§3. The third bounded cohomology of closed surfaces

Let $\mathcal{H}(\Sigma_g)$ be the set of equivalence classes (M, f) such that each M is an oriented hyperbolic 3-manifold and $f : \Sigma_g \longrightarrow M$ is a homotopy equivalent map (called a *marking*). Here, two such elements (M_0, f_0), (M_1, f_1) are said to be *equivalent*, i.e. $(M_0, f_0) = (M_1, f_1)$ in $\mathcal{H}(\Sigma_g)$, if there exists an orientation-preserving isometry $\alpha : M_0 \longrightarrow M_1$ such that $\alpha \circ f_0$ is homotopic to f_1. An element $(M, f) \in \mathcal{H}(\Sigma_g)$ is called *doubly-degenerate* if M contains no geometrically finite ends. We also consider the subset $\mathcal{H}_+(\Sigma_g)$ of $\mathcal{H}(\Sigma_g)$ consisting of elements (M, f) such that the injectivity radius $\text{inj}(M) = \inf\{\text{inj}_M(x); x \in M\} > 0$.

In [11], Minsky proved that the equivalence class $(M, f) \in \mathcal{H}_+(\Sigma_g)$ is determined only by the ending invariants. This result raises the question what condition on elements of $\mathcal{H}_+(\Sigma_g)$ gives the same ending invariants. In this section, we present such a condition in terms of the (induced) fundamental classes $[f^*(\omega_M)]$ in $H_b^3(\Sigma_g;\mathbf{R})$ for $(M, f) \in \mathcal{H}_+(\Sigma_g)$.

Theorem 3.1 ([15]). *For each $i_0 > 0$, let (M_0, f_0) be any doubly-degenerate element of $\mathcal{H}_+(\Sigma_g)$ with $\text{inj}(M_0) \geq i_0$. There exists a constant $\varepsilon(g, i_0) > 0$ depending only on g and i_0 such that, for any $(M_1, f_1) \in \mathcal{H}_+(\Sigma_g)$, if $\|[f_0^*(\omega_{M_0})] - [f_1^*(\omega_{M_1})]\| < \varepsilon$ in $H_b^3(\Sigma_g; \mathbf{R})$, then $(M_0, f_0) = (M_1, f_1)$ in $\mathcal{H}_+(\Sigma_g)$.*

The proof of Theorem 3.1 given in [15] is rather long. So, we will give only a rough outline of the proof, and refer to [15] for details.

Outline of Proof. First, we consider how to compare M_0 with M_1. Let Γ_i be the Kleinian group with $M_i = \mathbf{H}^3/\Gamma_i$ for $i = 0, 1$. For the Fuchsian group Π with $\Sigma_g = \mathbf{H}^2/\Pi$, let $\rho_i : \Pi \longrightarrow \Gamma_i$ be the isomorphism induced from $f_i : \Sigma_g \longrightarrow M_i$. According to Minsky [11] (see also Cannon-Thurston [5]), there exists a ρ_i-equivariant, continuous map $F_i : \mathbf{B}^2 \longrightarrow \mathbf{B}^3$ with $F^{-1}(S_\infty^2) = S_\infty^1$. Since M_0 is doubly-degenerated, the limit set L_{Γ_0} of Γ_0 is equal to S_∞^2. This implies that $F_0|_{S_\infty^1} : S_\infty^1 \longrightarrow S_\infty^2$ is surjective. By [14], there exists a Π-invariant, zero-measure set Λ^1 in S_∞^1 and a Γ_0-invariant, zero-measure set Λ^2 in S_∞^2 with $\Lambda^1 = F^{-1}(\Lambda^2)$ such that the restriction $F_0|_{S_\infty^1 - \Lambda^1} : S_\infty^1 - \Lambda^1 \longrightarrow S_\infty^2 - \Lambda^2$ is a homeomorphism. For the inverse $\varphi_0 = (F_0|_{S_\infty^2 - \Lambda^2})^{-1}$, the composition

$$h = (F_1|_{S_\infty^1}) \circ \varphi_0 : S_\infty^2 - \Lambda_0 \longrightarrow S_\infty^2$$

is continuous and equivariant with respect to the isomorphism $\rho_1 \circ \rho_0^{-1} : \Gamma_0 \longrightarrow \Gamma_1$. We say that h is a *comparing map* for M_0 and M_1.

It is shown that φ_0 is extended to a ρ_0^{-1}-equivariant, continuous map $\Phi : \mathbf{B}^3 - \Lambda^2 \longrightarrow \mathbf{B}^2 - \Lambda^1$. Consider the subspace $\widehat{S}(M_0) = (\Pi^4(\mathbf{B}^3 - \Lambda^2) - \nabla^4(S_\infty^2 - \Lambda^2))/\Gamma_0$ of $S(M_0)$. Since φ_0 is injective, the continuous map $\Psi : \widehat{S}(M_0) \longrightarrow S(\Sigma_g)$ covered by $\Pi^4\Phi$ is well-defined. Take a sequence of blocks $\{B_n\}$ in the $(+)$-end of $\widetilde{M}$ as in §1. Since $inj(M_0) \geq i_0 > 0$, we may choose $\{B_n\}$ so that $V \leq \text{vol}(B_n) \leq 2V$ for all $n \in \mathbf{N}$ and for some constant $V = V(g, i_0)$. For the Borel measure $z(n, n') = z(L(n, n')) \in \mathcal{B}(M_0)$, we set $u(n, n') = \Psi_*(z(n, n')|_{\widehat{S}(M_0)})$. Then, we have $\|u(n, n')\| = \|z(n, n')\|$ and $u(n, n')$ is an element of $\mathcal{B}(\Sigma_g)$. Let $c_i : \mathcal{B}(\Sigma_g) \longrightarrow \mathbf{R}$ be the $\mathbf{R}$-homomorphism continuously extending $f_i^*(\omega_{M_i})|S(\Sigma_g)$.

Let $q : \mathbf{H}^3 \longrightarrow M_0$ be the universal covering. For a fixed point x_n of each B_n, choose a point $\tilde{x}_n$ of $q^{-1}(x_n)$. There exists a constant $R = R(g, i_0) > 0$ such that $\mathcal{N}_R(\tilde{x}_n)$ meets a geodesic line $\overline{L}_n$ which covers a closed geodesic $\overline{l}_n$ in M_0 with $i_0 \leq \text{length}(\overline{l}_n) \leq R$. Let $l_n \in \Gamma_0$ be an element with $l_n(\overline{L}_n) = \overline{L}_n$ and such that the translation length of l_n is equal to $\text{length}(\overline{l}_n)$. Let $\tilde{x}_0 = (\sqrt{-1}, 0) \in \mathbf{C} \times \mathbf{R}_+$ be the center of the upper half space model $\mathbf{U}$ of $\mathbf{H}^3 \cup S_\infty^2$. For any element $\gamma_n \in \text{PSL}_2(\mathbf{C})$ with $\gamma_n(\mathcal{N}_{i_0/2}(\tilde{x}_n)) \ni \tilde{x}_0$, there exits $k = k(g, i_0) \in \mathbf{N}$ such that the 2-disk in $\mathbf{C}$ of radius 2 centered at one of end points of $\gamma_n(\overline{L}_n)$ is contained in any regular triangle T_k in $\mathbf{C}$ of edge length 2^k and with the origin $0 \in \mathbf{C}$ as its inner center.

For a fixed $\varepsilon > 0$, we suppose that $\|[c_0] - [c_1]\| < \varepsilon$. Then, there exists a bounded 3-cocycle $e \in Z_b^3(\Sigma_g)$ with $\|e\| < \varepsilon$, and a bounded 2-cochain $d \in C_b^2(\Sigma_g)$ with $(c_0 - c_1 + e)(\sigma) = \delta(d)(\sigma)$ for any straight simplex $\sigma : \Delta^3 \longrightarrow \Sigma_g$.

For any $\varepsilon_0 > 0$, we say that the bounded 3-cocycle c_1 is ε_0-*bad* in $B_n = L(n, n+1)$ with respect to $u(n, n+1)$ if $\|u(n, n+1)|_{\mathcal{B}(n, \sqrt{\varepsilon_0})}\| \geq \sqrt{\varepsilon_0}\text{vol}(B_n)$, where $\mathcal{B}(n, \sqrt{\varepsilon_0})$

is the subset of $\mathrm{supp}(u(n, n+1))$ consisting of straight simplices σ with $|c_1(\sigma)| \leq \mathbf{v}_3 - \sqrt{\varepsilon_0}$. Otherwise, c_1 is said to be ε_0-*good* in B_n. If c_1 is ε_0-bad in B_n, then we have

$$c_1(u(n, n+1)) \leq (\mathbf{v}_3 - \varepsilon_0)\mathrm{vol}(B_n).$$

Let $B_{n_1}, \cdots, B_{n_a}$ be the blocks in $L(n, n')$ in which c_1 is ε_0-bad. Then,

$$
\begin{aligned}
c_1(u(n, n')) &\leq \mathbf{v}_3(\mathrm{vol}(L(n, n')) - \sum_{i=1}^{a} \mathrm{vol}(B_{n_i})) + (\mathbf{v}_3 - \varepsilon_0)\sum_{i=1}^{a} \mathrm{vol}(B_{n_i}) \\
&= \mathbf{v}_3\mathrm{vol}(L(n, n')) - \varepsilon_0 \sum_{i=1}^{a} \mathrm{vol}(B_{n_i}).
\end{aligned}
$$

Since $\|\partial z(n, n')\| \leq K_1$ as seen in §1, $(c_0 - c_1)(u(n, n')) \leq K_1\|d\| + \varepsilon\|u(n, n')\|$. Since $c_0(u(n, n')) = \mathbf{v}_3\|u(n, n')\| = \mathbf{v}_3\|z(n, n')\| = \mathbf{v}_3\mathrm{vol}(L(n, n'))$,

$$(\mathbf{v}_3 - \varepsilon)\mathrm{vol}(L(n, n')) - \mathbf{v}_3\mathrm{vol}(L(n, n')) + \varepsilon_0 \sum_{i=1}^{a} \mathrm{vol}(B_{n_i}) \leq K_1\|d\|.$$

This implies that $\varepsilon_0 \sum_{i=1}^{a} \mathrm{vol}(B_{n_i}) \leq \varepsilon\mathrm{vol}(L(n, n')) + K_1\|d\|$. Since $V \leq \mathrm{vol}(B_m) \leq 2V$ for all $m \in \mathbf{N}$, we have $a\varepsilon_0 V \leq 2(n' - n)\varepsilon V + K_1\|d\|$. This gives the following inequality.

$$\frac{a}{n' - n} \leq \frac{2\varepsilon}{\varepsilon_0} + \frac{K\|d\|}{\varepsilon_0 V(n' - n)}.$$

This suggests that, for any arbitrarily small $\varepsilon_0 > 0$, one can choose $\varepsilon > 0$ and $n' - n \in \mathbf{N}$ so that ε_0-good blocks occupy a great part of $L(n, n')$. Hence, for any $s \in \mathbf{N}$, there exist $\varepsilon = \varepsilon(s) > 0$ and $n_0 \in \mathbf{N}$ such that, for each $m \in \mathbf{N}$, $L(mn_0, (m+1)n_0)$ contains a sub-bank $L(n_m - s, n_m + s + 1)$ such that, in each block B_n of $L(n_m - s, n_m + s + 1)$, c_1 is 2^{-2s}-good with respect to $u(n, n+1)$. Hence, for a lift $\tilde{\sigma} : \Delta^3 \longrightarrow \mathbf{U}$ of $\sigma \in \mathrm{supp}(z(n_m - s, n_m + s + 1)) - $ (small part), the straight simplex $h(\tilde{\sigma}) : \Delta^3 \longrightarrow \mathbf{U}$ with ideal vertices $h(\tilde{\sigma}(v_0)), \cdots, h(\tilde{\sigma}(v_3))$ is nearly regular. Thus, if we set $\tilde{\sigma}(v_0) = h(\tilde{\sigma}(v_0)) = \infty \in \partial\mathbf{U}$, then $\{\tilde{\sigma}(v_1), \tilde{\sigma}(v_2), \tilde{\sigma}(v_3)\}$ spans a regular triangle in $\mathbf{C}$ and $\{h(\tilde{\sigma}(v_1)), h(\tilde{\sigma}(v_2)), h(\tilde{\sigma}(v_3))\}$ does a nearly regular triangle.

In the proof of Theorem A in [15], we also defined the Borel measure $\mu_r(n, n') \in \mathcal{B}(M_0)$ and $v_r(n, n') = \Psi_*(\mu_r(n, n')|_{\widehat{S}(M_0)})$ for all sufficiently small $r > 0$ such that $\mathrm{supp}(\mu_r(n, n'))$ cosisting of all elements in $S(M_0)$ each of whose vertices is ideal, whose inner centers are contained in $L(n, n')$ and whose absolute volume is greater than $\mathbf{v}_3 - a(r)$, where $a(r)$ is a positive function with $\lim_{r \to 0} a(r) = 0$. In fact, by using these $\mu_r(n_m - s, n_m + s + 1)$, we showed that, for any mutually close $\sigma, \tau \in \mathrm{supp}(z(n_m - s, n_m + s + 1)) - $ (small part), $h(\tilde{\sigma})$ and $h(\tilde{\tau})$ are aslo close to each other, see [15, §4] for details. These facts imply that there exists $\gamma'_m \in \mathrm{PSL}_2(\mathbf{C})$ such that the measurable set $\bar{h}_m|_{T_k} : T_k \longrightarrow \mathbf{C} \subset \partial\mathbf{U}$ is well approximated by the inclusion $i : T_k \longrightarrow \mathbf{C}$, where $\bar{h}_m = (\gamma'_m|_{S_\infty^2}) \circ h \circ (\gamma_{n_m}|_{S_\infty^2})^{-1} : S_\infty^2 \longrightarrow S_\infty^2$. Since h is $\rho_1 \circ \rho_0^{-1}$-equivariant and since $i_0 \leq \mathrm{length}(\bar{l}_{n_m}) \leq R$, one can apply this approximation to show that $i_0/2 \leq \mathrm{length}(\bar{l}'_{n_m}) \leq 2R$, where $\bar{l}'_{n_m}$ is the closed geodesic in M_1 covered by the geodesic axis of $\rho_1 \circ \rho_0^{-1}(l_{n_m})$ in $\mathbf{U}$.

Since $\sup_m \{\text{length}(\vec{l}'_{n_m})\} < \infty$, $\{\vec{l}'_{n_m}\}_{m=1}^{\infty}$ does not remain in any compact subset of M_1. Then, by using some arguments in Bonahon [1], it can be shown that the $(+)$-ending lamination for M_0 coincides with that for M_1. And similarly, the $(-)$-ending lamination for M_0 coincides with that for M_1. Hence, by Minsky's Ending Lamination Theorem [11], there exists an orientation-preserving isometry $\alpha : M_0 \longrightarrow M_1$ such that $\alpha \circ f_0$ is homotopic to f_1. $\square$

Let $E_+(g)$, $E(g)$ be the smallest Banach subspaces of $HB^3(\Sigma_g; \mathbf{R})$ containing respectively $\{[f^*(\omega_M)]_B; (M,f) \in \mathcal{H}_+(\Sigma_g)\}$ and $\{[f^*(\omega_M)]_B; (M,f) \in \mathcal{H}(\Sigma_g)\}$. By the definitions of these vector spaces and Corollary 1.2, we have

$$\dim E_+(g) \leq \dim E(g) \leq \dim HB^3(\Sigma_g; \mathbf{R}) \leq \dim H_b^3(\Sigma_g; \mathbf{R}) \leq \#\mathbf{R}.$$

As an application of Theorem 3.1, it is shown that $\dim E_+(g) \geq \#\mathbf{R}$ and hence:

Corollary 3.2 ([15]). $\dim E_+(g) = \dim E(g) = \dim HB^3(\Sigma_g; \mathbf{R}) = \dim H_b^3(\Sigma_g; \mathbf{R})$ $= \#\mathbf{R}$.

By the argument similar to that in the proof of Theorem 3.1, one can also prove the following rigidity theorem for certain hyperbolic 3-manifolds of infinite volume.

Theorem 3.3 ([15]). *Let W be a boundary-irreducible Haken manifold with* genus $(\partial W) = g > 1$, *and let M_0, M_1 be oriented hyperbolic 3-manifolds without geometrically finite ends and with* $\text{inj}(M_0), \text{inj}(M_1) \geq i_0 > 0$. *Suppose that there exist homotopy equivalent maps $f_i : W \longrightarrow M_i$ for $i = 0, 1$. Then, the following three conditions are equivalent.*

(i) *There exists an orientation-preserving isometry $\alpha : M_0 \longrightarrow M_1$ such that $\alpha \circ f_0$ is homotopic to f_1.*

(ii) $[f_0^*(\omega_{M_0})] = [f_1^*(\omega_{M_1})]$ *in* $H_b^3(W; \mathbf{R})$.

(iii) $\|[f_0^*(\omega_{M_0})] - [f_1^*(\omega_{M_1})]\| < \varepsilon(g, i_0)$ *in* $H_b^3(W; \mathbf{R})$.

§4. Topologically tame Kleinian groups

As mentioned in Introduction, our results in this section are based on thorems concerning topologically tame Kleinian groups in Canary [2], [3], [4]. First, we will sketch the proof of the following theorem.

Theorem 4.1 ([16]). *Suppose that Γ is a topologically tame Kleinian group such that the volume of M_Γ is infinite. Then, $[\omega_\Gamma] = 0$ in $H_b^3(M_\Gamma; \mathbf{R})$ if and only if Γ is either elementary or geometrically finite.*

Sketch of Proof. For any differential 2-form η on an oriented Riemannian surface F with the area form Ω_F, there exists the smooth function $h : F \longrightarrow \mathbf{R}$ with $\eta = h\Omega_F$ on F. Consider a positive orthonormal basis $\mathbf{e}_1(p)$, $\mathbf{e}_2(p)$ for the tangent space to F at $p \in F$. Since $\eta_p(\mathbf{e}_1(p), \mathbf{e}_2(p)) = h(p)\Omega_F(\mathbf{e}_1(p), \mathbf{e}_2(p)) = h(p)/2$, the value $\eta_p(\mathbf{e}_1(p), \mathbf{e}_2(p))$ does not depend on the choice of a positive orthonormal basis. We set $\|\eta\| = \sup\{2|\eta_p(\mathbf{e}_1(p), \mathbf{e}_2(p))| ; p \in F\}$. If $\|\eta\| < \infty$, then

$$(4.1) \qquad \left| \int_F \eta \right| \leq \|\eta\| \text{Area}(F).$$

By Theorem 1.4, if Γ is topologically tame and geometrically infinite, then $\|[\omega_\Gamma]\| = v_3$ and hence $[\omega_\Gamma] \neq 0$ in $H_b^3(G; \mathbf{R})$. This shows the "only if" part of this theorem.

Now, we will sketch the proof of the "if" part. For simplicity, we only consider the case where Γ contains no parabolic elements. If Γ is elementary, then Γ has an abelian subgroup of finite index. By Gromov [7, §3.0], $H_b^3(M_\Gamma; \mathbf{R})$ is trivial and hence $[\omega_\Gamma] = 0$. We suppose that Γ is non-elementary and geometrically finite. Let Ω_Γ be the volume form on M_Γ. Since M_Γ is non-compact, the usual third cohomology $H^3(M_\Gamma; \mathbf{R})$ is trivial. By the de Rham theory, there exists a differential 2-form ζ on M_Γ with $\Omega_\Gamma = d\zeta$. For any 3-simplex $\sigma : \Delta^3 \longrightarrow M_\Gamma$, the Stokes Theorem implies that $\omega_\Gamma(\sigma) = \Omega_\Gamma(\mathrm{straight}(\sigma)) = d\zeta(\mathrm{straight}(\sigma)) = \zeta \circ \partial(\mathrm{straight}(\sigma)) = \zeta(\mathrm{straight}(\partial\sigma)) = \delta(\zeta \circ \mathrm{straight})(\sigma)$, where $\delta : C^2(M_\Gamma) \longrightarrow C^3(M_\Gamma)$ is the coboundary operator. This shows that $\omega_\Gamma = \delta(\zeta \circ \mathrm{straight})$. Since the homomorphism $j^* : H_b^3(M_\Gamma; \mathbf{R}) \longrightarrow H_b^3(C_\Gamma; \mathbf{R})$ induced by the inclusion $j : C_\Gamma \longrightarrow M_\Gamma$ is isomorphic, $[\omega_\Gamma] = 0$ if and only if $[j^*(\omega_\Gamma)] = 0$. Since M_Γ has no parabolic cusps, C_Γ is compact. Hence, there exists a constant $K > 0$ such that, for any oriented, (small) open disk D totally geodesically embedded in C_Γ by the inclusion $i_D : D \longrightarrow C_\Gamma$, $\|i_D^*(\zeta)\|$ is less than K with respect to the induced hyperbolic metric on D. Since C_Γ is convex in M_Γ, for any 2-simplex $\tau : \Delta^2 \longrightarrow C_\Gamma$, $\mathrm{straight}(\tau)$ is also a singular simplex in C_Γ. Since the absolute value of the area of $\mathrm{straight}(\tau)$ is less than π, $|\zeta \circ \mathrm{straight}(\tau)| < K\pi$ by (4.1). This shows that $j^*(\zeta \circ \mathrm{straight}) \in C_b^2(C_\Gamma)$ and hence $j^*(\omega_\Gamma) \in B_b^3(C_\Gamma)$. Thus, we have $[\omega_\Gamma] = 0$ in $H_b^3(M_\Gamma; \mathbf{R})$. $\square$

For an element g of a group G, let $C(g)$ be the set of elements of G conjugate to g, that is, $C(g) = \{hgh^{-1}; h \in G\}$, and set $C_\infty(g) = \cup_{n=1}^\infty C(g^n)$. One can prove the following theorem by using Theorem 4.1 together with [4, Corollary B].

Theorem 4.2 ([16]). *Let G be a group, and let $\Gamma_1, \cdots, \Gamma_n$ be topologically tame Kleinian groups admitting isomorphisms $\varphi_i : G \longrightarrow \Gamma_i$. Suppose that, for each M_{Γ_i}, only one end E_i of M_{Γ_i} is simply degenerate. Let Λ_i be a subgroup of Γ_i corresponding to E_i. If, for any integers i, j with $1 \leq i < j \leq n$, there exists $l_{ij} \in \Lambda_j$ such that $\varphi_i^{-1}(\Lambda_i) \cap C_\infty(\varphi_j^{-1}(l_{ij})) = \emptyset$, then $[\varphi_1^*(\omega_{\Gamma_1})]_B, \cdots, [\varphi_n^*(\omega_{\Gamma_n})]_B$ are linearly independent in $HB^3(G; \mathbf{R})$.*

There are various examples of sets $\{\Gamma_n\}$ of topologically tame Kleinian groups satisfying the conditions of Theorem 4.2. In fact, the following theorem is proved by applying Theorem 4.2 to the set of certain topologically tame Kleinian groups Γ_n ($n = 1, 2, \cdots$) each of which is isomorphic to $\mathbf{Z} * \mathbf{Z}$.

Theorem 4.3 ([16]). *For any finitely generated group G admitting a surjective homomorphism $f : G \longrightarrow \mathbf{Z} * \mathbf{Z}$, $\dim H_b^3(G; \mathbf{R}) = \dim HB^3(G; \mathbf{R}) = \#\mathbf{R}$.*

In Corollary 3.2, we have shown that $\dim E(g) = \dim HB^3(\Sigma_g; \mathbf{R}) = \#\mathbf{R}$. Now, it is a natural question whether $E(g) = HB^3(\Sigma_g; \mathbf{R})$ holds. However, Corollary 4.4 implies that $E(g)$ cannot occupy a large part of $HB^3(\Sigma_g; \mathbf{R})$.

Corollary 4.4 ([16]). $\dim HB^3(\Sigma_g; \mathbf{R})/E(g) = \#\mathbf{R}$.

References

1. F. Bonahon, *Bouts des variétés hyperboliques de dimension 3*, Ann. of Math. **124** (1986) 71-158.
2. R. Canary, *Ends of hyperbolic 3-manifolds*, J. Amer. Math. Soc. **6** (1993), 1-35.
3. R. Canary, *Covering theorems for hyperbolic 3-manifolds*, In: K. Johannson (ed.) Low-Dimensional Topology (Conference Proceedings and Lecture Notes in Geometry and Topology vol. III, pp. 21-30) International Press, Boston MA, 1994.
4. R. Canary, *A covering theorem for hyperbolic 3-manifolds and its applications*, preprint.
5. J. Cannon and W. Thurston, *Group invariant Peano curves*, preprint.
6. R. Grigorchuk, *Some results on bounded cohomology*, In: A. Duncan, N. Gilbert, J. Howie (ed.) Combinatorial and geometric group theory, (London Math. Soc. Lecture Note Series 204, pp. 111-163) Cambridge Univ. Press, Cambridge, 1995.
7. M. Gromov, *Volume and bounded cohomology*, Publ. Math. Inst. Huates Etud. Sci. **56** (1982) 5-100.
8. N. Ivanov, *Foundations of the theory of bounded cohomology*, J. Soviet Math. **37** (1987) 1090-1114.
9. N. Ivanov, *Second bounded cohomology group*, J. Soviet Math. **52** (1990) 2822-2824.
10. S. Matsumoto and S. Morita, *Bounded cohomology of certain groups of homeomorphisms*, Proc. Amer. Math. Soc. **94** (1985) 539-544.
11. Y. Minsky, *On rigidity, limit sets and end invariants of hyperbolic 3-manifolds*, J. Amer. Math. Soc. **7** (1994), 539-588.
12. Y. Minsky, *On Thurston's ending lamination conjecture*, In: K. Johannson (ed.) Low-Dimensional Topology (Conference Proceedings and Lecture Notes in Geometry and Topology vol. III, pp. 109-122) International Press, Boston MA, 1994.
13. Y. Mitsumatsu, *Bounded cohomology and l^1-homology of surfaces*, Topology **23** (1984) 465-471.
14. T. Soma, *Equivariant, almost homeomorphic maps between S^1 and S^2*, Proc. Amer. Math. Soc. **123** (1995) 2915-2920.
15. T. Soma, *Bounded cohomology of closed surfaces*, preprint.
16. T. Soma, *Bounded cohomology and topologically tame Kleinian groups*, preprint.
17. T. Soma, *Existence of non-Banach bounded cohomology*, preprint.
18. W. Thurston, *The geometry and topology of 3-manifolds*, Lect. Notes, Princeton Univ., 1978.
19. T. Yoshida, *On 3-dimensional bounded cohomology of surfaces*, In: H. Toda (ed.) Homotopy Theory and Related Topics (Advanced Studies in Pure Math. vol. 9, pp. 173-176) Kinokuniya, Tokyo, 1986.

DEPARTMENT OF MATHEMATICAL SCIENCES, COLLEGE OF SCIENCE AND ENGINEERING, TOKYO DENKI UNIVERSITY, HATOYAMA-MACHI, SAITAMA-KEN 350-03, JAPAN

E-mail address: soma@r.dendai.ac.jp

Proceedings of
THE 37TH TANIGUCHI SYMPOSIUM ON
TOPOLOGY AND TEICHMÜLLER SPACES
held in Finland, July 1995
ed. by Sadayoshi KOJIMA *et al.*
©1996 World Scientific Publishing Co.
pp. 279–293

BLOCH TOPOLOGY OF THE UNIVERSAL TEICHMÜLLER SPACE

MASAHIKO TANIGUCHI

Dedicated to Professor Yukio Kusunoki on his 70th birthday

(Received November 27, 1995)

The universal Teichmüller space $T(1)$ is a deformation space, suitable in a sense, of a simple closed curve in the complex plane. Hence, it may be natural to use uniform convergence topology (L^∞ topology on the circle). But if we come to discuss parametrization of planar fractals appearing as boundaries of simply connected domains, this topology is too strong. (In fact, uniform convergence can not fit on the case of 'blow-ups' of prime ends.) On the other hand, local uniform topology for Riemann maps of the interiors of curves, or equivalently Carathéodory topology marked at a point for simply connected domains, is a kind of weak topology. Hence it does not fit so well for the purpose.

Among other topologies of the universal Teichmüller space, Bers topology is famous and useful in many situations. But the closure of $T(1)$ in the Banach space $B(1)$ of (hyperbolically) bounded quadratic differentials on the unit disk U with respect to this topology is still mysterious (cf. for example, [Su]). Also note that the representation of $T(1)$ as quasisymmetric functions on the real axis can be embedded naturally into the space $Homeo^+(S^1)$ of all orientation-preserving self-homeomorphisms of the circle S^1, which in turn has a nice structure (cf. [Pe]). But correspondence, outside of $T(1)$, between complex weldings (represented by simple closed curves) and real welding functions are not so clear (cf. [Ha], [KNS], [MS]).

Here we consider another topology induced by Bloch norm. (Bloch norm has been used to define topologies of $T(1)$. An example is the one introduced by Astala and Zinsmeister [AZ], which is defined in a different way from ours. Also see [Gol].) Then the closure of $T(1)$ with respect to this topology is just the totality of normalized univalent functions on the unit disk, and we show that Bloch convergence of normalized Riemann maps is equivalent to (marked point free) Carathéodory convergence of the images (Theorem 2). Moreover, these topologies are the same as BMO topology. Also in the case of Kleinian groups with simply connected invariant components, we show that Carathéodory convergence of the ordinary sets implies Bloch convergence of Riemann maps of simply connected invariant components under a reasonably weak condition (Proposition 5), and give a quantitative condition for Bloch convergence of such Riemann maps (Theorem 3).

The author would like to thank Y. Gotoh, K. Matsuzaki and T. Sugawa for their many helpful conversations.

1. Injection into the Bloch space

We consider the set

$$\Sigma = \{f \mid f \text{ is univalent on } U \text{ and has a form } = \frac{1}{z} + \sum_{n=1}^{\infty} c_n z^n \text{ near } z = 0\}$$

of all normalized meromorphic univalent functions on the unit disk U, and equip Σ with Bers topology, or equivalently, identify Σ with it's image under the Schwarzian operator in the space of (hyperbolically) bounded quadratic differentials on U. It is well-known that Σ contains a subset $\Sigma(1)$ naturally identified with the universal Teichmüller space $T(1)$.

We have shown in [Ta] that Σ can be mapped injectively in the Banach space $VMO = VMO(S^1)$ consisting of functions of vanishing mean oscillation on the circle $S^1 = \partial U$. Moreover, identifying Σ with

$$z\Sigma = \{zf(z) \mid f \in \Sigma\},$$

we may consider Σ as a subset of the Banach space

$$\mathfrak{B}_0 = \{f \in \mathfrak{B} \mid \lim_{r \to 1} \sup_{\{r < |z| < 1\}} (1 - |z|^2)|f'(z)| = 0\}$$

of little Bloch functions on U, where the Bloch space $\mathfrak{B}$ consists of all holomorphic functions f on U with

$$\|f\| = \sup_U (1 - |z|^2)|f'(z)| < +\infty$$

Namely, we have the following

Proposition 1 ([Ta]). *$z\Sigma$ is contained in $VMO \subset \mathfrak{B}_0$.*

Proof. We give here a proof slightly different from that in [Ta]. First note that $z\Sigma$ is contained in the Dirichlet space $D(S^1)$ of the unit disk. Actually, fix a compact set K containing 0, and we have

$$\iint_{U-K} |\{zf(z)\}'|^2 dxdy \leq 2 \iint_{U-K} |f|^2 dxdy + 2 \iint_{U-K} |f'|^2 dxdy < +\infty,$$

for every $f \in \Sigma$. Hence Yamashita's theorem [Ya] implies that

$$z\Sigma \subset VMO.$$

(Also see [Ax] and [Au].) $\square$

The space VMO is just the closure of the set $C(S^1)$ of all continuous functions on the circle in the Banach space BMO, and hence contains a natural completion of the set of all simple closed curves. An element of VMO is not necessarily continuous, but can be represented as a sum of a continuous function and the Hilbert transformation of such one. Also recall that the famous Fefferman-Stein theorem implies that, for every Riemann map f, the Connes' quantized derivative $d_Q f \, (= [H, f]$ with the Hilbert transformation H) is a compact operator if and only if $f \in VMO$. (In this context, the Hilbert-Schmidt class of quantized derivatives is nothing but the Dirichlet-Sobolev space of harmonic functions in $D(S^1)$, and hence with the boundary value in $H^{1/2}$ (which Nag [NS] used to give another representation of $T(1)$).

It is well-known that the Bloch space $\mathfrak{B}$ is coincident with the 2-dimensional Euclidean BMO space $BMO(U)$, and hence Bloch norm measures certain local distortion of an element of Σ on U. On the other hand, BMO norm measures similar distortion on the boundary by definition (though BMO is nothing but the 2-dimensional hyperbolic BMO space on U). Combining these two norms, we can estimate wildness of elements on the closure of U. Thus the assertion of Proposition 1 will provide fruitful consequences.

Remark. For $f \in \Sigma$, the condition that $\hat{f}(z) = zf(z)$ is bounded and belongs to $\mathfrak{B}$, namely that

$$\sup_U (1 - |z|^2)|\hat{f}'(z)| < \infty, \quad \sup_U |\hat{f}| < \infty,$$

is equivalent to saying that

$$\sup_{U-K} (1 - |z|^2)|f'(z)| < \infty, \quad \sup_{U-K} |f| < \infty$$

for every compact set K in U containing 0.

Further, $zf(z) \in \mathfrak{B}_0$ means that

$$\lim_{|z| \to 1} (1 - |z|^2)|f'(z)| = 0.$$

Now Bers topology is very sensitive against the boundary behavior of univalent maps. In particular, even L^∞-convergence does not necessarily imply convergence with respect to Bers topology. On the other hand, BMO topology is obviously weaker than L^∞-topology.

Remark. The set Σ itself is contained in VMO, since multiplication by z is clearly an invertible BMO-multiplier which preserves VMO (cf. [Got]).

Proposition 2 ([Ta]). *The identical injection of Σ, equipped with Bers topology, into VMO is continuous on $\Sigma(1)$.*

Proof. (For the sake of convenience, we include a proof.) Suppose that $f_n \in \Sigma(1)$ converges to $f_\infty \in \Sigma(1)$. Then each $f = f_n$ or f_∞ can be extended to a quasiconformal automorphism of $\hat{\mathbb{C}}$ with the complex dilatation $\mu = \mu_n$ or μ_∞, which vanishes on U. Let f^μ be the quasiconformal automorphism of $\hat{\mathbb{C}}$ fixing $0, 1, \infty$. Then f^{μ_n} converges to f^{μ_∞} uniformly on $\hat{\mathbb{C}}$ with respect to the spherical metric.

282   MASAHIKO TANIGUCHI

Write, on U,

$$\frac{1}{f^{\mu}(z)} = \frac{a}{z} + b + \text{holomorphic terms}$$

Then

$$a = \frac{1}{2\pi} \int_{|z|=1/2} \frac{1}{f^{\mu}(z)} dz$$

and

$$b = \frac{1}{2\pi} \int_{|z|=1/2} \left(\frac{1}{f^{\mu}(z)} - \frac{a}{z} \right) \frac{dz}{z}$$

Hence a_n and b_n converges to a_∞ and b_∞, which implies that f_n converges to f_∞ uniformly on ∂U. $\square$

Recall that BMO topology is not weaker than Bloch topology and that they are actually different even on $D(S^1)$. But they are mutually equivalent to each other, at least, on $z\Sigma$. See Corollary after the proof of Theorem 2. The proof of this corollary is due to Y. Gotoh [Got'].

In the sequel, we concentrate on Bloch topology. And, to distinguish Bloch topology from Bers topology, we assume that Σ is equipped with Bers topology, and when we consider Bloch topology, we always use the image under the above injection ι of Σ into $\mathfrak{B}_0$. Namely, we set $\iota(f) = \hat{f}$ for every $f \in \Sigma$. Then Proposition 2 asserts that, at least on $\Sigma(1)$, Bers topology is not weaker than Bloch topology.

Now, almost all important function spaces satisfy the density condition of smooth elements. $\mathfrak{B}$ is not an exception. We next show that this property inherits to the subset $\iota(\Sigma)$, which reflects naturality of the subset $\iota(\Sigma)$ in $\mathfrak{B}_0$.

Theorem 1. *Let Σ_{smooth} be the subset of Σ consisting of elements having smooth boundary value, that is, $\Sigma_{smooth} = \Sigma \cap C^\infty(\partial U)$. Then $\iota(\Sigma_{smooth})$ is dense in $\iota(\Sigma)$.*

The classical theorem due to Warshawski implies that Σ_{smooth} actually equals to $\Sigma \cap C^\infty(\overline{U})$. Namely, smooth complex weldings give smooth structures on the whole sphere.

Proof. Fix $f \in \Sigma$ and set $f_r(z) = rf(rz)$. Then Proposition 1 and Remark following it imply that for every $\epsilon > 0$, there is an $\eta > 0$ such that

$$\sup_{1-\eta<|z|<1} (1 - |z|^2)|f'(z)| < \epsilon,$$

and hence

$$\sup_{1/(1+\eta)<|z|<1} (1 - |z|^2)|f_r'(z)| < \epsilon$$

for every $r > 1 - \eta^2$. Since f_r converges to f uniformly on compact sets in U, it is easy to see that $\hat{f}_r$ converges to $\hat{f}$ in $\mathfrak{B}_0$. $\square$

Remark. $\iota(\Sigma_{smooth})$ is clearly contained in $\iota(\Sigma(1))$. Hence, $\iota(\Sigma_{smooth})$ is dense also in $\iota(\Sigma(1))$.

Corollary. *$\iota(\Sigma(1))$ is dense in $\iota(\Sigma)$*

Remark. Bers topology on $T(1)$ is, in a sense, local Hölder topology on boundary curves, while local uniform topology on U seems too weak to have connection with boundary behavior. Bloch topology is clearly stronger than local uniform topology, but smooth curves are still dense in $\iota(\Sigma(1))$.

Next we note the following

Proposition 3. $\iota(\Sigma)$ *is closed.*

Proof. Suppose that $\hat{f}_n \in \iota(\Sigma)$ converges to $\hat{f} \in \mathfrak{B}_0$. Then $f_n \in \Sigma$ converges to $f = \hat{f}/z$ uniformly on compact sets. Hence Hurwitz's theorem implies the assertion. $\square$

Corollary (No Corona Theorem).

$$\iota(\Sigma) = \overline{\iota(\Sigma(1))}$$

Namely, $\iota(\Sigma)$ is a compactification of $\iota(\Sigma(1))$.

Proof. Since $\iota(\Sigma(1))$ is dense in $\iota(\Sigma)$, the assertion follows by Proposition 3. $\square$

Recall that, with respect to Bers topology, smooth elements are 'transversal' to the Teichmüller spaces of compact surfaces (cf. [NS]), and that several examples of points in the corona of $T(1)$ have been constructed by, for instance, Gehring, Thurston and Astala. Thus, we may say that Bers topology preserves some fractal pattern of curves. Anyway, Bloch topology is definitely different from Bers topology.

2. Set theoretic convergence

Definition. A sequence of open sets $\{\Omega_n\}$ in $\hat{\mathbf{C}}$ is said to converge in the sense of Carathéodory to an open set Ω if

(1) every compact subset K of Ω lies in Ω_n for every sufficiently large n, and
(2) every open set which lies in Ω_n for infinitely many n also lies in Ω.

The classical convergence theorem due to Carathéodory states that, when $f_n \in \Sigma$ converges to some f_0 uniformly on compact sets in U, $f_0(U)$ is the component, which contains ∞, of (but not necessarily equal to) the open set $\cup_k \mathrm{Int}(\cap_{n=k}^{\infty} f_n(U))$. In other words, local uniform convergence is equivalent to Carathéodory convergence marked a point, while the above definition is marked point free Carathéodory convergence.

Proposition 4. *Suppose that $f_n \in \Sigma$ converges to f_0 uniformly on compact sets in U and $f_0(U) = D_0$ is dense in $\hat{\mathbf{C}}$. Then $D_n = f_n(U)$ and $D'_n = \hat{\mathbf{C}} - \overline{f_n(U)}$ converge to D_0 and the empty set, respectively, in the sense of Carathéodory.*

Proof. The classical Carathéodory convergence theorem implies the condition (1). Hence $\cup_k \mathrm{Int}(\cap_{n=k}^{\infty} D_n)$ contains D_0.

Next, suppose that there is an open set B such that $B \subset D_n \cup D'_n$ for infinitely many n. We further assume that B is connected. By the assumption, $B \cap D_0 \neq \emptyset$. Hence $B \cap D_n \neq \emptyset$,which implies that $B \subset D_n$, for every sufficiently large n. Thus we conclude that $B \subset D_0$, which implies the assertion. $\square$

Theorem 2. *Let $\{f_n\}$ be a sequence in Σ. Then $\hat{f}_n$ converges to $\hat{f}_0$ in $\mathfrak{B}$ if and only if $D_n = f_n(U)$ converges to $D_0 = f_0(U)$ in the sense of Carathéodory.*

To show Theorem 2, first we recall that Carathéodory convergence of open sets is the same as Hausdorff convergence of the complements, where in general, a sequence of compact sets A_n converges to A_0 in the sense of Hausdorff if

$$\lim_{n \to \infty} \inf\{\epsilon > 0 \mid A_n \subset U_\epsilon(A_0), A_0 \subset U_\epsilon(A_n)\} = 0.$$

(Here $U_\epsilon(X)$ means the ϵ-neighborhood of a set X.)

Thus the distance to the boundary from a point behaves tamely. We denote by $d_f(z)$ the distance from $f(z)$ to the boundary of $f(U)$ for every $f \in \Sigma$ and $z \in U$. Then the following estimates are essentially well-known.

Comparison Theorem. *Fix $f \in \Sigma$ and set $D = f(U)$. Then we have*

$$\frac{1}{d_f(z)} \geq \lambda(f(z)) \geq \frac{M}{d_f(z)}$$

for every z in $\{1 > |z| \geq 1/2\}$ with a universal constant M. Here $\lambda(z)^2|dz|^2$ is the Poincaré metric on D.

Here recall that $f(\{|z| < 1/2\})$ contains $\{|z| > 8\}$ for every $f \in \Sigma$. Since, by definition, $\lambda(f(z))|f'(z)| = 1/(1 - |z|^2)$, we conclude the following

Corollary. *Let f and D be as above. Then*

$$d_f(z) \leq (1 - |z|^2)|f'(z)| \leq M \cdot d_f(z)$$

for every z in $\{1 > |z| \geq 1/2\}$.

Proof of Theorem 2. First to show the only-if part, suppose that D_n converges to D_0 in the sense of Carathéodory and fix $\epsilon > 0$ arbitrarily. Then as noted before, there is an n_0 such that $\mathbf{C} - D_n \subset U_\epsilon(\mathbf{C} - D_0)$ and $\mathbf{C} - D_0 \subset U_\epsilon(\mathbf{C} - D_n)$ for every $n \geq n_0$.

Since $f_0 \in \mathfrak{B}_0$, there is a compact set $K \subset U$ such that $d_{f_0}(z) \leq \epsilon$ for every $z \in U - K$. This implies that $f_0(U - K) \subset U_\epsilon(\partial D_0)$. And since f_n converges to f_0 uniformly on K, we may assume that $f_n(U - K) \subset U_\epsilon(\mathbf{C} - f_0(K)) \subset U_{2\epsilon}(\mathbf{C} - D_0)$ for every n.

Now fix $z \in U - K$. When $f_n(z)$ is not contained in $U_\epsilon(f_0(U - K))$, which implies, in particular, that $f_n(z)$ belongs to $(\mathbf{C} - D_0) \cap D_n \subset U_\epsilon(\mathbf{C} - D_n) \cap D_n$, then $f_n(z)$ is contained in $U_\epsilon(\partial D_n)$. Hence we have that $d_n(z) \leq \epsilon$. And when $f_n(z)$ is contained in $U_\epsilon(f_0(U - K))$, then $f_n(z)$ belongs to $U_{2\epsilon}(\mathbf{C} - D_0) \subset U_{3\epsilon}(\mathbf{C} - D_n)$. Hence we conclude that $f_n(z) \in U_{3\epsilon}(\partial D_n)$, or equivalently that $d_n(z) \leq 3\epsilon$.

Thus by Corollary, we have

$$(1 - |z|^2)|f_n'(z)| \leq 3M\epsilon$$

for every $z \in U - K$ and every n. Since $\hat{f}_n$ converges to $\hat{f}_0$ uniformly on compact sets in U and $|f_n(z) - f_0(z)| < 16$ for every $z \in U$, we conclude the convergence of $\hat{f}_n$ to $\hat{f}_0$ in $\mathfrak{B}$.

Next to show the converse, assume that $\hat{f}_n$ converges to $\hat{f}_0$ in $\mathfrak{B}$, but that D_n does not converge to D_0 in the sense of Carathéodory. Then since Carathéodory theorem implies the condition (1), there should be an open set B not contained in D_0 such that $B \subset D_{n(m)}$ for a subsequence $\{n(m)\}$ with $\lim_m n(m) = +\infty$. Fix $w \in B - D_0$

and set $z_n = f_n^{-1}(w)$. Then $|z_{n(m)}|$ should tend to 1. And for such $n(m)$, $d_{f_{n(m)}}(z_{n(m)})$ is not less than a suitable positive constant, which contradicts to the assumption by the above Corollary. $\square$

Corollary (Bloch = BMO on $z\Sigma$). *Bloch topology and BMO topology are mutually equivalent to each other on $z\Sigma$.*

Since Bloch topology is not stronger than BMO topology, we need only to show that, under the same notations as in Theorem 2, Carathéodory convergence of D_n to D_0, which is equivalent to convergence of $\hat{f}_n$ to $\hat{f}_0$ in $\mathfrak{B}$ by Theorem 2, implies BMO convergence of $\hat{f}_n$ to $\hat{f}_0$. To show this, we recall the following well-known equality.

Lemma 1. *Let f be an analytic function on U. Then*

$$\frac{1}{2\pi}\int_0^{2\pi}|f(e^{i\theta}) - f(z)|^2 P_z(\theta)d\theta = \frac{2}{\pi}\iint_U |f'(\zeta)|^2 \log\left|\frac{1 - \bar{z}\zeta}{z - \zeta}\right| d\xi d\eta.$$

where $P_z(\theta)$ is the Poisson kernel on ∂U for the point $z \in U$.

Especially $f \in BMO$ if and only if

$$\sup_{z\in U}\iint_U |f'(\zeta)|^2 \log\left|\frac{1 - \bar{z}\zeta}{z - \zeta}\right| d\xi d\eta < \infty,$$

and $f \in VMO$ if and only if

$$\iint_U |f'(\zeta)|^2 \log\left|\frac{1 - \bar{z}\zeta}{z - \zeta}\right| d\xi d\eta \to 0$$

as z tends to ∂U. And the following lemma due to Y. Gotoh is crucial for our proof of the above corollary.

Lemma 2 ([Got']). *For every $f \in \Sigma$, we have*

$$\iint_U |\hat{f}'(\zeta)|^2 \log\left|\frac{1 - \bar{z}\zeta}{z - \zeta}\right| d\xi d\eta \leq \frac{M_1}{\log\frac{32}{d_f(z)}} + M_2(1 - |z|),$$

on $\{|z| > 3/4\}$, where M_1 and M_2 are absolute constants independent of f.

Proof. Set $U_1 = \{|z| < 1/2\}$ and $U_2 = U - U_1$. Divide

$$\iint_U |\hat{f}'(\zeta)|^2 \log\left|\frac{1 - \bar{z}\zeta}{z - \zeta}\right| d\xi d\eta$$

into integrals I_k on U_k, $k = 1, 2$. Since $|\hat{f}'(\zeta)| \leq M_3$ for every $\zeta \in U_1$, where and in the sequel, M_j are absolute constants, we have

$$I_1 \leq \left(\max_{\zeta\in U_1}\log\left|\frac{1 - \bar{z}\zeta}{z - \zeta}\right|\right)\iint_{U_1}|(\hat{f}'(\zeta))|^2 d\xi d\eta < M_4(1 - |z|)$$

And

$$I_2 \leq 2\iint_{U_2}|f(\zeta)|^2 \log\left|\frac{1 - \bar{z}\zeta}{z - \zeta}\right| d\xi d\eta + 2\iint_{U_2}|f'(\zeta)|^2 \log\left|\frac{1 - \bar{z}\zeta}{z - \zeta}\right| d\xi d\eta,$$

which we set $2I_3 + 2I_4$. Then $I_3 \leq M_5(1 - |z|)$. And as for I_4, since $\iint_{U_2} |f'(\zeta)|^2 d\xi d\eta < M_6$, we have

$$I_4 = \iint_{U_2} |f'(\zeta)|^2 g_U(z, \zeta) d\xi d\eta \leq \iint_{U_2} |f'(\zeta)|^2 g_{U_2}(z, \zeta) d\xi d\eta + M_7(1 - |z|),$$

where and in the sequel, $g_\Omega(w, \cdot)$ is the Green function on Ω with the pole w. And, setting $B_1 = B(f(z), \frac{1}{2} d_f(z))$, we divide

$$\iint_{U_2} |f'(\zeta)|^2 g_{U_2}(z, \zeta) d\xi d\eta = \iint_{f(U_2)} g_{f(U_2)}(f(z), \zeta) d\xi d\eta$$

into integrals I_5 and I_6 on B_1 and on $f(U_2) - B_1$, respectively. Then by Koebe's distortion theorem, we can see that $f(B_0) \subset B_1$, where we set $B_0 = B(z, \frac{1}{10}(1 - |z|))$. Moreover

$$\max_{\zeta \in \partial B_1} g_{f(U_2)}(f(z), \zeta) \leq \max_{\zeta \in \partial B_0} g_U(z, \zeta) < M_8.$$

Hence

$$I_5 \leq \iint_{B_1} \left(\log \left| \frac{d_f(z)}{2(\zeta - f(z))} \right| + M_8 \right) d\xi d\eta < M_9 d_f(z)^2.$$

And since $f(U_2) \subset B(f(z), 16)$ bounding $g_{U_2}(f(z), \zeta)$ by a linear function of $\log \frac{1}{|\zeta - f(z)|}$, we can show that

$$I_6 \leq \frac{M_{10}}{\log \frac{32}{d_f(z)}}.$$

Thus we have the assertion. $\square$

Proof of Corollary (Bloch = BMO on $z\Sigma$). Suppose that D_n converges to D_0 in the sense of Carathéodory. And fix a positive ϵ arbitrarily. Then as in the proof of Theorem 2, we can find a compact set $K \subset U$ such that $\{d_{f_n}(z)\}_{n=0}^\infty$ are uniformly small on $U - K$. Hence by Lemma 2, we may assme that

$$\iint_U |\hat{f}'(\zeta) - \hat{f}_n'(\zeta)|^2 \log \left| \frac{1 - \bar{z}\zeta}{z - \zeta} \right| d\xi d\eta < \epsilon$$

for every $z \in U - K$ and every n.

On the other hand, since $\hat{f}_n$ converges to $\hat{f}_0$ uniformly on compact set of U, $\iint_U |\hat{f}'(\zeta) - \hat{f}_n'(\zeta)|^2 d\xi d\eta$ are uniformly bounded, and $|f_n - f_0| < 16$ on U, we can conclude that

$$\iint_U |\hat{f}'(\zeta) - \hat{f}_n'(\zeta)|^2 \log \left| \frac{1 - \bar{z}\zeta}{z - \zeta} \right| d\xi d\eta$$

converges to 0 uniformly on K. Thus by Lemma 1, we conclude that $\hat{f}_n$ converges to $\hat{f}_0$ in BMO. $\square$

Actually, we have obtained also the following assertion for the class S of holomorphic univalent functions on U.

Corollary. *Suppose that $f_n \in S$ are uniformly bounded. The following assertions are equivalent to each other.*

1) *f_n converges to f in BMO.*
2) *f_n converges to f in $\mathfrak{B}$.*

3) $f_n(U)$ *converges to* $f(U)$ *in the sense of Carathéodory.*

Remark. Consider the classes B_p $(p > 0)$ of holomorphic functions on U with

$$\|f\|_{B_p}^2 = \sup_{z \in U} \iint_U |f'(w)|^2 g_U^p(z, w) du\, dv < \infty,$$

which are investigated by Aulaskari and others. Note that $B_1 = BMO$ and $B_0 = AD$. Also we know (cf. [ASX]) that, for $0 < p < 1$, we have the strict inclusion relation

$$AD < B_p < BMO.$$

Now, the same arguments as in the above proof shows that Carathéodory convergence of images is equivalent to B_p convergence of univalent functions.

Finally, it is easy to see that Carathéodory topology marked at a point is not stronger than, and not equal to base point free Carathéodory topology, which in turn, is different in general from Carathéodory topology for the complements of the boundaries. We stress this fact by giving the following

Example. There is a sequence $\{f_n\} \subset T(1)$ such that

(1) f_n converges to *id* in Bloch topology, and hence by Theorem 2, $f_n(U)$ converges to U in the sense of Carathéodory, but

(2) $\mathbf{C} - \overline{f_n(U)}$ becomes narrower and narrower everywhere, and finally converges to the empty set in the sense of Carathéodory.

This example also shows that the Carathéodory topology on Σ allows certain geometric discontinuity of the boundaries.

3. The case of Kleinian groups

Now set-theoretic convergence as in the previous section plays important roles in various situations. Here we discuss the case of Kleinian groups.

Definition. A sequence $\{G_n = \langle g_{1n}, g_{2n}, \cdots \rangle\}$ of Kleinian groups isomorphic to a group G converges *algebraically* to G_0 if $\lim g_{in} = g_i$ exists for every i and $G_0 = \langle g_1, g_2, \cdots \rangle$.

Next, the envelop of a sequence $\{G_n\}$ is defined by

$$Env\{G_n\} = \{g \in PSL(2, \mathbf{C}) \mid g = \lim g_n, \, g_n \in G_n\}.$$

A sequence $\{G_n\}$ converges *geometrically* to $H = Env\{G_n\}$ if and only if the envelop of every infinite subsequence of $\{G_n\}$ is also coincident with H.

Lemma 3 (cf.[JM]). *Suppose that* G_n *converges to* G_0 *algebraically, then*

$$\cap_k \overline{\cup_{n=k}^\infty \Lambda(G_n)} \supset \Lambda(G_0),$$

or equivalently,

$$\cup_k Int(\cap_{n=k}^\infty \Omega(G_n)) \subset \Omega(G_0),$$

where and in the sequel, we denote by $\Lambda(G)$ *and* $\Omega(G)$, *respectively, the limit set and the ordinary set of a Kleinian group* G.

Proof. Fix $z_0 \in \Lambda(G_0)$ arbitrarily. Then there is a sequence $\{g_{n,0}\} \subset G_0$ such that a fixed point $a_{n,0}$ of $g_{n,0}$ converges to z_0. Since $g_{n,m}$ corresponding to $g_{n,0}$ in G_m converges to $g_{n,0}$, a fixed point $a_{n,m}$ of $g_{n,m}$ converges to $a_{n,0}$. Hence $z_0 \in \overline{\cup_{n=k}^{\infty} \Lambda(G_n)}$ for every k, which implies the assertion. $\quad\Box$

Remark ([JM]). The Carahtéodory condition (1) means that

$$\cup_k \mathrm{Int}(\cap_{n=k}^{\infty} \Omega(G_n)) \supset \Omega(G_0).$$

Hence, $\Omega(G_n)$ converges to $\Omega(G_0)$ in the sense of Carathéodory if and only if the condition (1) holds.

We here remark again that the Carathéodory convergence of $\Omega(G_n)$ to $\Omega(G_0)$ does not necessarily imply that of each component-wise convergence.

Problem. Find any condition which implies component-wise Carathéodory convergence.

For a sequence $\{G_n\}$ converging algebraically to G_0, $\Omega(G_n)$ does not necessarily converge to $\Omega(G_0)$ in the sense of Carathéodory. In the case of finitely generated Kleinian groups, the following theorem ([JM]) is fundamental.

Jorgensen-Marden theorem. *Suppose that $\{G_n\}_{n=1}^{\infty}$ is a sequence of finitely generated Kleinian groups, isomorphic to a group G, which converges algebraically to G_0. Suppose that G_0 is geometrically finite with $\Omega(G_0) \neq \emptyset$. Then $\{G_n\}$ converges geometrically to G_0 if and only if the ordinary sets $\Omega(G_n)$ converges to $\Omega(G_0)$ in the sense of Carathéodory.*

Remark. Even if G is not geometrically finite, Carathéodory convergence always implies the geometrical convergence ([JM, Proposition 4.2]).

In the case of a finitely generated Fuchsian group G of the first kind, suppose that G_n, which corresponds to a point in $T(G)$ and hence is isomorphic to G, tends to a quasi-Fuchsian group G_0 algebraically. Then since G_0 is qc stable, G_n should converge geometrically to G_0. Hence $\Omega(G_n)$ converges to $\Omega(G_0)$ in the sense of Carathéodory.

Since algebraic convergence is equivalent to Bers topology in this case, also Bloch topology is equivalent to Bers topology. Namely, we have the following

Corollary. *Suppose that the Teichmüller space $T(G)$ is finite dimensional. Then The natural injection of $T(G)$ into Σ is a homeomorphism onto its image.*

Let G be as in Corollary, and suppose that G_n isomorphic to G converges to G_0 with non-empty ordinary set in $\mathfrak{B}$. Since Riemann maps converge uniformly on compact sets, G_n converges to G_0 algebraically. Hence even on the set of b-groups defined below, Bloch convergence implies Bers convergence in this case.

Recall that a famous unsolved conjecture due to L. Bers asserts that every b-group corresponds to a boundary point of the Teichmüller space $T(G)$ with respect to Bers topology.

Now we turn to general cases.

Definition. We call a (not necessarily finitely generated) Kleinian group G a *generalized b-group* if $\Omega(G)$ has a simply connected invariant component Δ_G. If G is finitely generated, then G is usually called a b-group.

For a generalized b-group G, we always assume (by taking a conjugation if necessary) that a Riemann map f_G of U onto Δ_G belongs to Σ. Also, we call the complement of Δ_G the *filled limit set* of G and denote it by $K(G)$.

Corollary to Theorem 2. *Suppose that every G_n is a generalized b-group. Then $\hat{f}_{G_n}$ converges to $\hat{f}_{G_0}$ in $\mathfrak{B}_0$ if and only if $K(G_n)$ converges to $K(G_0)$ in the sense of Hausdorff.*

In general, Carathéodory convergence of the ordinary sets has no relation with Bloch convergence as noted before. In fact, some other components may vanish even if the invariant components converges as is seen by Example, and invariant components may produce some non-invariant components even if the total ordinary sets may converge in the sense of Carathéodory. (Also cf. [MS].) But in some cases, Carathéodory convergence of the ordinary sets implies Bloch convergence of Riemann maps.

Proposition 5. *Let G be a Fuchsian group. Let $\{G_n\}_{n=1}^{\infty}$ be a sequence of generalized b-groups isomorphic to G converging to G_0 and $f_{G_n} \in \Sigma$ the normalized Riemann map for G_n as before for every n. Suppose that the stabilizer of every component of G_0 is non-elementary. If $\Omega(G_n)$ converges to $\Omega(G_0)$ in the sense of Carathéodory, then $\hat{f}_{G_n}$ converges to $\hat{f}_{G_0}$ in $\mathfrak{B}$.*

Proof. If not, then by Theorem 2 there is a component D_0 of $\Omega(G_0) - \Delta_0$ which is contained in $\cup_k \mathrm{Int}(\cap_{n=k}^{\infty} \Delta_n)$, where we set $\Delta_n = \Delta_{G_n}$. By assumption, the stabilizer subgroup H_0 of G_0 for D_0 is non-elementary, and hence containing a loxodromic element g_0. Note that, since Δ_0 is invariant and simply connected, H_0 contains no APTs and $g_0(\Delta_0) = \Delta_0$.

Fix a point w on the axes of g_0 in D_0. Let ℓ_w be the geodesic segments from w to $g_0(w)$. Next fix a relatively compact, open neighborhood K_0 of ℓ_w in D_0. Then by the assumption, K_0 is contained in Δ_n for every sufficiently large n.

Now let $g_n \in G_n$ correspond to g_0 for every n. Then since f_{G_n} converges to f_{G_0} uniformly on compact sets of U, g_n converges to g_0, and hence $g_n(w)$ converges to $g_0(w)$. Considering the Poincaré metric on K_0, we can see that there are loops, say $\ell_{w,n}$, connecting w to $g_n(w)$, in Δ_n which have uniformly bounded hyperbolic lengths for every sufficiently large n. Hence we can show by hyperbolic geometry that every $\ell_{w,n}$ is in an M-neighborhood $(L_n)_M$ of the axis L_n of g_n in Δ_n with M independent of n. The local uniform convergence of f_n in U further implies that of the Poincaré density λ_n of Δ_n in Δ_0. Hence L_n converges to L_0 in a sense. To state this more precisely, recall that the attractive and repelling fixed points and the complex multiplier of g_n converges to those of g_0. Hence we can take a sufficiently small open neighborhood U and V of the attractive and repelling fixed points of g_0, respectively, so that $g_n(U) \subset U$ and $g_n^{-1}(V) \subset V$ for every sufficiently large n. Then the compact set $(L_0)_N - (U \cup V)$ is contained in Δ_n for every N and every sufficiently large n.

Thus we can show by local uniform convergence of λ_n that, taking a sufficiently large N, $W = U \cup (L_0)_N \cup V$ contains $(L_n)_M$ for every sufficiently large n.

Since $w \in D_0$, we can assume that w is not contained in W. But then, w could not be on $(L_n)_M$ for every sufficiently large n, which is a contradiction. $\square$

Here we recall that Proposition 4 and Theorem 2 immediately imply the assertion of Proposition 5 in some special cases.

Corollary. *Let $\{G_n\}$ and $f_{G_n} \in \Sigma$ be as in Proposition 5. Suppose that f_{G_n} converges to f_{G_0} uniformly on compact sets of U. Further suppose that G_0 has the connected and simply connected ordinary set without the exterior, then $\hat{f}_{G_n}$ converges to $\hat{f}_{G_0}$ in $\mathfrak{B}$.*

Proof. First by Proposition 4, both of $\Omega(G_n)$ and Δ_{G_n} converge to $\Omega(G_0) = \Delta_{G_0}$. Then Theorem 2 implies the assertions. $\square$

Corollary (cf.[Oh]). *Let $\{G_n\}$ and $f_{G_n} \in \Sigma$ be as in Proposition 5. Suppose that G is finitely generated and that f_{G_n} converges to f_{G_0} uniformly on compact sets of U (or equivalently, G_n converges to G_0 algebraically). Further suppose that G_0 is totally degenerate, then $\hat{f}_{G_n}$ converges to $\hat{f}_{G_0}$ in $\mathfrak{B}$ and G_n converges to G_0 geometrically.*

Proof. First recall that the limit set of a totally degenerate group has no interior point. Hence by Proposition 4, $\Omega(G_n)$ and Δ_{G_n} converge to $\Omega(G_0)$ and Δ_{G_0}, respectively. Then Theorem 2 and Remark after Jorgensen-Marden Theorem imply the assertions. $\square$

Also, Jorgensen-Marden Theorem implies the following

Corollary. *Let G_n and $f_{G_n} \in \Sigma$ be as in Proposition 5. Suppose that G is finitely generated. If G_n converges geometrically to a geometrically finite G_0, then $\hat{f}_{G_n}$ converges to $\hat{f}_{G_0}$ in $\mathfrak{B}$.*

Remark. Several other conditions for Carathéodory convergence, see for instance, [AC]. Also cf. [MT].

Now the main issue is how to characterize convergent sequences of Riemann maps in $\mathfrak{B}$. Finally, we give such a quantitative sufficient condition for convergence in $\mathfrak{B}$.

Definition. Let $\{G_n\}$ be an algebraically convergent sequence of generalized b-groups. Then we say that $\{G_n\}$ are *uniformly damping* on compact sets in U if, for every $\epsilon > 0$ and every compact set K in U,

$$|g_n'(f_{G_n}(z))| < \epsilon$$

for every $z \in K$, every n and every but a finite number of g in G, where we represent G_n as

$$G_n = \{g_n \mid f_{G_n} \circ g = g_n \circ f_{G_n}, \, g \in G\}.$$

A geometrical meaning of uniform damping-ness of a sequence of Kleinian groups G_n is the following: for every but a finite number of element in G, the Euclidean areas of the image balls under the actions by corresponding elements in G_n are uniformly small for every ball with a fixed radius in a compact set.

Theorem 3. *Fix a co-compact Fuchsian group G. Let $\{G_n\}$ be a sequence of b-groups isomorphic to G and $f_{G_n} \in \Sigma$ be the normalized Riemann maps as before. Suppose that f_{G_n} converges to f_{G_0} uniformly on compact sets of U. Further, if G_n are uniformly damping on compact sets in U, $\hat{f}_{G_n}$ converges to $\hat{f}_{G_0}$ in $\mathfrak{B}$.*

Remark. For a non co-compact G, uniform damping-ness on compact sets in U does not necessarily implies that $\hat{f}_{G_n}$ converges to $\hat{f}_{G_0}$ in $\mathfrak{B}$.

To prove Theorem 3, first set $f_n = f_{G_n}$ and recall the following elementary

Lemma 4. *Let G and $\{G_n\}_{n=0}^{\infty}$ be as in Theorem 3. Then there is a compact set E in U such that*

$$\|\hat{f}_n - \hat{f}_0\| \le \sup_{g \in G} \left(\sup_{g(E)} |\hat{f}_n'(z) - \hat{f}_0'(z)|(1 - |z|^2) \right)$$

Proof of Theorem 3. Fix E as in Lemma 4 and $\epsilon > 0$ arbitrarily. Then, since G_n are uniformly damping on compact sets in U, we can find a compact set F in U, containing E and $\{|z| \le 1 - \epsilon\}$, such that, if $g(E)$ is not contained in F, then $g(E)$ is contained in $\{|z| > 1/2\}$ and

$$(*) \qquad\qquad |g_n'(f_n(z))| < \epsilon$$

for every n and $z \in E$. Also we can find n_0 such that

$$(**) \qquad\qquad |f_n'(z) - f_0'(z)| < \epsilon$$

for every $z \in F$ and every $n \ge n_0$. (Hence, in particular, we have $|f_n'(z)| \le \epsilon + |f_0'(z)|$ for such n and z not equal to 0.)

Now fix n with $n \ge n_0$. If we show that

$$\hat{\omega}_g = \sup_{g(E)} |\hat{f}_n'(z) - \hat{f}_0'(z)|(1 - |z|^2) < M\epsilon(1 + \|\hat{f}_0\|)$$

with an absolute constant M, then the assertion follows by Lemma 4. For this purpose, since $|f_n(z) - f_0(z)| < 16$ for every $z \in U$ and f_n converges to f_0 uniformly on compact set, it suffices to show that

$$\omega_g = \sup_{g(E)} |f_n'(z) - f_0'(z)|(1 - |z|^2) < M\epsilon(1 + \|\hat{f}_0\|).$$

Here, if $g(E) \subset F$, then it is clear that $\omega_g < \epsilon$. And if $g(E)$ is not contained in F, then

$$\omega_g = \sup_{E} |f_n{}'(g(z)) - f_0{}'(g(z))|(1 - |g(z)|^2).$$

Here

$$|f_n{}'(g(z)) - f_0{}'(g(z))|(1 - |g(z)|^2)$$
$$= |f_n{}'(g(z)) - f_0{}'(g(z))||g'(z)|(1 - |z|^2)$$
$$= |(f_n \circ g)'(z) - (f_0 \circ g)'(z)|(1 - |z|^2)$$
$$= |g_n{}'(f_n(z))f_n{}'(z) - g_0{}'(f_0(z))f_0{}'(z)|(1 - |z|^2)$$
$$\le \{|g_n{}'(f_n(z))f_n{}'(z)| + |g_0{}'(f_0(z))f_0{}'(z)|\}(1 - |z|^2).$$

Hence by (*) and (**), we conclude that

$$\omega_g \leq \epsilon(1 + 2\sup_{g(E)} |f_0'(z)|(1 - |z|^2)) \leq M\epsilon(1 + \|\hat{f}_0\|)$$

with a constant M as desired. Thus we have the estimate even in this case. $\square$

Remark. Uniform damping property corresponds to uniform expanding property of complex dynamics near the boundary. A typical example is a simply connected attractive invariant component of a hyperbolic rational function.

Even for the case of finitely generated Kleinian groups, it seems not so easy to determine the class of Kleinian groups for which the assertion of Theorem 3 holds.

References

[Ab]   W. Abikoff, *Kleinian groups-geometrically finite and geometrically perverse*, Contemporary Math. **74** (1988), 1–50.

[AC]   J. A. Anderson and R. D. Canary, *Cores of hyperbolic 3-manifolds and limits of Kleinian groups*, preprint.

[AZ]   K. Astala and M. Zinsmeister, *Teichmüller spaces and BMOA*, Math. Ann. **289** (1991), 613–625.

[Au]   R. Aulaskali, *On VMOA for Riemann surfaces*, Can. J. Math. **XL** (1988), 1174–1185.

[ASX]   R. Aulaskari, D.A. Stegenga and J. Xiao, *Some subclass of BMOA and their characterization in terms of Carleson measures*, (to appear in Rocky Mountain J. Math.).

[Ax]   S. Axler, *The Bergman space, the Bloch space, and commutators of multiplication operators*, Duke Math. J. (1986), 315–332.

[Gol]   L. Goldberg, *On the shape of the unit sphere in $Q(\Delta)$*, Proc. AMS **118** (1993), 1179–1185.

[Got]   Y. Gotoh, *2-dimensional BMO spaces on general planar domains (Japanese)*, Topics in Complex Analysis 1991.

[Got'] Y. Gotoh, *(Oral communication)*.

[Ha]   D. H. Hamilton, *Simultaneous uniformization*, J. reine angew. Math. **455** (1994), 105–122.

[IT]   Y. Imayoshi and M. Taniguchi, *An introduction to Teichmüller spaces*, Springer-Verlag, 1992

[JM]   T. Jorgensen and A. Marden, *Algebraic and geometric convergence of Kleinian groups*, Math. Scand. **66** (1990), 47–72.

[KNS]   Y. Katznelson, S. Nag and D. Sullivan, *On conformal welding homeomorphisms associated to Jordan curves*, Ann. Acad. Sci. Fenn., Ser.A **15** (1990), 293–306.

[Le]   O. Lehto, *Univalent functions and Teichmüller spaces*, Springer-Verlag, 1986.

[MS]   K. Matsuzaki and H. Shiga, *Conformal conjugation of Fuchsian groups of the first kind to the second kind*, (to appear in J. reine angew. Math.).

[MT]   K. Matsuzaki and M. Taniguchi, *Hyperbolic manifolds and Kleinian groups*, Oxford Univ. Press

[NS]   S. Nag and D. Sullivan, *Teichmüller theory and the universal period mapping via quantum calculus and the $H^{1/2}$ space on the circle*, Osaka J. Math. (1995).

[Oh]   K. Ohshika, *Strong convergence of Kleinian groups and Carathéodory convergence of domains of discontinuity*, Math. Proc. Camb. Phil. Soc. **112** (1992), 297–307.

[Pe]   R. C. Penner, *Universal constructions in Teichmüller theory*, Adv. in Math. **98** (1993), 143–215.

[Po]   Ch. Pommerenke, *Schlichte Funktionen und analytische Funktionen von beschänkter mittelerer Oszillation*, Comment. Math. Helvetici **52** (1977), 591–602.

[Su]   T. Sugawa, *On the space of schlicht projective structures on compact Riemann surfaces with boundary*, J. Math. Kyoto Univ. **35** (1995), 697–732.

[Ta]   M. Taniguchi, *Quantized calculus and Teichmüller space*, RIMS (Koukyuroku) **882** (1994), 114–120.

[Ya] S. Yamashita, *Criteria for functions to be Bloch*, Bull. Australian Math. Soc. **21** (1980), 223–227.

DEPARTMENT OF MATHEMATICS, KYOTO UNIVERSITY
E-mail address: tanig@kusm.kyoto-u.ac.jp